# THE WILD HEART OF INDIA

# THE WILD HEART OF INDIA

nature and conservation in
the city, the country, and the wild

T.R. SHANKAR RAMAN

OXFORD
UNIVERSITY PRESS

OXFORD
UNIVERSITY PRESS

Oxford University Press is a department of the University of Oxford. It furthers the University's objective of excellence in research, scholarship, and education by publishing worldwide. Oxford is a registered trademark of Oxford University Press in the UK and in certain other countries.

Published in India by
Oxford University Press
2/11 Ground Floor, Ansari Road, Daryaganj, New Delhi 110 002, India



First Edition published in 2019



ISBN-13 (print edition): 978-0-19-949474-3
ISBN-10 (print edition): 0-19-949474-6

ISBN-13 (eBook): 978-0-19-909755-5
ISBN-10 (eBook): 0-19-909755-0

Typeset in Utopia Std 10.5/15
by Tranistics Data Technologies, Kolkata 700 091
Printed in India by Nutech Print Services India

*For*

Sitalakshmi Rajagopalan and T.R. Rajagopalan

Chandra Mudappa and (Late) B.P. Mudappa

# CONTENTS

## FIELD DAYS: AN ECOLOGICAL EDUCATION

## CONSERVATION: A WORLD OF WOUNDS

## REFLECTIONS: OUR PLACE IN NATURE

# Author's Note

The scientific names of plants and animals have generally not been included in the main text but are listed at the end of the book against their respective English or local names, the latter chosen as appropriate for the regional setting of the chapter. The scientific and English common names have been updated as per recent taxonomy and are based on sources included in the Select Bibliography. Place names, too, reflect current usage (such as Chennai for Madras and Bengaluru for Bangalore). Words derived from local languages now in common usage and with stable spelling in India (such as dosa, jhum) have not been italicized in the belief that if these are not already in the English dictionary, they should be. Sources of previously published work and co-authors are indicated for each chapter in the Publication Credits; the publishers are gratefully acknowledged for permission to use the material here. Some previously published work has been edited or revised to include relevant new scientific information or taxonomic changes, or for consistency in style, species names, and metric units, and for better coherence with the rest of the volume. In a couple of cases, essays were combined to provide greater depth or reduce overlap in content.

# Preface

> One of the penalties of an ecological education is that one lives alone in a world of wounds. Much of the damage inflicted on land is quite invisible to laymen. An ecologist must either harden his shell and make believe that the consequences of science are none of his business, or he must be the doctor who sees the marks of death in a community that believes itself well and does not want to be told otherwise.
>
> —*Round River*, Aldo Leopold

Nature, some people believe, is something *out there*, in forests or far wildernesses, separate from the dwelling or presence of humans. The book you hold in your hand is an attempt to dispel, or at least blur, such a sharp separation between people and nature. From city park to savanna, roadside to rainforest, and ocean bed to mountaintop, nature thrives and throbs around you, inseparable, omnipresent. It both envelops you and forms a part of who you are. You only have to open your senses to nature to perceive its eternal dance of life, death, and renewal, and participate in its shimmering wonder.

Such perception and participation carries both risk and responsibility. The risk, as Aldo Leopold noted presciently, comes when an ecological education opens your eyes to a wounded world: A world whose forests are stripped for timber and commodity plantations, whose earth is submerged by dams and gouged by mines. A world paved and polluted, increasingly defined and defiled by metal, mortar, and money. The sense of responsibility emerges from the awareness that the science of ecology (a word originating from the Greek *oikos*, dwelling place) concerns the science of home. As befitting a science of home, ecology's core concerns are relationships: life in relation to the environment, people in relation to planet. It is a science that traces connections, from one living organism to another, from sunlight through plant to predator, from humans to earth and landscape. Ecology, perhaps the most important science of the present century, also impels a curative response. It elicits a creative mission to retain the integrity of home, to heal the world of wounds, and restore relationships between people and place.

This collection of essays describes experiences and perspectives gained in wildly varied landscapes and waterscapes, from city and countryside to ocean and deep forest. Written over a period of 25 years, more than half my lifetime, these essays trace my own trajectory of learning about, engaging with, and reflecting on nature. They chart my journey from young student and naturalist in Chennai, through my training in wildlife science and ecology in Dehradun and Bengaluru, to my present life as a conservation scientist and writer working out of my home and field research station in the Anamalai Hills in southern India. Most of these essays emerged from journeys and field experiences with Divya Mudappa, wildlife biologist and fellow traveller into the wild heart of India. Journeying with Divya has always been an enriching experience of witnessing, photographing, and forming impressions, images, and ideas that finally found expression in these words. I co-authored 10 essays (or their earlier versions) with her, but most of the others, too, are from our time in the field together: out of a memorable encounter, an extended conversation, a close observation, a shared silence.

The essays in this collection are divided into three parts; each may appeal to readers whose inclinations and interests in nature take different shapes: those captivated by the lure and challenge of field observation and experience; those driven by a passion to conserve nature and wildness; and those trying to understand our place and role in nature in a rapidly changing world. I hope that these essays will resonate with you, reader, irrespective of where you live or why nature interests you.

The first part, Field Days—An Ecological Education, recounts things that I learnt or impressions I gained in the field that travellers, students of ecology, or observers of natural history may perhaps experience themselves. This part conveys varied experiences in the field that lasted a few hours or days to nearly three full years of research in forested mountains.

The second part, Conservation—A World of Wounds, carries pieces that stem from a growing awareness of the diverse threats to nature and what we can or must do to avert or reduce those threats. These essays try to link ecological knowledge to conservation policy and practice in the real world. As Leopold noted, in today's world an ecological awareness is inseparable from concerns over conservation, and the essays gathered in the first two parts retain an inevitable degree of overlap in linking field experience to motivations to heal the world of wounds.

The essays in the third part, Reflections—Our Place in Nature, represent attempts towards a deeper and more reflective perception of nature, landscape, and other species broadening into the domains of aesthetics, ethics, and the intangible. Few countries in the world offer opportunities for such perception more than India offers with its remarkable diversity of ecosystems, peoples, and ways of life that remain deeply connected to and enmeshed in nature. Other living beings, both plants and animals, have long been part of the spiritual and cultural worlds of Indians. They continue to inspire the imagination of poets and scholars, artists and musicians. They evoke appreciation and skilful uses among artisans and fishers, healers and herders, farmers and forest dwellers. In the city, the country, and the wild, they still find spaces in the hearts of people going about their busy and varied lives.

In field research or natural history, conservation or ethical reflection, there are people whose influence no fieldworker can fail to acknowledge. In my own life, as mentioned in the acknowledgements at the end of this book, there have been many such people: forest dwellers, field assistants, research guides, colleagues, scientists, historians, writers, friends, and family. It is to acknowledge this debt in a small way that I have included in this volume three essays corresponding to the three parts of this volume about three people who, although they are not alive today, remain an inspiration: R.K.G. 'Cutlet' Menon ('Lone Palm Tree, Sir!'), Ravi Sankaran ('A Life of Courage and Conviction'), and Rachel Carson ('An Enduring Relevance').

The essays in this volume track my own trajectory of personal and professional change—'growth' is perhaps too strong a word here—from my experiences as a field biologist, through forays into nature conservation, to deeper reflections on nature and place. Anyone who enters the field of natural history as a researcher or a conservation practitioner may begin where their inclinations reside but their trajectory often depends on the opportunities, learnings, and motivations that await them. Field research on wildlife and nature conservation in India faces many challenges, obstacles, limitations of funding, or priorities that may impel a researcher to work on particular topics or places. I have been fortunate to work mostly where I wanted, on subjects of deep interest to me, but my work, too, carried me to different places, brought me to consider different animals and plants, and spotlighted diverse conservation issues. Rather than a smooth narrative arc, I have presented these diverse encounters, experiences, and contemplations in a manner that mirrors my trajectory more realistically.

This book represents a personal journey rather than an attempt to provide a comprehensive overview of contemporary concerns in nature conservation in India. I believe there remains a role and space for such writing as is gathered here. Our world is awash with images and writings describing contrived adventures and imagined dangers in the wilderness. It is replete with news of peril and pillage of the world's vanishing species and wild places. Away from the spectacle and clamour, there remains a space for immersive and informed accounts that can provoke deeper engagement with nature and conservation, kindle greater sensitivity and personal transformation. It is a small but definite corner of that space that this volume seeks to occupy. And it is only you, reader, who can tell, as you turn the last page, whether this book does what it sets out to do, leads you to discover that the wild heart of India beats in your chest, too.

# Prologue

## *Where I had Always Wanted to Be*

There are times in your life, when, in an unexpected moment, you come face to face with yourself. It could happen anytime, to anyone. It could happen over breakfast as aroma and sound—hot coffee swirling in your cup and a dosa sizzling on the stove—suddenly release a cascade of recollections as history intersects happenstance. It could arise in a memory or dream where past and present merge into a fused and frozen time indistinct from the future. It could happen as you walk down a street and suddenly confront your own reflection in a shining, shop-front glass. In that moment, the person you were confronts the one you have become. Chances are, it might catch you unawares.

It happened to me like this. One morning, I was in Chennai, the city of my childhood, in southern India, at my parents' home in Adyar. The house, painted green and ringed by a small garden, stands along a line of homes in the quiet neighbourhood of Bakthavatsalam Nagar. It had been my home until two decades ago, when I left after my schooling and a Bachelor's degree in zoology from Loyola College for higher studies elsewhere, in Dehradun and Bengaluru, and then to other places where my work would take me. I was staying over that weekend, and my mother asked if I could take a look at some of my books and papers and clear out some of my old things from the room upstairs. With pending work and travel on my mind, the phones ringing every now and then, and various visitors coming and going, I was glad for an excuse to get away to my old room. I had a few hours to kill, so I told her I would take a look, not realizing that even in a half-distracted rummage through my stuff I would find something that would mark the day in my memory.

In the room upstairs, which my elder brother Sriram and I used to share, our mother had meticulously kept all our things in the glass-fronted wooden shelves nailed to the wall, protected behind two sliding glass doors with a ratchet lock tacked to the glass in the middle. I used the key she had had no trouble finding for me, heard the familiar sound of the glass grating open on the aluminium channels of the shelf. And there they all were. Rows of books, stacks of files, a welter of papers and envelopes, and even more of these stuffed behind the books. I pulled out and piled them all on the bed.

Textbooks I had saved from my high school days—biology, history, geography, English—slanted across the shelf from my brother's physics, chemistry, and maths textbooks from his school and engineering days at the Indian Institute of Technology in Chennai. A row full of my brother's notebooks filled with his emphatically neat, determined, cursive writing lay next to a stack of foolscap examination answer sheets: my brother's, impeccably scripted, organized under headers, keywords underlined for emphasis; mine, hastily scrawled, streaked with teacher's red, haphazardly heaped. Anthologies from our English classes, with some of our favourite essays, short stories, and poetry. A dog-eared *Wren & Martin*. Two of my scrapbooks on birds and mammals of the world. A file holding dot-matrix printouts of poems and puns and ribald jokes and cartoons. Another filled with yellowed newspaper clippings on events once recent and now remote; feature articles on subjects from garden plants to forest gibbons, on places from Central India to Antarctica, on people from Mahatma Gandhi to Sylvester Stallone. A tight envelope bursting with old postcards and letters from cousins and friends; another, more secretly wrapped and unopened, a passionate and poignant bunch of love notes and cards from a Muslim girl to a Catholic classmate of mine that had been too hot for him to even hide in his own home. Finally, rows of books: some dictionaries and reference, a trove of fiction from Charles Dickens to John le Carré, and poetry from Palgrave's *Golden Treasury* to Emily Dickinson. On the non-fiction shelf, Gerald Durrell and Cynthia Moss rubbed shoulders with Stephen Hawking and Richard Feynman. One full shelf held books on birds and natural history, alongside books on stars, planets, and amateur astronomy.

It was an archive: snippets, fragments, ornaments of a personal history already long past. In a quarter century, I had gone from schoolboy to scientist, become a wildlife biologist having secured a Masters degree from the Wildlife Institute of India in Dehradun after working in the mountains of Mizoram and a doctoral degree from the Indian Institute of Science in Bengaluru after field research on rainforest birds in the Western Ghats. It was in the Western Ghats that I now lived, in the Anamalai or elephant hills, named after the wild elephants that still roam the landscape. In 2000, Divya and I, a year after our marriage and a year before we completed our doctoral studies, had established a rainforest research station in the hill town of Valparai, about ten hours overland from Chennai. It is here that we work to conserve the dark rainforests—extensive tracts in the Anamalai Tiger Reserve and patches interspersed in the tea and coffee plantations on the undulating Valparai plateau.

Streams and rivers veined the landscape, fringed by lofty mountains draped with forests and grasslands, rock faces and cliffs, touched by the grandeur of great hornbills and wild elephants. Working in the hills and forests was enjoyable, of course, but what *wasn't* a walk in the park was trying to conserve wildlife in landscapes with people and plantations. To persuade people with business interests of profit and production from land to care for nature, to learn to live with wildlife, to help stave off conservation threats and crises as they arose, to prevent or minimize the destruction of beautiful places that one knew well: this was a full-time and often frustrating job. I vented my feelings writing essays that, according to friends, ranged from the lyrical to the depressing. I was now past forty and had little time for my family in Chennai.

That morning, I stared at the stuff on the shelves, then at the pile in front of me, rubbing my furrowed forehead as if it would bring some memory back, explain what I saw. As I picked up each item, I kept asking myself, what is this? Why have I kept it?

Frustrated at having to deal with the pile, I even asked my mother, 'Why have you kept all this?'

'Well, it was you and Sriram who kept them,' she answered, as she went about her work, 'and who knows when you may want your old things, or what you may want to make of them?'

Clearly, I was on my own here. Faced by a single room and its few shelves, I felt a sort of restless anxiety. I had choices to make: what do I keep, what discard?

It was as I began sorting that I found it in a pile of papers stashed in a dusty, flimsy file. I knew instantly what it was although a full twenty-seven years had passed during which it had faded to some innermost recess of my memory, so well hidden that it was effectively buried, forgotten. Until now. I took it out of the file, held it in my hands. Everything else began fading away.

Three foolscap sheets, now yellowing, held together at the top left by a rusted staple. In the unmistakable imprint of our old Remington typewriter—the one that had sat on a little desk in the dining room, wedged between the dining table, the puja room, and the path to the staircase—here was an essay from all those years ago. An essay written by me, or rather, by the boy I was then. A schoolboy, almost a stranger, known to close friends and family by a different name: Sridhar. A typewritten essay carrying his name.

I see him now at the typewriter with the bustle of the house around him. I hear his mother's energetic voice, his brother's footsteps, the quiet serenity of his father's unruffled presence. I see him, a gangly youth of fourteen years, his long, smooth limbs, his slim body. I see his dark eyes above darkening crescents, a knife-edge nose, a head of straight black hair falling on a forehead as yet untempered and uncluttered by life. There is a thin, shining patina of sweat pearling over his lips—lips that are full and fresh but do not move—as he sits slightly hunched at the typewriter. I see his index fingers stabbing, the keys clacking. And then, without pause, I read what he has written.

## WHERE I HAD ALWAYS WANTED TO BE

By T.R. SRIDHAR

It had been a tiring day. The exams were just a month away. The teachers had not finished even half of their portions. The realization came only in the morning, when the HM announced over the intercom, the dates for the exam and reminded the boys to start studying. The boys had taken it cool. But not so the teachers. There began a mad scramble for the text-books, note-books, and guides and they came laden to the class with more books than a poor, studious, all-book-carrying boy.

One teacher rattled off three lessons in a period. Then, another finished a chapter so fast that he left the cleverest boys blinking. Fortunately, it was biology, my easiest subject, and I managed to catch on something here and there.

We all ended up feeling famished, exhausted, defeated and dehydrated. I had never heard such a heavenly sound as the bell, when it rang. While returning home, the bus I was on blew a tyre. I waited an hour and clambered or rather crawled into the next one. It crashed into a motor-bike. My money was running out. The conductors didn't give me a full refund. So I decided to walk it.

It was a tremendously wearying walk. The buildings, trees and telephone poles that had dashed so quickly past me, when I was in the bus, seemed to now become super-phlegmatically lethargic and dragged painfully by. My shoe's soles scraped on the dusty gravel of the road and I looked down to find my shoes brown and covered with dry dust. I was nearing my house when my knees started buckling. I bent and pulled them sluggishly and stumbled clumsily into my house. I walked directly to the bathroom and flopped into the bath-tub which I had filled with cold water. I soon fell asleep.

I was woken up by my mother's shouts. After crying out to calm her, I donned my clothes and walked out—without even looking at her—and sat down on the seat of the chair beside the dining table.

"Whatever happened to you?" my mother asked wide-eyed.

"It was a tiring day," I said. She didn't ask me anything else and joined me while I had my supper.

It was three minutes more before I reached my bed. My head must have still been falling down to my pillow when I fell asleep again; for I don't remember having laid it down on the soft cushion.

Then I had the dream. It is, usually, very diffucult to say how a dream began, but I remember this one clearly.

At first it was dark. Very dark. The kind of darkness that seeps into you, clogging the very recesses of your being. Then, there developed a haze. A thin greyish mantle that started spreading from the rightmost corner of my right eye. The haze spread throughout and then shrunk until it was just a ... sort of doorway through which bright light entered. It turned out to be a tunnel. The tunnel seemed to flicker and move. I realized that it was I who was moving out of the tunnel.

I came out. The chill morning mist hit me like a sledge-hammer. I was suddenly feeling free ... there was no weight on my legs and the path before me lead into a lush, green jungle. I looked up to see the Blue Hills in the distance. I was in the Annamalai woods.

My passion for ornithology had still not left me. The rising sun was directly in front of me. I soaked up its warmth greedily and experienced a state of quixotic euphoria. A Magpie-robin sang its melodious song from somewhere deep in the forest. I heard a tittering, musical cry from my left and turned to spot a beautiful Yellow-backed sunbird in its glossy yellow, green and crimson plumage diving into the thick undergrowth. A group of Orange-headed ground-thrushes and Slaty-headed babblers landed in front of me making a cacophony of gurgling calls. The whole forest came alive. I listened to the calls of a million birds, the harsh chatter of the nocturnal owls in quest of a roosting hole in some gnarled branch, to while away the day; the raucous cries of the macaques and, the faint trumpet of a wild elephant.

'It was absolute peace. I had been in the heat and dust that had made me so weary. But now I was in the Western Ghats—at the Annamalai jungles at the foot of the awesome Nilgiris. I was where I had always wanted to be.

THE END.

After reading the typescript, I am elated and confused, at once. Falling asleep in a bathtub after a tiring day at school? Really? Thoughts and emotions aswirl, I laugh at the dream, cringe at the use of language. Super-phlegmatically lethargic? Where did he even find such words, leave aside the horror of using such an expression? (The answer stares back at me from the bookshelf: the well-thumbed pages of *How to Build a Better Vocabulary* within its bright blue cover, tacked alongside its white sequel, *All about Words*, by Maxwell Nurnberg and Morris Rosenblum.) I note with satisfaction his attempts to proofread and correct the typescript and typos with a pen, but itch to confront him, to correct the errors that remain. You have identified the sunbird wrong, your punctuation is awry, and go easy on the adverbs and hyphens will you? Also, it is Anamalai, not Annamalai, I want to tell him. And the Nilgiris is a different hill range over 50 kilometres to the north.

But most of all, I am incredulous. Incredulous at the boy imagining himself as an ornithologist in the Anamalai, someone he has no assurance of becoming. How could he? My first field research on wildlife was a study on deer and antelope in Guindy National Park in Chennai. My Masters fieldwork took me to tropical rainforests in remote northeastern India, studying the effects of shifting cultivation on birds and primates in Dampa Tiger Reserve in Mizoram. Then, I had scouted widely for topics and sites for my doctoral research, before electing to work on rainforest birds in the southern Western Ghats, in Kalakad Mundanthurai Tiger Reserve in the extreme south and in the Anamalai Hills. Surely, the 14-year-old boy had no way of knowing that—after my doctoral degree in ornithology—I would remain to work in the Anamalai, would rediscover as a 41-year-old his words from over a quarter of a century ago.

I'm confused: what is this typescript on its yellowing paper saying? Is it prescient prophecy, plain fact, or fiction?

The words seem prophetic. I do live and work in the Anamalai Hills now, in a landscape where, on any day, we need to only step out of home or research station to be assured of seeing wildlife: great hornbills whooshing over the canopy, stately gaur moving through the plantations, creatures of all sorts from fireflies to frogs and earthworms to elephants, amid great trees festooned with orchids and ferns. A landscape where I can take that walk in the woods he writes about, hear the conversation of macaques and the sounds of elephants.

Yet, that is not what I do most of the time, not what I have become. Instead, we work as a team in a landscape where extensive plantations have historically replaced and now lie between forest patches, where land is managed not so much for conservation as for commodities and cash: tea and eucalyptus, coffee and cardamom. In land intensively used by people, we work to restore degraded rainforests by raising native plants in a nursery and planting them out in degraded sites and coaxing private landowners to protect the forest remnants. We work to reduce conflicts with wildlife like elephants and leopards, studying the ecology of these animals, informing local people of elephant movements to prevent unexpected encounters, helping planters and the Forest Department implement appropriate measures to reduce or avoid conflict, all to build a landscape of coexistence with wildlife. To keep our research, field station, and conservation efforts going, we raise grants to support our work, write proposals and reports, meet all sorts of people from tribals to tourists, make presentations to planters and policymakers, try to start dialogues and bridge gaps: this is what takes up much of our time in the field. This is not the story of a boy who forgets his worldly cares when placed in the forest of his dream. This is about what it means to care, deeply and all the time, for the world one is in, the real world. A world where the forest is only part of a landscape that also includes the human.

Could the child have known what he would become later in life? Clearly, when he wrote this, he already loved biology and birdwatching. He had been birdwatching since he was eleven. ('My passion for ornithology had still not left me,' he writes, at fourteen!) That summer, he had gone on a memorable trip with his family to Mudumalai Wildlife Sanctuary in the Nilgiris and to the farm at the foot of the forested Anamalai, which had been his great-grandfather's. The same farm that his grandfather, after laughing at the boy's offer to study agriculture and take to farming when he grew up, sold and distributed the proceeds of across more than a dozen heirs. The farm is inaccessible under new owners, but the forests could still be visited. Perhaps the boy had only projected a subject he liked onto a place he loved. Perhaps the dream had gone deep and dormant, working surreptitiously, like autosuggestion or astrology, towards an eventuality that seemed inevitable. Yet, that very year, the boy had also taken a course in journalism as training for a career as a writer. What of that? *Two roads diverged in a wood, and I—I took the one less travelled by*? No, he could never have known I would end up in the Anamalai Hills. And besides, these are words put down on paper, and words taken literally must mean what they say. I know the place he wrote about, in the forests around Sarkarpathi, for it is a place in my memory, too, and he has been there, but I, in all the years since, have not. This was no prophecy.

Perhaps it is fact, then. A dry, reasonable, factual narrative of a day at school, and then an account of coming home, falling asleep, dreaming his dream. But the dream, the tunnel: is it not a classic artifice to enter another reality? Anyhow, what he says cannot be true: the bathtub is a dead giveaway. He has never lived in a house with a bathtub, not to mention that, in perpetually water-starved Chennai, using so much water for a bath was unthinkable.

*Unthinkable.* Yet, here it is: thought, articulated, punched on paper in black on white. It is all made up. It is fiction. After all, does the boy not describe his euphoria as *quixotic*? A word conveying an imagined and fanciful idealism, a quest for the unattainable, made immortal by Cervantes' *Don Quixote*, a founding work of literary fiction? It is a concocted world in which the boy has placed himself, not me. I am not in it, I am not it. It is just a teenage boy, with his straight black hair falling over his unfurrowed brow, sitting at a typewriter, dreaming up a world out of his imagination.

It is then I recall that his first published story was a work of fiction, a short story that appeared in the *Indian Express*. The newspaper had paid three hundred rupees, and his mother had opened a bank account to deposit his first honest earnings. Then, he published a poem, even began work on a novel. Short story, poem, novel: I have no copies of those now, no trace of their existence except in memory. And no, I am not making all this up. This is not fiction. This is true.

What then is the dream and what the fiction? And who, ultimately, is he, and who am I? Shankar, the birdwatcher-scientist walking the woods, or Sridhar, spontaneous writer of fiction? Or do the two roads that diverged in a wood now converge, or connect by myriad streets, to create a scientist who is better able to say why he cares so much for all the life in the real world, and why others should care, because he may be a writer, too? Perhaps one is mistaken in thinking of the road taken in terms of origin and destination, as diverging, as separating past and future from the present when all it is, is a trail that turns into itself, a closed loop walked once where each point exists and connects eventually to every other. Who among us has not imagined or yearned for alternate lives that may have been? And yet, what if those alternate lives are only distractions, suitable for fictional worlds or for the life of imagination itself perhaps, but not for a life lived truly and well in the world we inhabit? The subtle seduction of imagined other-lives may be subsumed in a love for the life that is palpably real.

The only thing I end up discarding that day is a skeleton. Yes, really, a skeleton in the cupboard! Actually, a parcel of deer bones: lower jaws, skull, ribs, hip bones, femur, vertebrae, a couple of small antlers. These, collected during my field project on deer and antelope in Guindy, were meant to be handed over to a scientist who studied animal diets through isotopic analysis of bones. Clearly, I had never got around to doing that. My parents, steadfast vegetarians who could not have liked having old bones lying around, even if they were bleached white with no trace of flesh or putrefaction and wrapped in a parcel and placed deep in my shelf, had nevertheless tolerated this for over two decades. Long enough, I thought. The bones need to go. I shall keep the typescript with me and place everything else, for now, back in the shelf. I shall tell my mother: I need more time.

Weeks later, I find myself taking courses in writing: in creative nonfiction, fiction, and poetry, writing for newspapers and magazines. I try my hand again at fiction and poetry, at an occasional essay. I carry the typescript around with me and I still wonder what that boy was doing. I wonder if he was making up a world not because he wanted to live in his dream, but only because he was already *there. There,* in front of a blank sheet of paper open to the imprint of human imagination, *where he had always wanted to be.* And I wonder at how he dragged me into it.

# FIELD DAYS

## an ecological education

### Earth's Bounty

Bless you, earth:

field,
forest,
valley,
or hill,

you are only
as good
as the good young men
in each place.

—*Purananuru* 187, Auvaiyar,
Translated by A. K. Ramanujan, *Poems of Love and War*

Let the path become where I choose to walk, and not otherwise established.

—*Blue Horses*, Mary Oliver

# Six Seasons in the City

In the early light of dawn, I park my bicycle beside a tree near the forest check-gate and walk in shouldering a small backpack, a pair of binoculars slung on a strap around my neck, a spiral notepad and pen tucked into my shirt pocket. Shoes crunching on the rust-red gravel road, I enter the forest, a woodland with a luxuriant understorey of evergreen shrubs amid umbrella thorn and other trees whose branches arch overhead. Suddenly, the fresh morning air resounds with lusty bellows: a deep-throated series of six notes, part roar, part groan, punctuated by short pauses and ending on a softer, hoarse, almost exhausted note. Quietly, I step along the trail and find him a little ahead. His dark brown coat spotted with white, a stalwart pair of antlers on his head, he stretches his thickened neck and bellows again. A chital stag in rut. Dark glands glisten wetly before his eyes as the taut-bodied stag, his swollen testes dangling, walks to a nearby bush and thrashes it with his antlers. Minutes later, as if in answer to his vocal challenge, another stag bellows in the distance. In the lull that follows, I hear the chuckling conversation of white-browed bulbuls from the undergrowth, the loud proclamation of a grey francolin from the thorn forest ahead, and the distant horns and hum of traffic in the city.

Yes, the city. It is just a short walk from the streets, pollution, and bustle of the city into Guindy National Park, home to the dainty chital—the only deer species found here—and herds of handsome blackbuck—the only antelope. In the heart of the teeming metropolis of Chennai, the state capital of Tamil Nadu in south India, this small wilderness—with free-ranging herds of deer and antelope, with francolins and bulbuls—somehow survived the onslaught of urbanization that devoured the surrounding countryside. Sprawling over 270 hectares, the park contains a range of habitats, including tropical dry evergreen forest, thorn forest or scrubland, and open grassland or meadow, besides small ponds. A 5-minute walk takes you out of the congested traffic on Sardar Patel Road, from city noise and pollution to forest birdsong and greenery.

My experience in Guindy National Park began in 1983, when I first visited as an 11-year-old birdwatcher in the company of V. Santharam, a quietly cheerful and exceptional ornithologist with the Madras Naturalists' Society. Santharam, over years of watching and studying the birds of Guindy and other thorn forests around Chennai, has documented more than 150 bird species in the park. In an eventful morning's birding, we tallied over 40 species of birds, including several 'lifers,' species that I had never seen before in my life: chestnut-shouldered petronia, chestnut-tailed starling, and common woodshrike. I was hooked. I began to eagerly await each weekend and the possibility of visiting Guindy again.

Guindy, a small slice of the natural world in the city, was a source of wonder and captivating discoveries to me as a young naturalist. I spent many memorable hours peering at spittle bugs and lynx spiders under the leaves of shrubs, watching bonnet macaques and birds in the trees, tracking golden jackals and deer through the scrub, picking up feathers and bones, savouring fallen fruits and the aromas of wet earth and the fragrance of wild flowers, napping in the shade of banyan trees, and walking mile upon mile seeing, smelling, listening, feeling, and gaining a sense of place. But it was my field research on deer and antelope from mid-1990 to the summer of 1993 when I was in college that really drew me to the science of ecology, to understanding that in nature what we see and when we see it depends intimately on the cycle of the seasons.

## Early Summer

The antelopes, harassed unceasingly
By the terrific glare of sunshine,
With thirsty palates and parching tongues,
Can scarce espy the firmament
Dark like the powdered collyrium;
They think 'tis a sheet of water
In the midst of an alien wood,
And, bounding, lightly break from the earth.

—*Ritusamhara* or *The Pageant of the Seasons*,
Kalidasa (Canto I: 11)

In the month of May, walking farther into the woodland that morning, I encounter more chital: solitary animals or small herds of two to four individuals. The woodland, distributed as three patches of dry evergreen forest occupying about 60% of the park area, remains true to its name, dry yet green, casting welcome shade in the heat. Despite the recent pre-monsoon afternoon showers, fresh green grass is scarce in the scrubland that occupies a little over a third of the park area, and in the small patch of grassland.

The chital roam across all habitats in summer but crowd the woodland in greater numbers, browsing on leaves of woody vegetation and eating fallen fruit of banyan, *Albizia*, tamarind, and other trees. Even the blackbuck, despite their love of open spaces and loathing of dense forests, are sometimes forced to enter the woodland in the summer seeking browse and fallen fruit—something they almost never do at other times of the year.

Later, I find another adult stag with a few does and their young under a tamarind tree in which a troop of bonnet macaques is foraging. I watch the deer through my binoculars and scribble observations in my notepad. The stag sports a pair of hard antlers nearly three feet long, corrugated and russet brown at the base and tapering to bone-white tips. He follows the females, checking if they are receptive; when one urinates, he sniffs at her rear and holds his snout in a peculiar lip-curl display for half a minute. The females seem to show no interest in him. They browse on fallen leaves or nibble the pods of tamarind the monkeys drop. In the summer heat, the deer are at fever pitch in their breeding ardour. After a few minutes, the male manages to draw a female away from the rest of the herd, into the woodland.

Farther down the trail, a dark *mahua* tree in flower gives out a sweet, intoxicating fragrance. Sitting on its boughs, a male koel who has found his voice of the breeding season pours out his incessant cooing. The cuckoo's song is the soulful, shimmering sound of summer.

## Late Summer

Green, like fragments of jade,
The grass rises on tiptoe,
Stretching its blades to catch the raindrops;
And a mass of the blithe new foliage
Bursts from the kandali plants,
And indragopaka insects make a crimson riot;
With patches of green and purple and gold,
The good earth is decked with many coloured jewels ...
—*Ritusamhara* (Canto II: 5)

The sea breeze blows in from the Bay of Bengal, barely four kilometres to the east, wafting a welcome coolness over the forest in the still-hot afternoon. The skies are blue with a few scattered dark clouds, and the air carries delicate aromas of wet earth and small flowers. Now the thorn scrub does not have the grim, crackling dryness and heat of early summer; the first rains of the southwest monsoon, meagre as they are, have enlivened the land with green. On the soil, red velvet mites have surfaced with the rains and forage like sedate giants. The thorny *Randia* and *Carissa* bushes have flowered and the scattered palmyra palm trees tower over the savanna-like vegetation. As they sweep through the air, the sleek palm swifts' shrill chitter accompanies the arrhythmic clack of dry palmyra leaves clapping in the wind.

I walk due south on my line transect on my monthly survey of the herds. I note the herds, counting the individuals and measuring their distance from the transect line with a rangefinder: gathering data that I hope will help unravel the grouping behaviour and estimate population density of deer and antelope. The first herd, nearly sixty metres away, comprises four chital: a large hard-antlered male with a broken left antler, a female, a sub-adult female, and a large fawn. A bit further, at just over twenty-five metres, I note five females, a sub-adult female, a large fawn, and a 'spiker', a juvenile male with thin spike-like hard antlers. I spook two more chital herds, one right ahead of me on the transect, before I can get a full count. A lone blackbuck male eyes me warily before walking away. Finally, a little over twenty metres away, I record a chital female and her large fawn just before the transect ends, at the wall of the neighbouring campus of the Indian Institute of Technology.

The chital now fill the thorn scrub in numbers, cropping the flush of grasses and herbs, nibbling only occasionally on the leaves of shrubs and trees. Their density has dropped to a third in the woodland, and more than doubled in the scrubland. Larger herds form in the scrubland possibly because more forage is now available and more individuals can forage together without competition, or because greater numbers and the more open habitat allow herds to meet and mingle. Other compulsions, too, could be at work: an instinct of safety in numbers or better opportunities for behavioural interactions in these social animals.

As June gives way to July, past the peak rut where bigger stags have courted and mated with receptive does—some with weaned fawns—younger males become invigorated. Pairs of young males spar on the sidelines, their hard antlers locking and clashing, testing their strength against each other as if preparing for greater battles ahead. Many two- to three-year-old juvenile males, their sensitive, growing antlers still covered in a furry skin layer called velvet, avoid confrontations. In time, they, too, will come into their own.

## The Rains

Amidst the kovidara trees,
In whose waving boughs
And tender young foliage
Mixed with the bursting flower-buds,
The breeze skims merrily,
There is a soft whispering turbulence of leaves;
And there the delirious bees
Are greedily sucking the flowing honey;
Whose heart does not swell with joy
At the sight of these lovely trees?

—*Ritusamhara* (Canto III: 6)

Entering the park one afternoon late in August, I find few deer in the woodland even though I hear bellows in the distance. In a herd of three, a young stag freezes, neck bent down, staring at me, his eyes glowing a mysterious emerald green in the dense, dark undergrowth. In the scrubland and grassland, several herds graze quietly, including one with eight males, young and old, and a single adult female. Below a palmyra tree, a young male whose bony antlers are scarcely a foot long gnaws at the ripe fallen fruit, biting into the succulent, creamy-yellow pulp.

In August and September, as the antlers of many juvenile chital mature, the velvet peeling off or rubbed away to reveal bony, hard antlers, the southwest monsoon transforms the land. This is also when the larger males begin to cast off their antlers, and become more vulnerable. Juvenile males attaining hard antler and full breeding capability may now improve their chances of finding a late-breeding female. In this season, there's a lesser chance of being supplanted by larger males. Still, by the end of the season, many of the males, young and old, will abandon the company of females altogether. They will form all-male herds and roam together for much of the cool, wet season to follow.

Later in the afternoon, I take the track heading north, along the edge of the Kathan Kollai tank in which the rains have so far formed only two low puddles, each about ten metres across. Along the edge of this shallow water body, *jamun* trees are in fruit, their luscious purple-black berries sprinkling the branches and the ground below. I run into five people who have collected as many kilograms of the fruit heading out to sell them in the neighbourhood of Velachery just outside the Park. I pick up a few myself, rub the dust off on my shirt, and pop them into my mouth, enjoying their watery sweetness and the astringent tang. My mouth turns purple as I eat and spit out the seeds. Further down, on the track heading to the grassland ahead, I find the spoor of a jackal: dog-like paw prints in the sand and purplish scat packed with a cluster of *jamun* seeds.

At dusk, the drone of cicadas resounds from the trees. Darkness settles over the park. Walking back, I hear the rhythmic *chuck-chuck-chuck-chuckrrrr* refrain of an Indian nightjar and see the dim shape of the bird sitting on the darkening trail at the edge of the grassland.

## Cool Season, Cyclones

> The lakes, where the decorative lily is blooming,
> And the amorous kalahamsa blends its grace,
> The limpid waters and verdant weeds,
> Steal the hearts of men.
>
> —*Ritusamhara* (Canto IV: 9)

Dark, spent rain clouds twist away in the skies over the park. On the open grassland called Polo Field, the sun casts a warm glow over the evening through a gap in the clouds, like light through amber.

The skies bear little trace now of the depression over the Bay of Bengal that had intensified into a cyclonic storm and swept across the east coast of Tamil Nadu, lashing rain and wind into the land, marking these months of the cool season as the wettest in the year. In the city outside, flood waters recede in low-lying areas leaving dirty sidewalks and hard tarmac. Shrinking puddles, blue skies, a touch of coolness in the air, the renewed bustle of people on the street: the city breathes its relief at the cyclone's passing and settles into its routines. But in Guindy, in November, the rains have filled out the ponds and green turf carpets the grassland. In this season of whirling winds of the northeast monsoon, the blackbuck reach their yearly turmoil of rut.

On Polo Field, two blackbuck males, dark-pelaged, long horns twisted into five full spirals, occupy territories on either side of the meadow. One male half-squats over a dark circle of dung on his meadow, marking it with his pellets and his odours again. On the other territory, the male snorts and follows a female, jerking his head up at her in a peculiar display, circling around and trying to cull her out from the rest of the herd. The female looks up from her grazing, breaks away from the herd, eyes alert, ears flapping, tail twitching. He follows.

The blackbuck and large herds of chital have trimmed Polo Field down to a meadow of prostrate herbs and flowering grasses. As part of my fieldwork in the grassland and scrub I try to document the herbaceous plants, grasses, and sedges, annuals and perennials that now carpet the wet earth. I am astonished at their resurgence and diversity following the rains. The tiny *Justicia* flowers in mauve or white, the creeping *Commelina* herbs with deep blue flowers, the *Aristida* and *Chrysopogon* grasses waving their spiky spindles and branched inflorescences in the breeze, the delicate *Evolvulus* creeper with its blue and white flowers, the stiff little *Kyllinga* and *Cyperus* sedges topped with white-heads or prickly brown flower clusters: a world of intricate beauty lies at my feet. Plants proliferate around and in the ponds. In shallow water, the peculiar marshweed (*Limnophila heterophylla* in Latin, or a lake-lover of varied leaf) emerges from beneath an inch or two of water, its tiny leaves above and below water shaped differently, making the best of both worlds. Before the end of the season, with a little effort of my own and the help of botanists from the Madras Christian College at Tambaram, my list of the smaller plants in Guindy grows to over a hundred species.

The grassland forms the centre of activities for the rutting blackbuck during the northeast monsoon, although a few herds also use some of the more open areas in the adjoining scrubland. The males stay within their territories through the day as herds of females and young pass through or stay to forage and rest or play. A younger male, carrying fewer spirals on his horns and a coat not as black as the males on the grassland, has established a territory in a patch of open scrub nearby.

This evening, the two males of Polo Field do not interact, but I have seen them fight over their territories. Clashing and locking horns, I saw one twist and pin his rival's head down, pushing and kicking up a swirl of dust. The outcome of that tussle: an invisible line of control now runs across the small patch of grassland that the two males often patrol, walking close and parallel to each other. A measured walk, this, carrying meaning, heavy with intent, indicating territories that I can sense but not see, spaces that they tacitly acknowledge and respect. The line separating their territories extends from the southern edge of the grassland, near the road, to its northern edge, where it ends at a fence of chain-link and barbed wire, the fence of Raj Bhavan, separating Guindy National Park from the residence of the State Governor, patrolled by armed sentries with guns.

## Season of Evening Dew

> The breath of Winter comes
> Stealing into the lush greenery of fields of sugarcane,
> While yet the earth is spread
> With the well-grown paddy ...
>
> —*Ritusamhara* (Canto IV: 1)

It is early January. I walk into the forest, my feet crunching on the red gravel track. I pass under umbrella-thorn acacias lit gold by the morning sun, *nila palai* and wood apple trees in dark greens. Farther ahead, a large banyan forms a dome-like canopy over the trail. The winter song of a male koel—an explosive crescendo of notes—pummels the morning chorus of bird calls. The dense and inviting undergrowth of *Clausena* and *Glycosmis*—shrubs of the citrus family—draws me in. I decide to look for birds before I begin my day's observations on deer and the antelope.

In a little clearing in the forest, I see a sparrow-sized bird on the ground dressed in dark slate-grey, a flush of red in the underparts: black redstart. Another small bird, in brown with black-and-white markings on his breast, takes off from the ground with a soft *pink ... pink* call and perches on a low branch some distance away, eyeing me and swinging his tail from side to side: forest wagtail. On low branches festooned with the flaming, flamboyant blooms of the creeping lily skulks a pair of blue-faced malkohas, their dark, long-tailed bodies in counterpoint to their blue-ringed eyes. A soft *churr* guides me to a drab Asian brown flycatcher in the trees while a nervous ticking from the shrubs leads me to her colourful cousin, a blue-throated flycatcher. Both birds, like the redstart and the wagtail, are winter migrants that have fled cold northern winters to enjoy the warmer climate here in the city forest. Here, they will spend several months, in the company of year-round residents like the koel, malkoha, babblers, and many others before they return north with the summer.

After that short birding interlude, I crawl out of the undergrowth to the edge of the track and freeze. A little distance ahead, under the arch of the banyan's branches, stands a chital doe with a tiny fawn. Scarcely a month old, the fawn passes between the doe's forelegs with spry, almost playful steps and reaches for her teats at the back of her abdomen. His tail, curved under the doe, is a raised flag of white hair. Another young male emerges from behind the doe, perhaps her offspring from the previous year. A few inches shorter than the doe, his snout and body still have the lithe, graceful contours of the young. On his head bulge two stud-like buttons, the developing pedicels from which he will sprout his antlers.

The new year brings a crop of chital fawns to Guindy. The dry weather that begins now portends a season of promise, colour, and vitality. By mid-January, when the harvest festival of Pongal is celebrated across the state, with propitiation of the Sun God,

the weather has turned, with the skies clear and bright blue, and a drop in humidity augurs dry days ahead. The land carries an anticipation of spring, a presentiment of the hot summer to come.

## Season of Morning Dew

> The Spring has adorned the earth, in a trice,
> With the groves of palasha trees aflower
> Swinging in the breeze,
> Bowed with the load of blossom
> Resembling flaming fire,
> The earth looks like a newly-wed bride
> In lovely red attire.
>
> —*Ritusamhara* (Canto V: 19)

One February morning, I take the road through the scrubland towards the Polo Field meadow. In the scrub, the flame-of-the-forest trees have burst into an incandescence of orange flowers. At half past seven in the morning, the line of Indian ash trees drips copious dew onto me as I pass. In the grassland, swathes of *Chrysopogon* grasses hold delicate, branched flower heads, while *Aristida setacea* grasses with long, thin leaves spike upwards carrying bristly inflorescences. Now, the meadow—cropped by deer and antelope and the occasional broom grass harvester who managed to slip past the forest guards—begins to show a few patches of dry earth.

In the scrubland nearby graze several chital herds. There are does with small fawns and yearlings, juvenile chital males, their heads sporting bony spikes a few inches long, as well as adult males whose rounded branching antlers are still growing, covered in soft velvet. A couple of males in velvet, to avoid injury to their sensitive antlers, rise on their hind legs and box with their forelegs.

In the grassland, the territorial blackbuck males watch over a slowly moving herd of eighteen females and young. The females impregnated in the rut late last year will soon drop their fawns. Then, the males will strut around in their territories in a second, minor rut, once again looking for females that are in season.

Soon, the days will begin to stretch a little longer, the sun will climb a tad higher in the sky, the grasses and herbs will dry up and thin out, and the chital will again head into the dry evergreen forest and break up into smaller herds. By the end of the dry season, many adult stags will strip the velvet off and come into hard antler beating and rubbing them on bushes and saplings. In mid-April, when much of the country, including Tamil Nadu, celebrates the New Year, golden, pendent inflorescences burst forth from Indian laburnum trees. With the approach of summer, the air once again resounds with the bellows of rutting chital males.

Poring over my field notes a quarter of a century later, I am struck by how each visit to Guindy enriched me with field experiences. Observing chital and blackbuck; distinguishing birds; identifying countless plants; marvelling at red velvet mites, at strange water scorpions lurking in shallow ponds; watching dragonflies zipping about, or mating and laying eggs in Bogy pond; or just sitting on the banks of the Kathan Kollai tank in the shade of banyan trees—these were all wonderfully refreshing, intimately connected experiences that left a lasting impression.

In the technical and scientific publications of my research in Guindy I had grouped my data and observations into four seasons for analyses and description: the dry season at the beginning of the year, the summer, the southwest monsoon, and northeast monsoon. Each season was marked by distinctive changes in the vegetation, in the movements of chital and their annual cycles of antler formation, herding, and breeding, in the rut and rhythms of the blackbuck, in sounds and colours and smells.

Still, the four seasons were, in a sense, arbitrary markers of continuous time. One could have lumped the months into just two seasons: a dry season from the beginning of the year through the heat of summer in May, and a wet season encompassing the rest of the year and the two monsoons. One could, more justifiably, characterize the year into the traditional six seasons as poets and writers have long done. One could have wielded a sharper knife, made a finer cut, going by the events of every month, by the rhythms of flowering and fruiting of different trees and herbs, by the emergence of the red velvet mites or water plants. One could have defined the seasons by the arrival and departure of migrant birds, or the seasonal pulse of rut or fawns of deer and antelope. Or, watching a female dragonfly, flying in tandem with a male, repeatedly dipping the tip of her abdomen in the ephemeral water at the edge of Bogy pond, I could even imagine the seasonal clock ticking from the instant she laid her first egg.

In Chennai, people joke even now that there are only three seasons: hot, hotter, and hottest. Yet, even that distinction seems to have blurred gradually, especially among the growing middle and upper classes. Air coolers and air conditioners have proliferated in homes, offices, shops, and cars, providing an equable temperature and humidity year-round. Supermarkets and malls deliver vegetables and fruits from afar throughout the year. The harvest festivities of Pongal or Onam, the celebrations of Navarathri, Diwali, Christmas, or the Tamil New Year are no longer markers of renewal and thanksgiving, or propitiation and prayer, as much as dates that separate the holidays from the workweek, when we may break the monotony of the everyday. At other times, when rains flood the streets or disrupt schedules in the monsoon, or when power outages in the summer make people swelter and sweat, the season becomes mere annoyance. We ignore the natural cycle of the seasons more and more as we sacrifice the pulse of seasonal rhythms for the familiar comfort of aseasonal routines.

In Guindy, from chital and blackbuck, from plants and birds and smaller creatures, I learnt to discern and savour the seasons in nature. Here, one can see, year after year, habitats transform with the seasons in ways seamless and spontaneous. Here, all life cycles annually through death and decay, renewal and growth, showing what lives lived in rhythm with nature are like. Each season brings the expected and the familiar, but also novelty and serendipity. Each season plays its part in an endless symphony of ecological renewal, manifesting the eternal cycles of nature in a small corner of a bustling metropolis.

# Night Life in Chennai

The moon played hide-and-seek in the cloudy night sky; the stars winked like fireflies. Far, very far below, a tight group of about 1,500 music lovers sat awestruck as Ustad Zakir Hussain's fingers shimmered over his tabla. The rhythms merged with the melody of Pandit Shivkumar Sharma's santoor to create an other-worldly ambience in IIT Madras' open-air theatre. Spellbound by great music, no one heard the raucous squawking of a night heron flying high in the sky. Neither did they notice a grey owl flying overhead on hushed wings. Barely visible in the darkness, little bats fluttered silently over the listeners. In the floodlights illuminating the stage, to the tunes of the maestros, a moth danced. The night was alive.

As I stumbled through the darkness a bit later towards IIT lake, I was painfully aware of the limitations on my ability to see. Yet, all around me were creatures evolved by millions of years of natural selection to survive in their night-time environments. Among mammals, the group most spectacularly successful in exploiting this are bats. At the lake, dark and swift, a number of bats flew back and forth, skimming the water's surface. Millions of years before our ancestors came down from the trees, bats, like other early mammals, were forced to become active at night, for the world was dominated during the daytime by the great reptiles—the dinosaurs. Bats solved the problem of finding their way about in the dark, or even in the absence of light, with a solution that even the huge crowd of engineers, technologists, and others—entranced now by the sounds of music—would acknowledge as remarkable. To understand this, how some bats use not sight but sound to find their way around, one has to delve into the extraordinary sensory world of insect-eating bats.

It was around the time of the Second World War that Donald Griffin, the famous Harvard scientist, and his colleagues studied and described the remarkable system of navigation of bats. Griffin, one of the pioneers of the study of animal behaviour and cognition, also coined the term 'echolocation' to describe the phenomenon. Bats use high-pitched sounds, sometimes audible as clicks to the human ear, but mostly as ultrasound at frequencies above the range of human hearing, to fly around without bumping into things. The sound waves that echolocating bats emit bounce off objects in their path, producing echoes that they detect and use to navigate and find prey. In some bats, this ability is so precise that it can help them locate a tiny fruit-fly from about a metre away, allowing them to close-in while estimating the distance to the prey to a fraction of a millimetre.

Using echoes to pinpoint objects is something both engineers and bats have perfected. It was during the World Wars that engineers originally devised and applied systems that used sound waves to detect underwater objects such as submarines. Sonar, originally an acronym for sound navigation and ranging, became a key feature on oceanic vessels. Radar, again originally an acronym for radio detection and ranging, and now in widespread use, is also based on the same principle, except of course that it uses radio waves instead of sound waves. Yet, the World War engineers' technological triumph had been developed and perfected *over 50 million years ago* by bats.

Bats emit their high-pitched sounds in discrete pulses. The pulse and echo travel at the speed of sound, around 340 metres per second. In order for a bat to distinguish it from the rapidly arriving echo bouncing off nearby objects, each pulse has to be of a really short duration. The duration of an individual pulse or 'click' may thus be just a thousandth of a second. Bats merely cruising along may emit around 10 pulses per second, speeding up to around 200 pulses per second as they approach a detected object such as an insect. The echoes return faster as the bats get closer to their destination, so they emit shorter, more rapid pulses to avoid overlap between pulse and echo. This increases the precision as well as 'resolution' of their perception. To put this in perspective: we watch movies—essentially a rapid sequence of images—at 24 frames per second.

Using sound comes with a specific problem, one also familiar to engineers: attenuation. The sound waves of a bat's click radiate away in all directions in an ever-increasing sphere. As the surface area of the sphere is proportional to the square of the radius, the intensity of sound decreases proportional to square of the distance travelled. This also happens to the sound waves that impinge upon an insect and rebound, so the intensity of the echo reduces drastically—to about the fourth power of the distance—by the time it returns to the bat. A pretty hushed-up echo, indeed.

The physics of detecting faint echoes carries two consequences. First, the bats need to make really loud sounds and/or use large and sensitive receivers (ears). In radar, engineers have devised transmitters that produce high intensity radio waves and receivers that are very sensitive. Likewise, many insect-eating bats can be astonishingly loud and extremely sensitive. At 10 centimetres, bat sounds can be as loud as 140 decibels—louder than a thunderclap or a jet engine at take-off a hundred metres away. Fortunately, the audience at IIT Madras was undisturbed by the bats flying above, as bats echolocate using call frequencies above the range of human hearing. The human ear can hear sounds at frequencies between 0.02 kilohertz and 20 kilohertz, whereas bat echolocation spans a range from 11 kilohertz to 212 kilohertz. Most bats emit sounds through the mouth, but in many species, sound is transmitted from the larynx through a weird flap of skin with special resonating structures mounted on the bat's nose: the nose-leaf.

Second, bats need to avoid being deafened by their own noise. The bat's ears, typically large and sensitive, ingeniously switches off hearing when the pulse is emitted and re-opens it for the echo. The switch is a small muscle that contracts to pull apart the bones in the middle ear to momentarily disconnect the hearing circuit. This switching is done in perfect synchrony: a bat emitting 100 pulses a second, opens and shuts the hearing circuitry 100 times a second. Evolutionary biologist Richard Dawkins, who discusses these remarkable adaptations in his book *The Blind Watchmaker*, compares this ability to World War I aircraft gunners, whose machine guns were timed with gun synchronizers to shoot between the rotating propellers of their fighter planes. To shield sensitive radar antennae from the outgoing pulse, engineers use a similar send–receive switch, just as bats do.

The other challenges for any engineer toying with radar or sonar is to be able to detect not just the location and distance of an object but its size and movement. Once again, bats can give engineers a run for their money in the solutions they have evolved. To detect the velocity of moving prey, some bats, like the horseshoe bats belonging to the genus *Rhinolophus*, emit clicks of a constant frequency and make use of the Doppler effect. This is analogous to the radar guns used by the police to check the speed of moving vehicles. The clicks are long by bat standards but still under one-tenth of a second in duration. Bats moving towards a stationary object will have their sound clicks and returning echoes shifted upward in pitch (increased frequency). The bat's brain, which Dawkins likens to an 'onboard computer', processes the signals, interpreting the inherent Doppler shift to measure the distance. Some bats go one step further. They continuously alter the frequency of their outgoing call so that the returning Doppler-shifted echo always arrives at a particular frequency—the frequency at which the bat's ears are most sensitive.

To determine the distance of an object, such as an insect, bats use another ingenious technique. Most bats, except horseshoe and leaf-nosed bats, send out frequency modulated (FM) chirps—once again, analogous to the 'chirp-radar' used by the engineers. These bat calls are not constant frequency; instead, the frequency usually drops by an octave or more from the beginning to the end of the pulse. This helps the bat assess objects at different distances and can help separate prey from background clutter. Some bats, such as the horseshoe bats, use a combination of these methods, emitting constant frequency clicks with descending or ascending terminal FM squeaks. The constant frequency pulse helps to detect and classify insect prey (even sensing the tiny flicker in echoes caused by insect wing beats), while the broadband frequency sweep helps localize the target.

Some bats even catch fish, but as ultrasonic calls are almost fully damped, so to speak, by the water, these bats detect the fish by the ripples they make on the water's surface. The extraordinary adaptations of bats that use echolocation do not end with this and are indeed far more intricate and astonishing. More insights have emerged and continue to emerge from research that bridges the sciences of animal physiology, neurobiology, animal behaviour, and ecology. The science is clear: it is time to reconsider the old idiom *as blind as a bat*.

Back at the IIT lake, as I watched the bats skimming the water, I couldn't quite make out what they were doing. Perhaps they were just drinking water, using the echolocation signature and reflectivity of the smooth surface to recognize the water body, or just skimming insects off the surface. Tiny pipistrelles, hardly three inches long, fluttered like moths in the air. A frequent sight in Chennai, this small insect-eating bat sometimes hunts for insects and roosts in houses and dilapidated buildings.

Given the sensory world that they inhabit, insect-eating bats have large ears, prominent nose-leaves in many species (such as the mouse-tailed bats), while their eyes tend to be small and insignificant. Most bats are skilled and strong fliers. Their wings are extensions of translucent skin stretched between the extended digits of their forearm and often between their thighs and tail as well.

Insect-eating bats are not the only bats that one can see in Chennai city.

At dusk, one only has to look up to see the dark shapes of large bats flying sedately across the skies. These are fruit bats called flying foxes, the largest of the Indian bats, with a wingspan of nearly 5 feet. One night, I tried to bring a powerful torch to bear on a flying fox winging past 20 to 30 metres overhead. Each time I did that, the bat turned away from the beam. Even as the city lit up with artificial lights, he turned away into a more comfortable darkness. Clearly, the bat sensed the light as a visual distraction or disturbance, so I never repeated the trial after that. Compared to human eyes, the eyes of the flying foxes are said to be ten times better at seeing in the dark. Fruit bats also usually have an elongated muzzle and a good sense of smell that helps them find ripe fruit and flowers with nectar. By day, large numbers of flying foxes often roost together in large trees, squabbling noisily. They can also see well in daytime, but do not venture out, preferring to roost, suspended upside-down on the branches of trees. As dusk falls, they set off on wide forays to fruiting trees, forests, and groves in quest of ripe fruit. They chew the pulp, drink the juices, spitting out the inedible parts.

Another smaller fruit bat found here and in the neighbouring Guindy National Park is the short-nosed fruit bat. Fruit bats are unfortunately much-maligned species. They have been categorized as 'vermin' even in the Wildlife Protection Act, ostensibly because of the damage they cause to orchards. Yet, in forest ecosystems, they are valuable players in a key process: the dispersal of seeds of trees and other plants. While many seeds are scattered widely and are not obviously noticed as the bats spit or defecate after eating fruit on their forays, one may find small piles or middens of seeds under favourite feeding spots—perches to which bats carry fruit to consume at leisure. Many insect-eating and fruit-eating bats also visit flowers for nectar, during which time they help in cross-pollination of the plants. Across India every day, flying foxes and other bats must be helping pollinate flowers and disperse seeds in the millions over huge swathes of land.

Other bats, too, have had their share of bad press. Vampire bats, despised and feared by humans for centuries, have spawned many Dracula-type tales. This is despite the fact that true vampire bats rarely ever attack humans, feeding primarily on the blood of large mammals such as horses and cattle, locating them by smell, while some feed on the blood of birds. Three species exist, found only in Central and South America. These vampire bats slice a neat cut almost painlessly with razor-sharp teeth and lap up the blood. Their saliva contains anticoagulants to keep the blood flowing. Yet, even these maligned bats show tenderness to their young and altruism towards their companions: a bat may regurgitate a blood meal to feed another bat that is hungry or starving. India has no true vampire bats; but there are the so-called false vampire bats, such as the Indian false vampire, also known to occur in the vicinity of Chennai. This species is an insect-eating bat, but it is also reported to feed on rats, small birds, and even other bats.

Insect-eating bats were earlier classified as one group called the microchiroptera or micro-bats, separate from the fruit-eating bats, called megachiropterans or megabats. Recent research, including molecular and genetic studies, has recast the bats of the world into two large suborders called the Yinpterochiroptera and Yangochiroptera, which for brevity's sake, we can call yin bats and yang bats. Yin bats include fruit bats and insect-eating bats such as horseshoe bats, Old World leaf-nosed bats, and false vampire bats. Yang bats include insect-eating bats such as evening bats, New World leaf-nosed bats, and several other families.

Bats have been around for millions of years before human beings arrived on the scene. Today, bats, after rodents, are the most diverse, numerous, and widely distributed of mammals. In many countries, including India, bats are being driven to extinction because of the loss of habitat and food trees; destruction of roosting areas like caves, old buildings, and large trees; and deaths from pesticide poisoning, hunting, wind turbines, and transmission power lines. Their plight is perhaps not as dire in India as elsewhere and we still share the land with a multitude of these fascinating creatures.

The raucous and familiar *cheeak-cheeak-chierr-chierr* calls announced the presence of a different nocturnal predator. Perched on a tree by the road, a spotted owlet watched me as I approached. Owls have the uncanny knack of spotting you before you spot them. They then freeze and give you a stolid stare. As I walked around, the squat, myna-sized owlet turned its head to follow me. The head turned nearly 180 degrees and then flipped back swiftly the other way to look at me over the other shoulder. This is perhaps the commonest owl on the campus and surrounds. One can see pairs even during the day, resting in some leafy boughs of a large tree, emerging at dusk with their loud chatter to announce the beginning of a busy night of hunting.

Bats are not the only flying predators out at night. Silent nocturnal hunters, owls are also up and about. In IIT and other parts of Chennai, even over the busy Anna Salai or Mount Road, one occasionally sees the rather peculiar barn owl. Larger than the owlet, this is a bird dressed in greys, browns, and whites, with a rounded head and a heart-shaped disc outlining a face that features a powerful hooked beak and large prominent eyes. Like humans, they have good binocular vision, enabling better depth perception and accurate location of prey. Their large eyes function even in low-light conditions—many owls can see in a hundred times less illumination than that needed for human vision. Owl wings are rounded, relatively large for their body size, with soft-edged feathers, allowing for silent flight. Adept at catching prey off-guard, they swoop down on their quarry on silent wings, seize it with powerful talons, and rip into and eat the flesh using their hooked beaks.

The barn owl may also use sound to locate prey. At the gentle rustle of a rat scurrying on dry leaves, the owl can strike in near-total darkness. Our human detection of the direction of sounds is based on the split-second difference in the arrival times of the same sound at our two ears, but barn owls have gone a step further. Their ears are not symmetrically placed on either side; the right ear is directed slightly upwards and the left ear slightly downward. With this, the owls can spatially locate sounds not only on a two-dimensional horizontal plane, but also in a vertical plane, giving it a three-dimensional sense of location, to an accuracy of one degree.

Two other owl species occur in the IIT campus. The handsome Indian scops-owl, a small owl with triangular ear-tufts, and the slightly larger and mysterious brown hawk-owl, whose soft and ghostly *oo-uk, oo-uk, oo-uk* can be heard on a quiet night.

Owls, too, are an important part of the city and its remnant forests and gardens. They are predators that help keep populations of rats and mice under check, including in agricultural fields. Yet, owls and other birds of prey are often slandered as poultry thieves and shot or poisoned all over India. They suffer from the loss of their roost and nest sites, or fall victim to accumulation of persistent pesticides in their body tissues. Rampant quarrying for stones to feed the burgeoning construction industry in Indian cities has destroyed many hills and cliffs that earlier held nests of Indian eagle owls in their rocky ledges.

Walking at dusk on the less-frequented roads and trails at IIT and in the neighbouring Guindy National Park, I would often encounter a drab, brown bird squatting on the path. More often, I would hear the bird first. The call—a repetitive *chuck-chuck-chuck-chuckrrrr*—announces the presence of the Indian nightjar. Nightjars are elusive birds, well camouflaged on the ground in their mottled brown plumages. Crepuscular and nocturnal, they spend the day resting or roosting well hidden among trees and shrubs or on a quiet spot on the ground. At dusk, one bird would alight on the road and begin calling. Soon, another would answer from some distance away. Often, one of the 'duetting' pair would fly to join the other for a busy night, hawking insects in the air.

Although dark, the night life is not without its lights. Often, I have encountered the stars of the insect world, tiny fireflies twinkling in the air against the dark undergrowth. It is pity that this subtle beauty of the fireflies is lost in the blaze of lights and concrete jungle in the surrounding city. Reflected lights would greet me when I walked around with a torch. A thousand tiny pin-pricks in the leaf litter on the ground, in the shrubs and trees: the eyes of spiders beaming back at me. The eyes of civets also reflect, but I never saw one in IIT by night. In Guindy, on a couple of early mornings I have seen the small Indian civet, a species with a spotted body and a prominent tail marked with white and black rings. The common palm civet is another species found here, a grey-black animal looking rather melancholic and resting by day in palmyra and other trees and in the nooks and crannies of buildings. By dawn or dusk, occasionally at night, one sometimes sees Indian hares, drab grey-brown mammals with a black patch between and behind their large ears.

Night is also the time when many other animals become active, like frogs and toads, crickets and myriad insects, and venomous creatures such as some snakes, scorpions, and spiders. The snakes I most often encountered while roaming the IIT campus and in Guindy were usually non-venomous ones, such as the familiar rat snake and the swift and sleek bronze-backed tree snake, seen mostly by day. I would often wander around through the forest and scrub vegetation in my slippers or even barefoot and never had an unpleasant encounter with anything except thorns. Yet, walking into a snake by night can be dangerous and one should be wary of the saw-scaled viper, a nocturnal snake less than a foot in length, whose venom is haemotoxic. During the day, groups of saw-scaled vipers may huddle in palmyra tree saplings, in the niches where the broad leaf stalks attach to the stem.

Even in the city, the life of the night abides for all—whether trained zoologist or engineer, curious naturalist or citizen—to discover. As a student of zoology at Loyola College, I felt quite distanced from my education when roaming in the woods of IIT and Guindy. Zoology was taught in classrooms like a dead subject to be dissected in the laboratory rather than as an exposition of living nature that we are all part of. In IIT itself, the behemoth of technology and the activities of campus life rolled on, like engines left running, largely oblivious to the other life and lives around, the institute spewing hundreds of students out every year, a majority leaving not just the campus, but the country, to work and live in distant places. India churned out its engineers, as it still does, with little effort to integrate environmental awareness or ecological understanding into their education. With every passing year, Aldo Leopold's words in *A Sand County Almanac*, 'Education, I fear, is learning to see one thing by going blind to another,' seemed to ring truer.

Even knowing the species that live around us, one often has to cultivate a greater sensitivity. As Henry Beston wrote in a famous passage in his book, *The Outermost House*:

> We need another and a wiser and perhaps a more mystical concept of animals. Remote from universal nature, and living by complicated artifice, man in civilization surveys the creature through the glass of his knowledge and sees thereby a feather magnified and the whole image in distortion. We patronize them for their incompleteness, for their tragic fate of having taken form so far below ourselves. And therein we err, and greatly err. For the animal shall not be measured by man. In a world older and more complete than ours they move finished and complete, gifted with extensions of the senses we have lost or never attained, living by voices we shall never hear. They are not brethren, they are not underlings; they are other nations, caught with ourselves in the net of life and time, fellow prisoners of the splendour and travail of the earth.

Unseen, unheard, the living beings around us occupy sensory worlds denied to us in our limitations. And yet they can be experienced, even understood, by cultivating perceptions open to new discoveries, imagination, and wonder. If one gains such a perception, perhaps the nights may not be so dark after all.

# Lone Palm Tree, Sir!

As 2008 drew to a close, he passed on from this world, almost unnoticed, unappreciated even. Not that he looked for attention. He looked for other things in his long and self-made life. Right until the end, there were things that would light up his eyes—a reminiscence of hours spent in the wilderness in years past, his younger biking days and his Calcutta, tinkering with binoculars and radio equipment, a good book, a new stock of interesting tobacco for his pipe, getting together with friends for a chat, and, of course, a good joke, the dirtier the better.

The name given him was R.K.G. Menon, but that wasn't what we called him. He had a nickname of long standing—60 years, no less—emerging from the hallowed corridors of Madras Christian College: Cutlet. He was always, to any of us who knew him, just 'Cutlet'.

Imagine a rugged man turning into his fifties carrying out, during 1977–9, a full-fledged field study of the behaviour of blackbuck at Guindy National Park and Point Calimere, initiating systematic waterbird counts in Vedanthangal, carrying out and publishing in 1982 what were perhaps the first population estimates for an ungulate in India using line transect techniques—and all of this years ahead of any similar effort by other Indian, university-trained and funded researchers and field biologists. Imagine a man without a formal college degree or training or affiliation, who yet kept pace with the advances in scientific thinking in animal behaviour and ethology and could not only discuss this with clarity but also apply it in his own work. Cutlet was this and more.

I first met Cutlet during meetings or field trips of the Madras Naturalists' Society (MNS), an organization he helped found. There was little close interaction of the sort that came later; in the initial days I was merely learning the ropes of basic birdwatching, interested in just getting outdoors, excited with every new species I saw, and had time for little else. Even then, at the Adyar estuary and other places, I remember him teaching me to use my binoculars properly, and telling me to take detailed field notes, to count the number of birds and not stop with just identification, and to observe their behaviour. 'Write it down. If you think it's all in your memory, it is not worth it. It's just *kaka-pee* [crow-shit],' he'd say—or something similar—and with more choice adjectives that I, unfortunately, cannot repeat here. All the while, clouds of smoke would billow from his pipe.

It was almost exactly twenty years ago, when he had crossed 60 years of age and I was dawdling through my late teens, that I got to watch him in the field. We were both part of a small group of nature enthusiasts from MNS trekking to Konalar in the Anamalai Hills. Although he kept company with us on the trek and in the evenings, he would take off on his own through the grassland during the day to sit quietly somewhere observing tahr or langur or whatever else caught his attention that day. He would not brook crowds, noisy or otherwise, even of nature enthusiasts, that came to see wildlife but did not observe—and he'd make no bones about it. One day, we came across a dead sambar. Cutlet observed the broken neck and patiently tracked the signs around, showing us signs and scats, inferring that this was a tiger kill. On that trip, I learnt from him some of the hallmarks of fieldwork, about good backpacks and footwear, field clothing and sleeping bags, the proper use of binoculars, the drawbacks of cameras, about silence and observation, not to mention a number of hilarious jokes, songs, and limericks.

Cutlet was a well-read man with a scientific temper, which immediately distinguished him from many other naturalists around him. I do not know of him ever missing a chance to immediately borrow and read any interesting book, whether it was field research or a serious scientific text or monograph on animal behaviour, ethology, and evolution. The list of books and authors that I was introduced to and read thanks to him is large, indeed. *The Mountain Gorilla* by George Schaller turned me towards field research in wildlife. Cutlet did not just tell me to read it—he helped locate what was perhaps the only accessible copy in Chennai, from the shelves of the library of IIT, where the book lay almost unnoticed. Not having the means himself to purchase many books or build a private collection, Cutlet was heavily dependent on libraries and friends for access to books or journals. He pointed me to the Connemara library for *Gorillas in the Mist* by Dian Fossey, or old volumes of the *Journal of the Bombay Natural History Society*. To R. Selvakumar's house to request copies of other books by Schaller. To the British Council Library for Niko Tinbergen's *The Study of Instinct*. And so on.

The authors and books I got to read and discuss threadbare with him in his one-room rented house in Gandhi Nagar, Adyar, are a revealing list, when I think of them now. He'd read all of them, and if I managed to get a copy, he often read them a second time. In ethology, books by Niko Tinbergen (*The Study of Instinct*, extracts from *The Herring Gull's World*) and Konrad Lorenz (*King Solomon's Ring, On Aggression*) topped the list. From these, we would chat about Tinbergen's four questions, and other aspects of animal behaviour—interpreting super-normal stimuli and intention movements, displacement activities and imprinting. More textbook-like among the books were McFarland's *Animal Behaviour* and Dimond's *The Social Behaviour of Animals*. Among field studies, we would discuss classics like Fraser Darling's *A Herd of Red Deer* and David Lack's *The Life of the Robin* and a whole host of more recent books from field research. George Schaller on lions, gorillas, deer, and tiger would recur. Hans Kruuk on hyenas, Douglas-Hamilton, Cynthia Moss, and Joyce Poole on African elephants, Clutton-Brock on red deer and primates, David Mech on wolves, and, of course, out of his special interest in blackbuck, Fritz Walther and Elizabeth Cary Mungall on gazelles and antelopes. It was through Cutlet that I first heard of many of these books, which I read later, thanks to the libraries at the Centre for Ecological Sciences (CES) and the Indian Institute of Science (IISc), and through the help of Raman Sukumar.

Cutlet was also deeply interested in the rapidly growing field of evolution and sociobiology. He'd read and could hold forth on E.O. Wilson's *Sociobiology*, Richard Dawkins' *The Selfish Gene*, and a slew of other books and ideas from the 1970s through the 1990s that were among the most interesting. I remember wading through arguments over *The Blind Watchmaker* and *The Extended Phenotype* with him. He appreciated and was serious and critical about ideas considered rather divergent at the time, such as Zahavi's concept of signal selection and the handicap principle and Wynne-Edwards's theory of group selection. In all of this, he would try to connect the concepts to his own observations of blackbuck and other animals: How the blackbuck pelage and behaviour linked to signal selection. How its territoriality could be understood in relation to ideas spanning Fraser Darling and Lack and Robert Ardrey to Walther and Mungall.

Books like Anthony R.E. Sinclair's *The African Buffalo* and George Schaller's *The Deer and the Tiger* linked behaviour and ecology. Cutlet was not too enthusiastic about the field of ecology per se and somewhat de-emphasized looking at plants. Still, the field of behavioural ecology interested him. When I got a copy of a new edition of the classic Krebs and Davies textbook on behavioural ecology, he read it and tried to see it in the context of blackbuck behavioural ecology. We'd repeatedly discuss the decline of blackbuck population in Guindy National Park: Cutlet saw how reduced numbers had profoundly changed the social behaviour and reduced interactions among males. He saw territorial and social interactions as key in stimulating reproduction, and believed that the population decline was an example of an Allee effect at work. Trying to bring to the attention of the Wildlife Warden pertinent aspects related to conservation of this blackbuck population, Cutlet explained the possibility of such an effect in simple terms in a letter written in January 1993—how higher blackbuck population density influenced better breeding success, or conversely, when blackbuck numbers dropped, how their breeding behaviour and success would break down. Once again, this was an idea ahead of its time or our own data, and which I found being discussed in leading journals only years later.

In retrospect, what made these bouts of reading and discussions a great learning experience for me and fascinating for Cutlet, was perhaps the fact that neither of us had anything to lose or anything material to gain from it. It was pure curiosity and personal interest. Cutlet was far removed from any academic or peer pressures to perform cutting-edge research, publish papers, or proclaim his scientific interest or ability. The bureaucracy and corridors of academia, that can stultify as often as it can stimulate, were not for him. He had no job on the line, no tenure to uphold, no defining seminar or workshop to commit to, no funding priority to meet, no deadline-driven reports to prepare (barring a few that he wrote for the Forest Department on management issues). My college coursework (BSc Zoology) was as archaic and lifeless as a beat-up tin can and the dead specimens we dissected in our labs. What I dabbled with in ecology or ethology and the books I read were totally unrelated to exams and grades and performance in courses. And so the reading and the discussions seemed to work, and they seemed worth it.

Although he had no formal training in quantitative aspects of the science, Cutlet still believed in repeated observation and quantification using proper sampling techniques. Years before I was formally (and in a more text-book fashion) introduced to behavioural sampling techniques, I got a thorough grounding in the basic methods from Cutlet. Out of his sundry collection of reprints, he yanked out a well-used photocopy of a paper that still remains a classic in the field: Jeanne Altmann's 1974 work on sampling methods for the observational study of behaviour, which has seen upwards of 6,000 citations till date, some of them Cutlet's. Cutlet spoke of the benefits of different kinds of sampling for different aspects of his study of blackbuck behaviour, and the terms and ideas slowly sank in: ad lib sampling, focal animal sampling (his favoured method, especially on identified individuals), scans, and other methods. He insisted that I make a copy of the paper and read it; we would march off to observe the behaviour of chital and blackbuck at Guindy. Cutlet taught me how to make an ethogram, identify and name individuals, code behavioural data, and watch animals unobtrusively. When

I thought I would start a study on chital behaviour in Guindy to complement his work on blackbuck, he gave me a copy of a 1981 paper by Shingo Miura on social behaviour of chital in Guindy that helped me get started. Cutlet would similarly urge other MNS members and students to add value to their field trips by doing systematic counts and observations. The number of younger people he helped in the field of wildlife studies is not a small one.

Cutlet also taught me the basics of field work, by example and demonstration rather than lecture. Besides behavioural observations, he trained me in the basic line transect method, which involved walking along straight lines through the forest and counting animals and measuring distances to animals on either side. Cutlet was perhaps the first person to apply line transect techniques to estimate population density for ungulates in India, publishing a paper in 1982 in the Indian journal *Cheetal* with estimates of chital populations. His work was based on one of the early publications, from 1970, that developed this survey method, by Anderson and Pospahala. Cutlet had also approached a statistics professor at Madras University to understand the method and then applied it in his work. By the time I began my work, the methods had developed further and a computer software called TRANSECT was available and I could easily learn how to use it with from Sukumar and others at CES.

Still, I had to learn the ropes in the field. Cutlet had an excellent liquid-filled magnetic compass (he always appreciated good equipment, particularly binoculars and telescopes, and would repair and maintain them in good condition himself) and he taught me to use it to walk the transect and maintain course through the forest. Find and hold to the bearing, use the mirror, and sight along the viewing slit at a distant tree or landmark and then march towards that. 'When an army marches through the desert, the guy holding the compass would have to direct the others. He needs to find a lone palm tree that can be a reference to navigate. And as he marches, he'd have to call out "Lone palm tree, Sir!" periodically,' Cutlet said, half in jest. As the compass-bearer while walking transects with Cutlet in Guindy and the nearby IIT campus, I would then often choose a palmyra as marker and say: 'Lone palm tree, Sir!'

Still, Cutlet was highly self-deprecating. He would reiterate his lack of formal qualifications and scientific training and tell me that he was no good and that if I wanted to really learn the ropes or make a mark in this field I should go see others, the real scientists, the professors. As I made a faltering start at my own field project on chital and blackbuck in Guindy National Park with his help, Cutlet repeatedly urged me to go to Sukumar at IISc for guidance: 'He is the elephant man who knows stuff about populations and ecology.' He also pointed me to Ajith Kumar, another person he held in great professional regard and personal affection: 'Ajith is a Cambridge man who's worked with David Chivers, go talk to him.'

Cutlet, the man and the ethologist-naturalist, was largely overlooked by most people who knew him. He was a no-nonsense man and would get rather irritated by those who chose to remain ignorant of science and ideas, who merely went for nature trips to picnic outdoors but still loudly spouted an entrenched opinion about why animals did this or that. He would not mince words when talking to them and, in his earlier years, would not baulk at using the most colourful language either. As expected, this put some people off. This, coupled with his lack of formal qualifications, his self-deprecating comments, and his solitary life, appeared to provide adequate reason to those who chose to turn away from him. And there were those who perhaps thought he was a mere curiosity, a loner better left to his pipe and his eccentricities. Although Cutlet was somewhat chauvinistic at times and could come out as strongly opinionated, he could and would be swayed by a well-substantiated and logical argument. Kavita Isvaran, now a leading scientist who has herself carried out detailed studies of blackbuck behaviour, speaks about how when she first met Cutlet to discuss the phenomenon of lekking in blackbuck he was very sceptical, almost dismissive. Through the course of a thorough discussion he eventually came round to recognize that his understanding, restricted as it was largely to one population, needed to be expanded to accommodate the findings of newer research.

Yet, it was his own fieldwork that really defined Cutlet. He made over a 1,000 hours of focal animal observations on blackbuck (often working from dawn to dusk in the field) and analysed and worked on several drafts and manuscripts on blackbuck behaviour. He carried out fortnightly water bird counts at Vedanthangal in 1981–2 using a block count technique from standard locations, a method that others from MNS were able to replicate in 1991 to compare with his data. When a collaborative opportunity arose (with very meagre but vital funds) to monitor chital antler cycles, Cutlet would pedal off on his bicycle to Guindy and IIT and walk all over to conduct a survey for up to 15 days every month for two years, meticulously classifying individuals by antler size, stage, and condition.

Cutlet was from a well-off family, and once the proud owner of a 1,000cc V-twin Vincent HRD Black Shadow (one of the fastest motorcycles of that period) among other bikes, though he lived his final years in a one-room house, getting around on a moped or a bicycle, but still remarkably content. He worked without funds and in his spare time from the various jobs—editing wildlife brochures and booklets, working with the Snake Park, short stints here and there—that he took on to make ends meet. He had no formal support for statistical analysis or preparation of results and graphics. He extracted numbers from his notes and punched them into a trusty calculator to calculate quantitative measures describing blackbuck behaviour: rates of aggression, time spent in various activities, and so on. He made charts and territory maps, drawing them with ruler and pencil on graph paper. He wrote drafts of manuscripts in flowing longhand on foolscap sheets, usually with an excellent fountain pen (Parker was a favourite brand, with Chelpark ink, as was Sheaffer). His English was old-style and excellent, but he would re-read and edit, and if major reorganization was required he would rewrite by hand. He often copied these by hand when he wanted to send them for comments. When he deemed it worthy of submitting for publication or for comments from a scientific colleague, he would march off to a nearby commercial typist and get it typed, proofed (especially to correct the typist's glaring errors of all biological and scientific terms, not to mention having a good laugh every time 'agonistic behaviour' was typed as 'agnostic behaviour'), and then typed again, so the few errors that crept into the published versions should perhaps not be laid at his doorstep.

Only a small part of Cutlet's studies has ever been published. I refer not only to his research on blackbuck, but his work on antler cycles that was meant as a collaborative study. While Cutlet wrote drafts on various aspects of blackbuck behaviour based on his field study, he'd laid much effort into analysing and writing about agonistic and territorial behaviour. He worked on detailed manuscripts on these aspects (the originals of which are unfortunately not available) and sent them to Dr Elizabeth Cary Mungall, the leading blackbuck researcher at the time, for comments and feedback. She responded with detailed comments on the text, tables, and figures exhorting him to publish it as it '... will be an interesting contribution to the literature.' In a letter dated 4 May 1983, she writes, 'You are doing good work and all of us who share your interest in blackbuck and their relatives thank you for your efforts in bringing your results to publication so that the rest of us can learn about your results also.'

Some of the drafts of his writings are available at blackbuck.wordpress.com for reference by biologists, naturalists, those interested in animal behaviour, and anyone else who would like to see and understand the fascinating world of blackbuck and other species through Cutlet's eye. These and his published writings provide a glimpse of the earnestness and range of interests of the man who wrote as R.K.G. Menon, but was still just Cutlet to everyone. Besides his own work on blackbuck and chital, he worked on scientific papers with G.U. Kurup (on the behaviour of blackbuck during a solar eclipse), with A. Rajaram on the microscopic study of hairs of Indian mammals, with V. Santharam on waterbird populations at Vedanthangal, and he guided and co-authored work with me and R. Sukumar on the decline of blackbuck and ecology of chital and blackbuck in Guindy. In more general articles, he wrote about crows and dogs, sambar and tiger, and of course, blackbuck. There are brief articles about crop-raiding elephants and man-eating leopard (he visited Suligiri, where a man-eating leopard was shot by government diktat). He wrote of rollers and lapwings, of cannibalism and protean behaviour, and of days spent in the jungles of his memories.

Cutlet had a number of other friends who he was fond of and had often had a rollicking good time with. My association with Cutlet largely related to our shared interest in animal behaviour and Guindy, and I know little about his other friends, his family, or his life beyond the blackbuck or prior to the 1990s. Still, among those naturalists and nature enthusiasts I knew, Cutlet stood apart for his efforts and his enthusiasm concerning animal behaviour. As Mungall wrote in her letter of 1983, 'You mention that you have no support from any group and yet you list yourself as a naturalist of the Range Rover Foundation, Adyar, Madras. Is this a volunteer group? Is it very active in wildlife conservation? If all its members are like you, it certainly is a wonderful organization for India.'

A couple of days before he died, he called his close friends, went out with them, had a meal at their home. He was happy, but in his talk his friends detected a poignant tone. On the morning of 26 December 2008, while walking back home after a regular meal at his usual restaurant, he collapsed and passed away on the streets of Chennai. He was 80 years old.

I have pondered over what can be a fitting memory of this remarkable man who sought no recognition or acclaim and always stayed out of the limelight. Perhaps a permanent record of his contributions, as we have tried to do in the website dedicated to him. Or if there is someone watching (and watching over) the blackbuck of Guindy that he loved so much, that would be apt. Perhaps by carrying on the bird counts that he initiated at Vedanthangal, the water bird populations would mark his memory in the trends of their numbers. Perhaps some effort at sustaining this locally through the MNS. Or perhaps, in a much broader sense, the very continuation of a free spirit of enquiry and passion for ethology that marked his life would be sufficient. In the final reckoning, Cutlet lived alone and cut his own swathe through this life. He was, in a very real way, like the lone palm tree he spoke of. A lone palm tree, serving as a benchmark in the wilderness to help us find direction in our own lives.

# The Tropicbirds of Memory

The arrival of the postman at our doorstep, just before noon, is always a welcome event; more so when he brings something interesting: a book or a magazine, or even the rare letter. Today, he's brought the latest issue of *Indian Birds*, volume 8 number 5 of our subscription to this excellent journal, which we've had in our little library from the very first issue. This latest one promises interesting reading over a quiet weekend indoors, tucked away from melancholy mists and monsoon rains.

Opening the issue, I'm delighted to see the first paper titled 'Notes on Indian rarities—1: Seabirds,' by three leading Indian ornithologists, J. Praveen, Rajah Jayapal, and Aasheesh Pittie. It is the first part of a very welcome series on Indian birds, a critical review of all bird species reported from India, especially rare birds with few records, as a prelude to publishing a systematic checklist of Indian birds. Their task is not an easy one, to sift through reports and publications, examine records and evidence, verify details and locations mentioned, to arrive at a reliable and complete list of India's birds. In their first paragraph, they highlight several difficulties and lacunae that are spot-on. I like their observation on species that have made their way into checklists without adequate basis:

> An unfortunate fallout is that several contentious species, with dubious provenance, have crept into such lists virtually unchallenged, often abetted by the professional standing of the observers and/or the periodicals they are published in.

The authors' emphasis on the necessity for critical review, on the need to establish an Indian Bird Records Committee, on verifiable evidence to support bird records, are all timely and pertinent. After explaining their intent and the thorough methodology they followed, they provide detailed species accounts summarizing their findings and conclusions.

Though I'm grateful to them for the initiative, for clearing up the occasionally murky waters of Indian ornithology, I immediately wonder what the authors have to say about a species that I'd seen and reported along with my brother (who, like me was also a compulsive birder) nearly three decades ago. My eyes gravitate automatically to the bird, the second species on their list, a pelagic bird of our oceans and islands: White-tailed tropicbird. In one paragraph, with an accompanying table, they summarize the records from across India on this beautiful bird.

At once, I note with appreciation yet unavoidable poignancy, their decision to mention, but discard as questionable, the records of white-tailed tropicbirds seen in the mid-1980s in Madras (or Chennai as it is now called), by a couple of schoolboys, Sriram and Sridhar. Having reviewed the methods and knowing how thorough the authors have been, I respect their decision, although even to this day I have no doubt whatsoever that the birds were indeed white-tailed tropicbirds.

We'd seen the tropicbirds not once or twice, but five times in 1984, and four times in the following year, sightings and observations that we reported as notes published in *Blackbuck*, the journal of the Madras Naturalists' Society. In 1986, I saw the tropicbird once more, in an extraordinary coincidence, on the very same date (20 July), time (morning), and place (near Santhome cathedral) that I had first seen it in 1984, and reported that in a note in the *Newsletter for Birdwatchers*. Of these records, the authors of the *Indian Birds* paper say:

> A set of ten June–August sight records of this species reported from Chennai during 1984–1986 (Sridhar and Sriram 1985; Sridhar 1987; Sriram and Sridhar 1985) are not considered here as no other observer has reported this species, before or after, from that area, and all Indian records are from January–April.

They are definitive in their decision and I feel I should agree with them, for they have taken a cold, hard look at the records, as they certainly should. How could I compel them otherwise? How can I convey the certainty of those records to them, when the person I was in 1984—a boy not yet 12, a birder not yet a year into birding—is almost as distant to me as he must be to them: an unfamiliar person from a nearly forgotten past?

Imagine then, if you will, that morning in July 1984, the boy (in white shirt, khaki shorts, black leather shoes, and white cotton socks) standing in a line with his classmates on the cement basketball court beside the rain tree in the morning assembly at St. Bede's. Imagine the magical moment when he looks up, not at the Principal pontificating from the podium, but *up*, at the sky, at the beautiful, white, fairy-like bird that has suddenly appeared near the high spire of the adjoining Santhome Basilica cathedral, flying on graceful tapering wings and extended tail streamers, over the school, over the upturned face of the open-mouthed child.

It was a bird he had never seen before. He did not know then what species it was. Yet, he saw it again, and his brother did, too—and again, and again. Yet, what reliable or verifiable evidence did they produce? None, perhaps. The two wrote a note for *Blackbuck*, probably punching it out on their old Remington typewriter, describing the bird and the details they recalled: white plumage and long tail streamers, black wing-bars and wing-tips, dark eye-stripe, and orange-yellow bill. The editors carried their observations, appending a critical note, humouring the 'enthusiastic student members' who wrote it, cautioning the reader on such errors as greenhorns may make, confusing terns for tropicbirds, and asking other members, possibly more reliable ones, to keep an eye out for it, anyhow, in Chennai and along the Tamil Nadu coast.

Yet, it seemed fated that only the boys would see and report it again the following year, a *pair* of tropicbirds this time, again only their eyes as evidence. They would admire the birds' flight, which they reported in their note as 'slower and steadier' than a pigeon's, with every flap of their wings seeming 'to raise and lower the bird in the air'. The schoolgoing boys would not carry their heavy pair of binoculars along with their heavy load of books. They had no camera, nor money to buy and develop photographic film. Whether they drew sketches or kept notes of those sightings, no one really knows, for there is no trace of that anymore. They could only talk or write about what they saw. Sadly, more reliable and experienced members of the Madras Naturalists' Society never did see the birds, all to *their* loss (felt the boys).

With no evidence, no corroboration, no photographs, it is only right that experts now reverse the written record in their recent review. And it is only appropriate of others, who report new sightings of white-tailed tropicbirds—as a person does in a note in the same issue of *Indian Birds*, a few pages past the recent review—that they make no mention of these dubious records from Chennai, the city once known as Madras. And so, some records, like some names, are consigned to the dustbin, as they say, of history.

What will remain then, of those boys' experience beyond the yellowing pages of journals and the white birds of memory? Will memory, too, fade away one day, arcing from surprise and certainty to obscurity and oblivion? I shall wonder at that over the weekend.

# Fording the Flood

Have you been stranded in the middle of nowhere, lost in a forest or in a featureless desert, or disoriented on a thickly overcast day somewhere far from human habitation, and wondered what would happen to you if you could not find your way back to base camp? If so, did you panic? Did you pray for deliverance? Did you atone and repent for your sins in a fervent and hysterical hope that you would be saved? And when you finally managed to deliver yourself from your plight, or someone rescued you, were you not ecstatic? Whatever the case, it would be an incident that you are unlikely to forget. And months or years later, when you look back on that day, you would perhaps find your panic replaced by poignancy and your memory brimming with nostalgia and new meaning.

The 27th of August 1994 was one such day. The morning was bright and sunny, auguring a promising day for our trip to Rajaji

National Park in the Sivalik Hills at the foothills of the Himalaya. There were six of us, students at the Wildlife Institute of India, Dehradun: M.D. Madhusudan, Rohan Arthur, Kavita Isvaran, Advait Edgaonkar, Divya Mudappa, and myself. We left the campus just after dawn with our teacher and head of the wildlife biology faculty, Dr A.J.T. Johnsingh. Dr Johnsingh, who would turn 50 in a few months, was legendary for walking in mountains and forests across India. As he had done many times before, he was leading us on a field trip to help us hone our skills and learn to observe plants and animals in the wild. In a few months, most of us would leave to carry out field research projects for our Masters theses. Later, finishing up as qualified wildlife scientists, who knew where the flow of our lives would take us?

Dr Johnsingh drove the 20 kilometres or so along a road washed clean and dry by rain and sun, until we reached the river's edge a kilometre before Mohand village. We parked the jeep by the side of the road and descended the litter-laden, slippery embankment onto the boulder-strewn riverbed or *rau*. There hadn't been much rain over the last few days, and there was little water in the Mohand *rau*. Across the river rose the somewhat steep Mohand ridge, clothed in ornate *sal* and pine forests, which we were to climb that morning. We cursorily scanned the upper slopes—the grassy openings amid the pine trees and the steep muddy cliff faces—for any animals, particularly Himalayan goral, a species of goat-antelope with a brownish coat. Sighting none, we forded the river, taking our footwear off to wade through the less-than-one-foot-deep gurgling waters across to the other side.

Soon we were climbing towards the ridge. For the most part, it was an easy path, and we stuck to the small trails winding their way through ferns, bushes, and dense clumps of grasses higher up. The wet soil held stagnant, algae-filled puddles of water in some places and was slippery, especially in the upper reaches where a treacherous mat of glistening brown and half-decayed pine needles shrouded the soil. In the clear and energizing morning air, the sun blended with the wet freshness of the lush and verdant vegetation to make the labours of our uphill trek scarcely noticeable. We stopped to see a few birds, listen to the hill warbler's throaty and cheery outpourings, and admire the fervid and effusive leafbirds, with their golden, deep purple, and green hues. On the grassy slopes of the opposite ridge, a couple of goral and a sambar doe graciously allowed us a look.

Out of breath after a trek that had lasted nearly two hours, we reached the ridge top and flopped onto the grass, drinking to the full the breathtaking view spread out before us along with deep lungfuls of the clear mountain air. Dr Johnsingh fished out a packet of biscuits and a bottle of water, which the rest of us had neglected to carry, and regaled us with a plethora of wildlife anecdotes, only stopping to point out the distinct signs on the grasses nearby that had been grazed by goral and sambar.

Then, someone had the prescience to turn around and look behind. Perhaps a mere kilometre away, like giant grey genies, loomed huge, dark, rain clouds pelting the dissected ridges and narrow valleys with millions of tiny sparkling-clear raindrops. After a minute's silence, someone portentously said, 'Looks like it's going to rain.' We lazed a few minutes on the grass, finishing off the biscuits and water, then rose to begin our descent back to the waiting jeep.

We slipped and slithered our way downhill, delighted with our day so far. Dr Johnsingh trudged on ahead, skilfully negotiating the treacherous portions of the often practically non-existent trail, only sometimes losing his foothold to give us occasion for concern or suppressed amusement. Over many such trips to

Rajaji all of us had become proficient in the 'bottom technique' of mountain-descending: the wet, muddy patches on our trousers and jeans the proud certificates of our expertise.

A fine drizzle started. As we tucked our binoculars and cameras into protective, waterproof covers and scarcely water-resistant shirts, the steady drizzle was a delicate reminder to Rohan, who'd had a bath that morning, and a mocking castigation to the rest of us, who hadn't. As if to emphasize its point, the drizzle turned into a downpour, drowning all noises and voices in its incessant beat. I don't know whether the rain hindered our progress or, since we were slipping so often, it abetted our descent to the river bed, but through the forests and the rain and the bushes and the ferns, reach the river we did. And there, an awesome sight awaited us.

What we'd left behind as a tolerable little stream of water had transmuted into a roaring, furious river. We laughed in surprise and amazement and stood quietly in the rain watching the thundering waters flash past, hauling branches and logs and burnishing the boulders, even as small rivulets coursed down our bodies. On the hill slopes, muddy rapids formed and grew, and hurtled down the rills and nullahs to join the river in its downstream wanderlust. When large boulders stayed even these seething waters, the river billowed and surged in anger, venting spray in exasperation. We watched Nature play her impulsive games, while our jeep beckoned us from the other side. There was no way we could cross the river.

We kept our eyes on the clouds and the torrents for any signs of abatement. Then, after many long minutes, thinking that the river had indeed abated, we decided to attempt a crossing. The seven of us formed a chain, gripping each other's hands tightly as we stepped carefully and firmly into the raging, muddy rapids. Mud swirled through our shoes and our clothes; the water's force threatening to sweep us off our feet. Moving forward, one step at a time, while staying upright, maintaining our balance on the smooth stones underfoot, took quite an effort. We still had nearly 20 metres left to cross. And, ahead, it only looked worse. We decided that the crossing was too risky and squelched and splashed our way back to the safety of the bank; we stood there watching the river helplessly, awed despite our fears. Even the trees seemed to be getting impatient as they fumed out their misty vapours to the clouds above, an exhilarating sight in itself.

Meanwhile, the rain just seemed to get worse and the waters continued rising, even where we stood. Dr Johnsingh told us how, a few years ago, a Land Rover with nine trainees from the Wildlife Institute was carried away for several hundred metres by a monsoon flash flood, the occupants barely managing to jump out and swim across to save themselves.

Would the rain ever stop? When would we ever get back to the warmth and safety of our rooms, to food that would quiet our rumbling stomachs? The jeep beckoned us from the other side. The river mocked us at our feet.

The clouds relented. Slowly, the downpour subsided into a soothing sprinkle. More slowly, the river appeared to abate its temper. When we were still remarking that the river was, even now, too difficult to cross on foot, a Gujjar on a bicycle appeared mysteriously on the road on the other side and urged us with gesticulations to move upstream to cross at a more favourable point. Glad to do something after being such mute witnesses to the river's caprices, we did as he suggested. We found what we thought was a suitable place to cross, reconstituted our human chain, and strode determinedly into the water.

The turbulent brown waters were still powerful and tugged at our feet and legs. We lumbered carefully ahead, laughing and grinning and exclaiming in excitement at each surprised step forward. Dr Johnsingh, again leading us, faltered on some slippery boulders, even as he turned to give us a beaming smile and a hearty, thrilled laugh. Three steps more, two steps, one step, and we had made it!

We stood on the banks of the river and laughed in delight, like a bunch of schoolkids playing hooky, with the thought, 'Wow! That was great!' echoing in each of our heads. Still laughing and chattering (Advait from the cold, the rest of us in glee), we headed back to the jeep, and I turned to give a last respectful glance at the river.

Back in the jeep as we headed back after a few cups of refreshing hot tea at Mohand, Dr Johnsingh had the last word: 'You know? If it hadn't been for the rain and the flood, this morning wouldn't have been so memorable, *illai ya* [is it not]?'

## Postscript

> Our lives, however dear,
> follow their own course,
>     rafts drifting
>     in the rapids of a great river
>     sounding and dashing over the rocks
>     after a downpour
>     from skies slashed by lightnings—

—from 'Every Town a Home Town,'
by Kaniyan Punkunran, *Purananuru* 192,
*Poems of Love and War*

A few months after that day, some of us who forded the river hand-in-hand, embarked on another journey together. A journey that began as conversations in the Wildlife Institute of India campus continued as ideas penned in letters written to each other from our scattered field sites—Rohan in the coral reefs off the coast of Gujarat, Divya in the rainforests of the Anamalai Hills, Madhu among tahr and grasslands and cliffs in Eravikulam, and I in the forests of Dampa Tiger Reserve in Mizoram. We conceived of working together to create an organization that would build on field research and scientific understanding to conserve India's wildlife. We roped in Charudutt Mishra, from the senior batch at WII, and used our pocket money to establish the Nature Conservation Foundation in 1996. Other friends and field biologists, too, later came on board. Aparajita Datta and Yashveer Bhatnagar, also seniors at WII, joined NCF, leading programmes in the Himalaya, while Anindya Sinha, a friend and faculty at the National Institute of Advanced Studies in Bengaluru spearheaded collaborative work on primates. Kavita, who, after her post-doctoral work became a professor at the Indian Institute of Science, joined us as an adjunct faculty. Suhel Quader, too, joined NCF after a stint as faculty elsewhere, even as we continued to connect with other classmates—Advait, who joined the Indian Institute of Forest Management, Bhopal, as a scientist; Saravanakumar, who became a remarkable independent wildlife filmmaker and photographer; and many others. The challenges of building an institution from scratch, of conserving a multitude of wild species in diverse landscapes, and of learning from and working with local people across India made for a remarkable journey, one that still continues. Over the years, I have wondered about that morning in Rajaji, about walking hand-in-hand with my friends, a bunch of biologists crossing the river, the flood seething below us. And I have wondered, too, about the Gujjar who appeared on the other side and pointed the way.

# Answering the Call of the Hoolock Gibbon

As a young birdwatcher, I often leafed through the pages of Sálim Ali's book on the birds of the eastern Himalaya. Fascinated by the colourful and diverse avifauna of the region, I harboured a secret desire to visit northeast India someday. Occasionally, I would read about the region in the papers and magazines—about its deforestation and shifting cultivation, logging and civil strife, and its rich biodiversity—my interest unabated. Yet, a visit to the Northeast remained a dream.

I took my first step towards realizing this dream after I enrolled for a Masters degree in wildlife science from the Wildlife Institute of India, a premier national institute established in the picturesque foothills of the Himalaya in Dehradun. As three semesters of coursework wore on, my mind was set on visiting northeast India and, if possible, doing my field project that was a course requirement there. I was interested in the tropical evergreen forests of northeast India and thought that the best candidate for a study would be the hoolock gibbon. The gibbon—India's only ape—is a frugivorous primate found in the evergreen and semi-evergreen forests of this region. I read whatever published material I could find about the primates of northeast India, particularly gibbons, and discussed it with many people who had been there. Only a handful of studies had been carried out on the gibbon, and many aspects including their population status, distribution, ecology, and behaviour, remained poorly known. Clearly, there was more work that needed to be done. Still, the actual visit seemed only a remote possibility given the challenges of logistics, funding, and designing a feasible project in a region that I had never seen or visited.

Then, in December 1993, opportunity knocked: The Mizoram Forest Department approached the Wildlife Institute of India for wildlife biologists to carry out a survey of their protected areas (wildlife sanctuaries and national parks). The head of the biology faculty at the Wildlife Institute, Dr A.J.T. Johnsingh, discussed with the Chief Wildlife Warden of Mizoram, C. Ramhluna, about sending two of his students for the survey, and the latter showed keen interest. And so, my friend Charudutt Mishra, who had just finished his MSc, and I were off to Mizoram. On the survey, Charu would focus on the serow and goral, two species of Indian goat-antelopes, and I would look for primates.

Mizoram is the southernmost state in northeast India, wedged between Myanmar and Bangladesh. In 1993, one could get there only by a Vayudoot flight from Kolkata directly to the small airport near the capital city of Aizawl, or by reaching Silchar in Assam overland or by air, followed by an eight-hour bus journey. Our flight from Kolkata to Silchar offered me my first breathtaking aerial view of this part of India. A patchwork of forests dissected by numerous rivers, presumably some part of the Sunderbans delta, went past the aircraft's wings. Soon we were flying over a magnificent series of parallel hill ranges running north to south with oceans of brilliant white clouds snuggling cosily in the valleys. These were the hills that we would soon be exploring on foot. Our flight took us further north and descended in Silchar. Silchar was a quaint replica of a typical small town in Kerala. It was so strangely similar—in topography, people, houses, tall trees, with crop fields at the outskirts—to places in Kerala, that I could barely believe that I was really in northeast India. Scattered among houses and seas of paddy, tea, bamboo, and teak, stood huge, ponderous evergreen trees—silent emblems of the kind of forests that must have clothed the land at one time.

A few hours later, we were in a small bus leaving behind the flat plains and lowlands and heading into hillier terrain. Around half past four in the afternoon, we reached the Assam–Mizoram border. Our special inner-line permits, which we had got in New Delhi, were checked and we were allowed to enter the state. We travelled along the road winding over the Lushai Hills. In half an hour, the sun had set, and the area plunged into darkness. The hills were black against a starlit sky. Scattered fires, or a sprinkling of lights against the dark backdrop, indicated the presence of people or a village.

Aizawl, the state capital, is a picturesque city about 1,100 metres above mean sea level on the hills. Other than a few multi-storey buildings in the hub of the city, there are numerous houses perched precariously on stilts on the slopes at the edge of the roads. The bamboo cottages and tranquillity of the surrounding countryside contrasted rather sharply with the cement houses, high-fashion clothes, and vehicular traffic of Aizawl. We received a warm welcome at the Mizoram Forest Department office and, after a restful day's stay, left for the first sanctuary on the list—Dampa. Dampa Wildlife Sanctuary, which became a Tiger Reserve later, lies southwest of Aizawl on the Bangladesh border.

In altitude, Dampa is not very high—the highest point is just over 1,000 metres. An earlier survey by a friend and senior at the Wildlife Institute of India, Nitin Rai, had revealed the presence of at least four species of primates there, including the hoolock gibbon. I was particularly interested in this sanctuary for the opportunity it offered to see several primate species in one area.

To explore the forests we needed a local guide who knew the trails. We found a wildlife guard, Kimthanga, who could also speak Hindi, and set off into the forest. Our destination was a watchtower at Dampatlang (*tlang* means 'mountain' in the Mizo language). Our 8-kilometre trek to the watchtower took us through dense bamboo jungle, secondary forests—forests regenerating after logging, cultivation, or other disturbances—and patches of evergreen forests that had tall trees and a dense canopy at about 35 metres. This patchwork of regenerating forests amid mature evergreen or primary forests was created by the practice of shifting cultivation or jhum, prevalent in these parts.

Suddenly, the silence of our trek was punctured by a commotion from the canopy. Glancing up, we saw a troop of capped langurs crashing away through the treetops, having seen us. Focusing my binoculars, I got a clear but brief glimpse of one individual. He was a handsome male langur, looking attentively around—the black cap, the puffed out hairs on the side of the face, and the golden-hued underparts giving him an endearing appearance. I remarked at their extreme shyness to Charu as we continued on our trek. Before long, we encountered a troop of Assamese macaques. These monkeys, too, disappeared with harsh, resounding alarm calls. One female, though, sat on a low branch, allowing me to take a good look at her. I noted her pale pink face, her tail carriage, and well-groomed fur, and the absence of any pinkish-red on her hindquarters, when, with another younger macaque, she too crashed down into the lower canopy and disappeared.

We went on, excited, carefully scanning the canopy for primates and birds, and the path for tracks, droppings, or other animal signs. As we turned a corner, another surprise was in store. When we first heard a series of very loud grunts, we froze in our tracks. Not knowing what it was, I advanced cautiously: less than 20 metres away a small bear cub clambered about two metres up a tree trunk, then paused, clinging. I pointed out the cub breathlessly to Charu, who was behind me. Even as we identified the cub as a sloth bear, we heard the heavy footsteps of a large animal advancing through the undergrowth in our direction. Fearing this was the mother, and because we were so close to her cub, we turned and fled. We halted our ignominious retreat almost a hundred metres away. Kimthanga, our wildlife guard, and Rai Bahadur, an experienced local tracker, also seemed very alarmed at our predicament. The bear and her cub were between us and the safety of our camp at the watchtower, less than half a kilometre away. Kimthanga and Rai immediately set about making a fire with dry bamboo culms. As the bamboo burnt, it snapped and crackled like tiny thunderclaps, which they hoped would be enough to drive the bear off our path. After about fifteen minutes, we mustered courage and strode carefully ahead towards the watchtower. Passing the location of the initial action, we saw the bear cub clinging tenaciously to the same tree. Luckily, we did not meet the mother.

Later that night, sitting around the campfire under the moonlit sky, Rai Bahadur told us stories of tigers and bears and the *zamphu* (binturong or bearcat), their personalities, and about the people who have had close encounters with them. As we went to bed, after drinking the scented tea brewed in bamboo cups and eating rice and dal, we heard loud grunts from close by. The next morning, on some loose sand further down the path, we saw pug marks and a scrape by the side of the trail: a tiger had passed that way.

The next day had more excitement in store. We climbed down the narrow path towards Tuichar lui (*lui* means 'river' in Mizo) in the valley. On the way we saw two species of squirrels, both new to us. One was the unmistakable black-and-white Malayan giant squirrel high up in the canopy in primary forest. The other sightings were of the relatively more common and smaller red-bellied Pallas's squirrel. This was a charcoal-black squirrel with rust-orange or red on the underparts from the breast downwards. The staccato calls of this squirrel were quite different from the more metallic calls of the Malayan giant squirrel. Later, we also saw the diminutive Himalayan striped squirrel in the forest and the hoary-bellied squirrel near habitation in villages at the edge of the sanctuary.

Along a ravine not far from Tuichar lui, we saw our first troop of Phayre's langur. This folivorous or leaf-eating primate has been reported in India only from Tripura, Mizoram, and Assam's southern Cachar. Outside India, this langur occurs in Bangladesh and Myanmar. Phayre's langurs are very striking primates, unmistakable for their white eyepatches that resemble spectacles and make them look as if they are perpetually amazed. Their upperparts are a dark slate-grey and the underparts silvery white. Besides a few brief published notes, our knowledge about this species comes from an intensive study in the neighbouring state of Tripura by Atul Kumar Gupta, a committed forest officer, and Ajith Kumar, the eminent Indian primatologist. Their study troop spent much time in secondary forests that had many important food plants of this primate.

All this was good, but I still hadn't seen the hoolock gibbon. We wondered whether there were any gibbons left in the patch of primary forest we could see from the watchtower. (Overall, the sanctuary, and indeed Mizoram itself, had very little primary forest. Much of the area was under dense bamboo jungle that resulted from a short jhum cycle of three to five years. In Dampa, the primary forests stretched from the valley of Tuichar lui all the way up the opposite slope of Chawrpialtlang.) We hoped to detect the presence of gibbons by the loud calls and duets they are known to make every forenoon. These calls can be heard even a few kilometres away. Our first morning at the watchtower, we did not hear any gibbon calls. Nevertheless, we decided to set off the next day to Chawrpialtlang to check.

The trek to Chawrpialtlang took us first to the small stream that was Tuichar lui. The forests here were just amazing and would fit any superlative description of tropical rainforests. We had already seen the two squirrel species and a number of birds such as the long-tailed broadbill, common green-magpie, orange-bellied leafbird, black-naped monarch, necklaced and white-crested laughingthrushes. We had also seen elephant footprints and dung all along the trail, though none of it appeared fresh. After resting near Tuichar lui and drinking some of the clear, cold water, we moved on for the climb up Chawrpialtlang. We had gone hardly 300 metres before the low rumble of elephants reached our ears.

It was 9.45 a.m. The sounds were directly ahead of us, and unwilling to take any risks, we decided to return.

In that instant the whole forest exploded with the cries of hoolock gibbons. The calls, resounding through the tall rainforest, left us spellbound. Starting with a high-pitched hoot, they went *hoo hue hua hoo hua hua hua hoo hua... Hooo haak... hoo hua* and on and on. We headed towards the sound, stopping whenever the calls paused. The gibbons' calls went on for nearly twenty minutes. By that time, we had reached a ravine, and the calls came from just across. We stopped and looked carefully through the dense canopy. And there he was! A male, in his jet-black pelage and contrasting white eyebrows, clinging onto the branches of an understorey tree. As we watched, he expertly manoeuvred his way, brachiating—swinging hand over hand—through the canopy. Nearby, a female in her brown coat also appeared on an open branch. I took a couple of photographs. The gibbons, however, were uneasy at our presence, and moved away. A subadult and a small infant that followed them brachiating in earnest were delightful to watch.

A fair bit of information is available on hoolock gibbons. They are mostly frugivorous—fruit-eaters—and a large part of their diet consists of *Ficus* (wild fig) fruits. They are monogamous and territorial, and each family group usually occupies a home range of 20–40 hectares. They are found almost exclusively in relatively undisturbed primary forests that currently exist in restricted pockets in most of northeast India. It has been suggested that gibbons may be a good flagship species of such forests and that, as frugivores, they may play an important role as seed dispersers. This aspect has received little attention. It would be useful to see how seed dispersal of fruiting trees that gibbons feed upon is affected in areas where gibbons have become locally extinct.

Other than habitat loss, another threat to primates in northeast India is hunting. In Mizoram, we saw hundreds of skulls adorning the walls of houses—primates, serow, goral, wild pig, bears, gaur, Indian muntjac, sambar, tiger, porcupines and other rodents, and even a clouded leopard, besides wreathed and great hornbill casques and heads. The Forest Department has taken many commendable steps to reduce and control hunting, and poachers have been caught occasionally. In Dampa, eleven villages were peaceably shifted out of the sanctuary and no shifting cultivation is now allowed inside. Still, poaching does go on. While this may not have a big effect on some species, slow breeders like great hornbills and hoolock gibbons may be considerably affected. Later on our survey, in Sangau village in southern Mizoram, we found a young female hoolock gibbon being kept as a pet. She had been seized by a wildlife guard from some Chakma tribals who had caught her as an infant. The gibbon, who they had named Janice, behaved very much like a child with her human family. She was agitated by my presence, but when the grandmother of the house took her into her arms and whispered a few soothing words, she immediately calmed down and clung to her like a human baby. Watching the young gibbon at close range, I thought she had the saddest eyes and face I had ever seen. I could not help wondering what the future has in store for these wonderful primates of northeast India.

The survey reinforced my interest in working in northeast India. The following year, back at the Wildlife Institute, I planned a study on the effects of shifting cultivation or jhum on wildlife, as this practice had a major influence on the landscape and was a significant conservation concern. After the survey, it felt only natural to select the Dampa landscape as my study site. I focused on forest birds to design a more feasible project, but included observations of squirrels and primates, especially the gibbon. The call of the gibbon rang in my ears as I returned to Dampa.

# Bamboo Bonfires and Biodiversity

The heat from the fire is intense, even from a hundred metres away. The entire slope is ablaze. Piles of slashed vegetation and tens of thousands of bamboo culms sun-dried for three months burn ferociously. The bamboo hisses, crackles, and explodes, audible a mile away. Hot gusts of wind scud the fire upslope, throwing branches and small trees ten metres into the air. High above, unmindful of the billowing fumes, swallows and drongos, in a frenzy of activity, hawk insects. Ash and smoke darken the sky, muddying the sun to a dull orange. In twenty minutes, almost as rapidly as it started, the spectacle ends; leaving only a blanket of smouldering ash and scorched tree trunks.

The fire was kindled by tribal farmers of Teirei, a remote village adjoining tropical rainforest in the Lushai Hills of Mizoram. The farmers practice traditional slash-and-burn shifting agriculture locally known as *jhum* or *lo*. Ash is an effective way to enrich poor soil with nutrients prior to cultivation. The burnt patch, significantly, was just within the border of the 500-square-kilometre Dampa Tiger Reserve. This reserve was established in 1989 to protect tigers and other wildlife species such as the hoolock gibbon, capped langur, clouded leopard, hornbills, great slaty woodpeckers, wren-babblers, and other endangered species, many of which are found only in the tropical rainforests of the Northeast within the country.

Jhum is a serious conservation issue in northeast India. Between 1989 and 1995, remote-sensing analyses estimated that more than a thousand square kilometres of forests were lost to jhum in the region. The effects on wildlife are largely unknown because few studies have been done in these often remote, insurgency-ridden parts of India. On the other hand, more than a hundred ethnic communities and well over a quarter of a million families depend upon jhum for livelihood and economy, frequently cultivating in or at the edge of protected areas, as in Dampa.

The conservation issues raised by jhum are many and controversial. Many conservationists claim that, by destroying forest cover, jhum causes wildlife declines and extinctions, soil erosion, and drastic environmental changes—most evident when tall, primary rainforest is replaced by crop fields. Others have argued that the effects of jhum may be relatively benign compared to those of terrace cultivation, tea plantations, and monoculture forestry. By maintaining a mosaic of fallows and regenerating forest, jhum may help increase biological diversity at the landscape level. Yet, the critical question is: do species of high conservation value—those that are rare or specialized, or have small geographical ranges—benefit or suffer from slash-and-burn shifting cultivation? To unravel the answer, one needs to first understand the cropping patterns and changes that occur in the forest vegetation as a result of jhum.

Although the timing of cultivation, types of crops, and agricultural practices of jhum vary in Indian communities, the broad pattern is remarkably similar to shifting cultivation in Southeast Asia and other tropical regions. Until recently, the enterprise in northeast India has been driven and regulated by the communities that control the land. Village Councils would allot each household a parcel of land between 1 and 4 hectares. Normally, this would be part of a slope of secondary forest that has been recovering—and regenerating—for 5–10 years since the end of the last bout of cropping. Tall, mature rainforest is also cleared, though rarely, owing to the scarcity of such forest and the difficulties of clearing.

After the cut, in January or February, the slash dries on the hills until April, when it is burnt just before the pre-monsoon rains. Farmers then sow several varieties of rice, their mainstay, along with more than a dozen other crops, including eggplants, beans, and tubers, as well as some cash crops such as tobacco and chillies. A busy season of weeding and multiple harvests follows until October, when the spent field is abandoned, or rather rested and allowed to recover. Fields are rarely cultivated for more than a year, because this one round of cultivation severely depletes the soil. The next year and in successive years, new areas are cleared, until the vegetation in the first site regenerates sufficiently—usually within ten years—to permit cultivation again. But is this a sufficient amount of time for native rainforest plants and wildlife to recover?

To observe a regenerating forest from the time it is cleared to when the original vegetation or a semblance of it recovers is practically impossible within the lifetime of a rainforest biologist. Field biologists therefore sometimes use a short-cut solution: they study various sites that were cleared and abandoned at different times, and which represent different ages and stages of forest regeneration, a method called 'space-for-time substitution'. Such an opportunity existed in Dampa. So, to study changes in vegetation and wildlife here, I surveyed sites that had regenerated for between 1 and 100 years and compared them to rainforest that had never been cleared. It was a special and awe-inspiring experience, like a virtual voyage through time, visualizing the birth, growth, death, and vicissitudes of a rainforest and its plants and animals—a fascinating subject for any rainforest biologist. After months of fieldwork, with the data from transects and plots in hand, the trajectory of changes could be pieced together.

Soon after a field is abandoned, weeds, grasses, surviving crop plants, and bamboos sprouting from underground rhizomes run amok, creating a dense and vigorous tangle that at first threatens to smother forest regeneration.

In these open fields with hardly any tree canopy, common and widespread wildlife proliferates. The ubiquitous red-vented bulbul, common tailorbird, white-rumped munia, and grey bushchat thrive in open land that has lain fallow for one year. Most rainforest species avoid these areas, although the occasional pigeon or woodpecker may briefly visit an isolated tree standing dry and forlorn in the field. The common hoary-bellied squirrel scurries on the ground, picking at choice bits of food. The grass looks good for ungulates, but the shy Indian muntjac and sambar seldom venture here, for they may be snared or shot.

Fortunately, this situation does not last long. The vegetation recovers with astonishing rapidity. The open, weedy fallows quickly give way to bamboo forests. In five years, the bamboo, along with pioneer trees—species such as *Macaranga* and *Trema* that are first to establish in open or disturbed areas—form dense stands that reach 10 feet and higher. Wildlife from the surrounding landscape begins to colonize. Understorey birds are among the first to appear in sizeable numbers: rainforest babblers, warblers, flycatchers, and bulbuls. If lucky, you might also see the bamboo-loving woodpeckers: the pale-headed woodpecker and the white-browed piculet, clinging to the smooth culms, searching for insects.

Bamboos reign supreme for many years. In Mizoram, the *mautak* bamboo (*Melocanna baccifera*) dominates regenerating fallows for at least the first 30 years. As time passes, more bird species appear, and the air is alive with their calls. Some arboreal mammals, too, venture into tall bamboo and secondary forests that have been allowed to regenerate for 10 years or more, particularly if they are near mature rainforests. Phayre's langurs forage in troops of a dozen or so individuals in the canopy. They feed on leaves of trees and climbers, often nibbling only at the leaf petiole and discarding the rest. The sprightly, dark-furred and red-bellied Pallas's squirrel, and even a few of the cautious black-and-white Malayan giant squirrel scamper through the canopy or pause to gaze suspiciously at observers. As bamboos and pioneer trees grow taller and larger, rainforest tree seedlings sprout and flourish in their shade.

If left undisturbed, the slow-growing saplings eventually take over after the bamboos flower *en masse* and die. One site—it had regenerated for a hundred years—contained mostly tall rainforest trees and lianas with little trace of bamboo. Here, and even more so in primary rainforest that has never been cleared, plants and animals achieve their highest diversity.

Camping in a cave by the Tuichar lui, deep in primary rainforest, I could experience this first-hand every day. Here were lofty rainforests with their profusion of life. In a single day's observation at a wild fig tree fruiting just above my camp, I saw four species of primates including a family of hoolock gibbons, five species of squirrels, three species of green pigeons in large flocks, great, oriental pied-, and wreathed hornbills, mountain imperial-pigeons, Asian fairy bluebirds, and, surprisingly, even a flock of laughingthrushes that had ascended into the canopy. In stark contrast, fig trees that were left standing alone and tall above a jhum fallow or a bamboo forest held only a vestige of these spectacular gatherings, fewer species, and mostly common ones.

When regenerating bamboo forests are cleared for cultivation within ten years, as usually happens in northeast India, rainforest recovery is interrupted and the land undergoes another of the endless cycles of bamboo. Owing to the spread of shifting cultivation in the region—which means short fallow cycles of fewer than ten years—huge areas are under this 'arrested succession' of dense, almost monotypic bamboo forests. Besides having fewer species, these bamboo forests are also prone to destructive fires after bamboos flower synchronously and dry up. Nearly fifty years after the last bamboo flowering around 1959, vast areas of Mizoram underwent a spectacular bout of flowering during 2006–7.

The wildlife species that suffer most because of jhum are often those that are most critical from a conservation point of view—those that are rare, specialized, or restricted to the northeast Indian rainforests. India's only ape, the hoolock gibbon, and other arboreal mammals such as the capped langur and the Malayan giant squirrel, occur only in mature rainforest and are locally extinct or very rare in jhum-altered landscapes. This pattern of change was also evident among bird species. The number of bird species increases with forest regeneration, rapidly at first, then slowly, to reach maximum diversity and abundance in the 100-year-old mature forest and undisturbed tropical rainforest. Moreover, the mix of bird species or bird community composition also changes with time, only achieving a high similarity with primary forest after a 100 years.

Where does this leave the claim that jhum increases biological diversity in the landscape? Obviously, more species can be accommodated in a tropical rainforest landscape when new habitats such as open fallows and dense bamboo forests are created by jhum. The additional species appearing in the landscape are, however, mostly common and widespread species, of open scrub or dry deciduous forest habitats. Many species considered more important for conservation—rare, endangered, restricted-range, and habitat-specialist species—decline or suffer from habitat alteration because of jhum. The increase in biological diversity in the landscape may thus come at the cost of such rainforest species.

Still, any assessment of the effects of jhum has to consider the livelihood needs and traditional rights of the people who practise it. Social scientists and activists have justifiably championed the cause of indigenous peoples. Yet, defining what is traditional and who is indigenous in communities undergoing rapid socioeconomic changes, market integration, and migration is difficult. In and around Dampa, as in other parts of northeast India, the human population has soared in recent years and includes many settlers from other parts of Mizoram, Tripura, Nepal, and Bangladesh, all with their own customs and traditions, and far removed from places where their traditions initially evolved. Over the last century, a large proportion of the people in Mizoram have converted from animism to Christianity. With road- and market-penetration, human societies do not stay static, but change dynamically.

Jhum is not the only problem for wildlife conservation in northeast India. Large-scale logging by the government, illegal timber poaching, and conversion of rainforests to monoculture plantations of tea and teak—widespread ecological ills caused by state and private interests, mostly non-tribal—consume precious land and forest. As a consequence, the burgeoning tribal populations, growing at among the fastest rates within India, are forced to clear remnant forest tracts and cultivate at shorter fallow periods. And so, the vicious cycle of arrested bamboo succession continues. If wildlife conservation in the Northeast is to be effective, all the forces of landscape change must be addressed, squarely and urgently.

Although it is easy to say that from a biological perspective one needs undisturbed, preferably large, tropical rainforests, it is not an easy conservation objective to achieve. Such areas are scarce, and one is often left with only various-sized, disturbed fragments of rainforest in a jhum-dominated landscape. There are, however, alternatives. In Meghalaya, tribal communities protect small, sacred groves. In Mizoram, thanks to state laws passed in the 1960s, villagers use a network of 'supply' forests under regulated use for biomass harvests. More infrequently, a few 'safety' forests exist, fringes around villages created to protect them from jhum fires. These areas are rapidly diminishing, even vanishing, as villages grow and lifestyles change. It is important to include these areas, along with agriculture and plantations, within the ambit of conservation planning for it to be effective at the landscape scale.

Conservation efforts in northeast India cannot proceed without due consideration of the legitimate needs of the millions of poor farmers, such as the people of Teirei, who depend on jhum for their livelihoods. Shifting cultivation is an organic system of multiple cropping well adapted to areas of high rainfall. Alteration of jhum to mechanized or terraced agriculture or monoculture plantations, even if possible, may be even worse for biological diversity and food security.

In the final reckoning, many forms of land use and forest types will be a part of the landscapes of the future. It is also evident that in parts of northeast India, intense, short-cycle jhum and wildlife conservation are largely incompatible. For wildlife conservation to be a reality, there is one type of land use and forest that is essential in this mix: protected sites with primary rainforest.

In Dampa Tiger Reserve, conservation efforts have been promising. After initial difficulties, eleven villages with nearly five hundred families located inside the sanctuary were resettled on the periphery in 1989. Today, jhum is mainly restricted to the buffer zone and areas outside sanctuary boundaries. A large project implemented through the local government and village councils has been underway in Mizoram to develop and sustain alternate livelihoods for the villagers, with the goal of finding alternatives to jhum, and ostensibly to minimize pressures on forests.

Meanwhile, scientific surveys continue to reveal the extraordinary diversity in these rainforests. Using camera traps, forest staff have obtained photographs of the rare and elusive marbled cat and the clouded leopard. A survey to catalogue resident reptiles and amphibians has revealed the presence of several rare and endemic species, including some that could be new to science. In many ways, Dampa represents a tantalizing pocket of hope for what is possible in these remarkable rainforests.

# In Clouded Leopard Country

In the rainforest, the rewards of silence sometimes exceed your wildest expectations. From where I sit quietly, I don't hear a single artificial sound. Unseen cicadas shrill and set the air ringing, woodpeckers cackle from the treetops, and frogs click and boom from the rock-pools alongside the singing river below. From somewhere in the undergrowth, a grey peacock-pheasant sounds an echoing, guttural laugh. In the distance rise great grey cliffs, home of serow (a forest goat-antelope) and bear, overlooking rainforests that pulsate with the hoots and songs of hoolock gibbons every morning. Around the steep rock slope where I'm stretched out on my back, the looming rainforest envelops me like an amphitheatre. I feel like a tiny flame steady in an evergreen sconce. As yet, I have no inkling of what we are about to witness.

The evening drops like a shade over Dampa Tiger Reserve, a 500-square-kilometre wildlife reserve set in the Lushai hill ranges of Mizoram, near the Bangladesh border. After setting afire the far cliffs and the tops of tall, emergent dipterocarp and *Tetrameles* trees, the sun dips behind the mountains. On the rock slope with me are Divya Mudappa, my partner, fellow biologist, and wilderness wanderer; Jaydev Mandal, friend and young ornithologist; and Zakhuma, intrepid wildlife photographer with the Mizoram Forest Department. As we wait, the humid heat of day yields to a cool silver dusk, presaging a starlit night.

Suddenly, the forest resounds with the clamour of alarm calls. Harsh shrieks rip through the air and the high branches and tall bamboos crash and sway. In the tree canopy, through binoculars, we see an entire troop of Assamese macaques in tumult.

The brown monkeys seem to have materialized out of the mysterious depths of the forest to roost above a rocky bluff overlooking our slope. Why are they alarmed? It must be us, we think. But it is not: the monkeys are looking away, over their shoulders into the forest. We sit, frozen, trying to become as immobile and silent as the rock. In the trees, the monkeys cry out, swivelling their heads and peering into the dense undergrowth. Down there, something is moving.

And then, a golden shape emerges from the shadows: Intense eyes, lean, feline torso, splotched and marked in black, long, flicking tail, and purposeful paws—he stalks forward. Softly, incredibly, he walks along a smooth, rounded bamboo culm as thick as an arm angled over the edge of the bluff. Up in the treetops, the monkeys erupt in alarm; below, the air fills our lungs and refuses to leave. In thrall, we watch the predator on his hunt, an animal emblematic of these rainforests: the clouded leopard.

As the monkeys leap away, the clouded leopard turns on the precarious bamboo, enters the shrubbery, and emerges again above the bluff. He looks up at the monkeys, yawns to reveal strikingly large and curved canines, wipes his face with his paws, and ignores us completely for the few precious minutes that he remains visible. And then, almost as suddenly as he appeared, he is gone. The falling darkness hides the direction of his passage. It leaves his hunt to our imagination, announced only by the terrified calls of the monkeys.

In India, clouded leopards are restricted to the country's northeast: the eastern Himalaya, the Assam valley, and the hills south of the Brahmaputra. This region falls within two global biodiversity hotspots, the Himalaya and Indo-Burma, renowned for supporting an astonishing diversity of life alongside an extraordinary diversity of peoples and cultures. From bamboos and babblers to pitcher plants and primates, Northeast India holds more species than any other part of the country.

Doubtless, many species await discovery and research, too. In Dampa itself, a new species of bamboo and two species of fish were described by scientists between 2010 and 2014. Field camera or 'camera trap' surveys by wildlife scientists and the Mizoram Forest Department indicate that clouded leopard density in Dampa is perhaps the highest among forests of South and South-East Asia, where the species is found. Yet, with sightings so rare and observations difficult in the dense rainforests, many aspects of clouded leopard ecology and behaviour are an abiding mystery.

Like the clouded leopard, much of the wildlife in India's Northeast remains elusive, poorly understood, and inadequately protected. While larger threatened mammals—tigers, elephants, and rhinos—and reserves like Assam's Kaziranga National Park garner conservation attention, innumerable equally significant species and areas get overlooked. Even as that changes slowly, rapid transformations in land use and other threats rise from river valleys to mountaintops across the region.

In upper Assam, the last vestiges of rainforests lie scattered, fragmented by pockets of agriculture and vast monoculture plantations of tea. Wildlife scientists have recorded rare and threatened wildlife such as hoolock gibbons—India's only ape species—stump-tailed macaques, and small cats in some of the remnant patches. In smaller fragments frayed by roads, timber logging, and other disturbances, the species are gone. A few forlorn forest plots hold lone gibbon families that, bereft of better prospects or others of their kind, have even stopped singing.

In Arunachal Pradesh and the hill states, dams, roads, mining, and plantations are transforming the landscape apace, creating more fragmented forests, replacing diversity with paltry monocultures. Once, shifting agriculture or jhum, practised by tribal people in the region, was considered a threat to forests. But recent research suggests that the diverse and dense mosaic of bamboo and regenerating forests that jhum creates is better for forests and wildlife than the expanding plantations of tea, teak, oil palm, and rubber.

Still, where communities, conservation organizations, and state authorities have come together, there are prospects of revival. In Assam, a remarkable reintroduction project has helped in the population recovery and reintroduction of over a hundred critically endangered pygmy hogs into the tall grasslands of the Brahmaputra Valley. In Arunachal Pradesh, innovative efforts have enabled tribal people to become protectors of endangered hornbills, birds that have suffered from hunting and forest loss. In Nagaland and Meghalaya, community reserves and initiatives protect migratory Amur falcons and other wildlife as well as remnant forests. Initiatives for forest recovery are also emerging; these use myriad native rainforest plant species to ecologically restore degraded areas and bring back a semblance of the human–nature connect that once marked this landscape.

Back on the rock slope the next morning, I gaze at the spot where we first saw the clouded leopard. It seems easy, too easy, to identify the place with this predator. Ironical, too, in a reserve that carries the name of its more imperial cousin, the tiger.

I swing my eyes around to see something different and from deeper memory.

There, on the slope, are the natural potholes that lead to a subterranean stream from where a romp of otters had once emerged wet and dripping almost at my feet, only to disappear downriver. Here spreads the buttressed roots of a great tree between mossy boulders cradling a crevice where a pair of wren-babblers had once made their tiny nest. Above, on a natural wooden ledge 30 metres up a tall rainforest tree, juts the platform from where a flying squirrel had launched herself in a great glide into the night. There, over the ridge, lie the burial mounds and old forests of people who once lived softly off the land.

Around me, swaying with the bamboos, rising with the trees, cushioned in the moss, glinting in the stream, and wafting across the skies, lives a raft of recollections of other living beings and the places they belonged to and that belonged to them.

# The Dance of the Bamboos

At first I thought it is the people of Mizoram who use bamboo to perform their celebrated dance, the Cheraw. After months of field research in remote forests of this small state in northeastern India, I know now it is the other way round. Through its intimate influence on the people, it is the bamboo that does its own dance on the mountains of Mizoram.

In March, Mizoram comes alive to the dance of the bamboos. Bamboos clap and clack to the rhythm of the Cheraw in the Chapchar Kut festival; bamboos are worked and woven into intricate handicrafts and other products in the state's Bamboo Day exhibition; bamboos are cut and laid out to dry on the hill slopes where fields are prepared for jhum cultivation. As bamboos are integral to jhum farming, and jhum forms the mainstay of agriculture across the state, almost everywhere you look, you find bamboos.

Yet, despite being an inextricable part of Mizo life and culture for centuries, bamboos face new peril as politicians in the state push hard for a new policy of land use that aims to cover the hills with settled agriculture and industrial plantations and end shifting agriculture for ever.

*7 March 2014.* On the first Friday of March, Mizo people across the state celebrate Chapchar Kut, before the farmers begin another spell of jhum. In 1995, I first watched the grand Cheraw performance at the Assam Rifles stadium in Aizawl, the state capital, as Mizoram celebrated its gospel centenary year, marking hundred years of Christianity in the tribal state. Although the state had seen great transformations in religion, traditions, and economy over the last century, the Cheraw itself had been retained as a deeper marker of culture.

Nearly two decades later, I travel to Aizawl from Dampa Tiger Reserve in western Mizoram to watch the bamboo dance during Chapchar Kut once again. I'd studied the effects of shifting cultivation on forests and wildlife in the mid-1990s in a number of sites in Dampa that I was now re-surveying. The sites include tropical rainforests with hoolock gibbons and hornbills as well as old jhum fields now covered with tall bamboo forests brimming with life. Leaving my research as a field biologist aside for a moment, I come to Aizawl for a glimpse into the cultural side of the bamboo story.

En route, on green, forest-covered hills, I see small jhum fields with slashed bamboos drying in the sun (these will be set alight later), even as smoke rises from fields fired early, the bamboos crackling and popping as they burn with consuming ferocity. Soon, in the ash-enriched soils, farmers will raise another season of crops, and when the spent fields are later abandoned for a new site the next year, the bamboos will rise again. The clearing of the forest is only temporary, the bamboo returns quickly and with vigour. My past research showed that in five years, over 10,000 bamboo culms would regenerate per hectare (mainly *mautak, Melocanna baccifera*) in the jhum fallows and the density of bamboo would increase even more if left uncut for longer.

On Chapchar Kut day in Aizawl, crowds pour into Lammual stadium, as young men and women who will perform the Cheraw stream onto the grounds below. The performers wear traditional dresses of bright red and green and black and white, striped and hatched with curiously bamboo-like patterns, the men with dark cloth headbands with perpendicular crossing stripes, the girls in bamboo-weave headbands topped by a ring of red feathers.

The brightly dressed youngsters carry stacks of green bamboo culms into the expansive grounds, placing them in sets of ten. Two culms about ten feet long placed in parallel a couple of metres apart set the bounds of the arena for the eight dancing girls; then, four pairs of culms are placed in perpendicular, to be held by eight crouching boys and clapped and beaten to the rhythm of the dance. As each group of Mizo girls and boys assembles at their placed bamboos, the sun-baked earth begins to bristle with colour and life.

It is nearly noon when the dance begins. *Clap-clap-slap-slap* the bamboo sets the rhythm; the boys alternately clap the hand-held pair of bamboos and slap it on the culms on the ground. The girls, in two rows of four each, have the four spaces between the paired bamboos and the spaces outside to move in. And with grace, elan, and joy, they begin to dance, their feet stepping in and out of the culms, in sync with the bamboo. They step and swirl and hop and turn, toss their heads and swing their arms, face each other or turn away, dancing to the incessant beat of the bamboo culms worked by the boys at their feet. Though they step and stomp and turn and toss, the girls inexorably return to the same spot where they started. The bamboo delimits the space they have to dance—first, between the bamboo culms, and then, before the boys slap it shut, hop over, onto the space between the next pair of culms, and out, and back again. In my photographs the girls are frozen, heads aloft, long hair swinging, feet in the air even as the bamboo on the earth makes space for where they will land now, keeps space for later, and will make space once more where they began.

Watching the Cheraw, I begin to think it is not unlike jhum itself, in which bamboo plays such a pivotal role. *Mautak* bamboos making space for this year's cultivation, reserving shifting spaces for the next few years, always with the prospect of return to place within the bounds of the bamboo.

In the hills of Mizoram, I tried to understand this cycle of shifting cultivation through field research and by talking to farmers. After a span of 5–10 years, when forest vegetation and bamboo have recovered sufficiently in old jhum fields, farmers return to the same site again. The bamboo forest that has sheltered the soil for years from sun and erosion is then cut, dried, burnt, and replaced by crops, forming the centuries-old cycle of cultivation and regrowth that has helped maintain extensive areas under bamboo and regenerating forests in Mizoram.

For every hectare of forest cut for jhum, at least 5–10 hectares are retained as forest in the landscape. Furthermore, jhum farmers also leave uncut many uncultivable strips of forest on ridges, in ravines and valleys, besides areas that form boundaries between fields. For local people as for forest plants and wildlife these uncut spaces serve as small but significant resource patches, natural buffers, and refugia—where species can survive even when the surroundings change—in the landscape. In these areas, besides *mautak*, other bamboos may be found: the stalwart *rawnal* (*Dendrocalamus longispathus*), the giant *phulrua* (*Dendrocalamus hamiltonii*), the sturdy *rawthing* (*Bambusa tulda*), and forest bamboos such as the elegant *sairil* (*Melocalamus compactiflorus*), and the beautiful, smooth *chalthe* (*Pseudostachyum polymorphum*).

Even as jhum fires consume bamboos, opening fields for cultivation and nourishing their soil with ashes, the bamboo springs up in fields of the year past, it endures in refugia and ravines, it leaps towards the sky in older abandoned fields. On the hill slopes, in the blowing breeze and whipping winds spurred by jhum fires, tall bamboo culms sway and clack and swing and dance with grace and beauty, not unlike the girls of the Cheraw. The bamboos step aside temporarily for a farming season of a few months, only to return later and reclaim the land. As farmers move from one bamboo patch to another every year, and return to each site after a few years, the bamboos first yield to farms, then reappear in the wake of the farmer, forming the perpetual cycle of field and fallow, of farmer and forest.

Now, the spectacular Cheraw at Chapchar Kut seems emblematic of jhum, symbolizing the life and spirit of Mizoram that shifting cultivation embodies.

*15 March 2014*. A week later, I am back at work in Dampa Tiger Reserve. Today, farmers of Serhmun village will start a fire on the hills near Tuilut, to meet a deadline set by the state government. From Damparengpui village, Lal Sanga takes me in his autorickshaw with a couple of others up the bumpy, winding hill road to Tuilut.

'Do you really want to go all the way to see that?' he asks.

It would turn out to be the loudest, hottest, most spectacular fire that I ever had witnessed at close range. A deliberate fire that would reduce to ashes what had been a dense bamboo forest. And yet, the fire did not signify destruction as much as it did a new beginning.

From the skies of Mizoram, the hills look like a green patchwork carpet, broken by small openings. The bamboo forests sway in light greens and fading yellows, amid straw-coloured fields of cut bamboo drying in the sun and more verdant strips with tall trees along ridges and ravines.

The windows of a bamboo hut, perched on the slopes of a steep hill near Tuilut, offer a view of fields, fallows, forests, and the valley of the Sengmatawk lui (*lui* means 'river'). All around, the forests lie in various stages of regeneration, marking the fields of the past, the present, and the future.

Almost everywhere you look in Mizoram, you see bamboos. Houses of bamboo perch on the steep slopes. Bamboo baskets and beds adorn the interiors, along with bamboo kitchen implements and pipes, bamboo walls, floors, ceilings, and mats—everything from sieves and lunch boxes to furniture and fish traps are made using bamboo. Bamboo vinegar is made as a byproduct of bamboo charcoal production. Bamboo coffins are promoted as a more affordable and accessible alternative to wooden coffins in this predominantly Christian state.

The range of crafted products illustrates bamboo's utility and versatility as well as human skill and ingenuity: bamboo hats, boards, curtains, flower vases, lampshades, bowls, jewellery, musical instruments, bags, baskets, intricately woven *thul* boxes with lids that snap tightly shut, and more.

In Mizoram, cradles are made of bamboo, as are the coffins.

At around half past two, six farmers walk briskly past the hut, down the slope, towards a large field of dry bamboo on another slope across the valley. I ask if we can take photographs of the fire; they agree without breaking their stride, in a hurry to get started.

They reach the field and fan out along the base. Timing is crucial, so the farmers shout to each other to co-ordinate their movements. They torch the dry vegetation at several points using matches and faggots. The fires catch immediately and grow rapidly. The flames lick their way in as roiling gusts, scudding across the fields of tinder-dry vegetation; the men back away to a safe distance. Higher up on another steep slope, a lone farmer sets his smaller two-hectare field on fire.

The parched bamboo culms crackle and burst in sputtering explosions, and a distant thunder of the conflagration carries over the hills.

Plumes of smoke billow over the landscape, sending ash and scorched leaves sailing in the wind. (Neighbouring villages may know of jhum from the descending ash, even if they cannot see the fire.) At the height of the blaze, all the way across on the opposite slope, we are buffeted by the wind and heat. It looks as if the great fire will consume everything in its path. And then, suddenly, the fire dies almost as quickly as it began. The two-hectare field burns in less than fifteen minutes; the larger field, over ten times as big, takes about an hour.

The fire, which is usually set to burn the fields from one demarcated edge to the other, or work its way inwards as a ring, generally remains well contained. When it escapes as ground fire, it is often because of surrounding land-use changes. As the fire subsides, the funereal colours and thunder give way to an ash-grey soil and to the silence of smoke rising from the earth.

Monoculture plantations that were absent two decades ago (oil palm, rubber) or were more limited in area (teak) increasingly occupy or abut traditional jhum lands. While bamboo forests retain their greenery even at the peak of the dry season, the teak plantations turn bare and brown. Below the leafless teak trees lies sparse undergrowth with paper-dry leaves and exposed patches of desiccated earth. Neatly burnt jhum fields form dark patches in the green bamboo forests, while occasional fires escape and spread through the dry understorey of teak plantations.

A few weeks after the fires, with the onset of pre-monsoon rains and thundershowers, the farmers will plant another season of crops. In each small field, they will cultivate around two-dozen varieties of crops including rice, vegetables, tubers, bananas, chillies, and cash crops.

With no external inputs such as chemical fertilizers or pesticides, jhum continues to remain an entirely organic system of multi-cropping, arguably a better form of land use than the intensive monoculture plantations being promoted by state and industrial interests in Mizoram. Still, the decks are stacked against the practice, regarded as primitive and destructive at least partly from the sheer visual impact of seeing forests cut and burn in such spectacular fires. Sadly, this is too narrow a perspective.

As we leave Tuilut in the evening, I look back at the burnt slope, the green forests, and the Sengmatawk lui flowing into the valley. To gain an appreciation of jhum, one needs to look beyond the burn and see the years of rapid regeneration that bring back dense bamboo forests to rested fields. More important, perhaps, one needs to learn to see the mosaic of fields, fallows, and forests, as the farmer does, through the windows of the bamboo hut.

Reflecting later on Chapchar Kut, I wonder at the subtle and intricate cultural connections and values that the Cheraw dance seemed to epitomize. The group dance celebrates the sense of community (*tlawmngaihna*) that Mizoram is famous for, the bamboo attests the connection to forest and land (*ram*), and the circular dance within the bounds of the bamboo appeared, to me, to reflect the rotational system of jhum cultivation itself. But today, as oil palm and rubber plantations begin to replace bamboo and jhum, an economy based on culture, diversity, and community is being replaced by one dependent on cash, permanent monocrops, and private interests.

What will happen then, in future, to the dance of the bamboos?

I imagine Mizo youth assembling at the Assam Rifles Grounds for the dance. But the boys aren't holding bamboo culms any more. They are at the gates in dark suits selling tickets, collecting cash. The girls, clad in monotonous green dresses patterned with spikes and needles, stand in the sun, alone, their arms aloft, their palms open and fingers splayed wide, their eyes staring, unblinking at the fierce sun. The girls' feet are fixed to the earth, unmoving. And through the gates, the spectators trickle in, to see the Mizos perform the oil palm dance.

# Bird by Bird in the Rainforest

*'Stop bouncing around like a ping-pong ball, you tailless piece of shit!'* I said.

And it worked! The tiny, truncated bird, smaller than a sparrow, hopped onto a thin twig—and paused. It was just a couple of seconds, but after fifteen futile and frustrating minutes trying to catch a glimpse in the dense rainforest undergrowth on a misty morning, this was enough to get one clear view. If you can call staring through a pair of Swarovski 8.5 × 42 binoculars at the briefly motionless bird 20 feet away—my breath held to avoid fogging the eyepiece, my cold fingers clenched around both barrels, my shoulders and elbows locked, my torso twisted like a snake curling up an imaginary tree, my knees bent so that I could look lower in the undergrowth, past ferns and herbs and leafy tangles and fallen branches, through a tiny gap—one clear view. A lot of effort, for one little bird.

I hadn't said the words out loud, of course, but this was no time to wonder at the power of silent abuse to pinion birds to their perch. I had details to note. A sprightly, dumpy bird, smaller than a sparrow. Underparts ashy grey; a rather big head coloured a dull olive green that continued onto his wings and back. A dark-tipped, pointed beak, like an elongated and sharpened pencil lead. Black eye stripe stretching past the eye, topped by a lighter yellowish supercilium—a stripe starting from the base of the beak, passing over the eye, to the side of his almost non-existent neck. Long legs, for a tiny bird, gripping the twig firmly, as he perched and gave me a slightly indignant eye over his shoulder. No tail. At least none that I could see. I would have been grateful for a few more seconds, but he had had enough. '*Chirririt!*' he exclaimed, loud for such a little bird. '*Chirririp!*' he let rip, again, before he bounced onto another twig, and plunged into the undergrowth. But I had seen enough. With a smug smile, I wrote his name down in my notebook: grey-bellied tesia. A touch darker grey below, a flash brighter yellow on crown and supercilium, and I would have had to put him down as slaty-bellied tesia, an entirely different species. Satisfied, I pocketed my notebook, left him to sulk and skulk in peace.

The morning had not begun this way, with me casting silent abuse at unsuspecting birds. I had started on a far more polite note. I'd walked into the regenerating rainforest at the edge of Dampa Tiger Reserve near Teirei in Mizoram. I was here to carry out a comprehensive survey of birds at the behest of the enthusiastic Field Director of the Reserve, Lalthanhlua Zathang, on behalf of the Mizoram Forest Department. The December dawn had just broken beyond the hills and the sun was yet to crest the ridge.

In the chill morning, the dark forest stood cloaked in a grey mist. From the canopy, dew fell like rain on the shrubs and onto the leaf litter covering the earth. Through the soft patter of falling dew, all around, I heard the calls of waking birds. The excited tweets of a grey-headed canary-flycatcher perched on some distant branch, the chatter of bulbuls flitting around some unseen tree, the piercing screech of hill mynas flying across invisible sky, and soft churrs, and metallic clicks, and nasal notes, all punctuated the quiet morning air even as I struggled to pinpoint their locations. And not a single bird showed.

Still, I was hopeful. And polite.

'Hey guys,' I said, 'I dropped by last evening, but it was all very quiet out here. I figured you were busy or tired and I shouldn't disturb you. Well, it is morning now, and here I am! Please come out. Say hello!'

'*Nyeaahh*,' said the white-throated bulbul, in a nasty, nasal drawl. Like all the others, he refused to show. The coward.

I tried again. 'Where are you guys?'

More patter of falling dew.

Minutes passed. I heard soft rasping notes alternating with a loud, continuous *chaunk chaunk chaunk* as a pin-striped tit babbler called from low branches. Elusive and ventriloquial, she frustrated my efforts to spot her, the soft notes seeming to come from nearer, the loud notes from farther away than where she really was. Farther ahead, I stopped near a *khuangthli* (*Bischofia javanica*) tree fruiting copiously: there were birds busily feasting on the high branches. It was perhaps just a tad hasty, a little too jerky, the way I raised my binoculars to my eyes; before I could focus, a large flock flared off the canopy in a great flurry of wings. Green pigeons, but which species? Pin-tailed, thick-billed, ashy-headed? No telling now, however crisply I focused on the still-quivering twigs. Later still, a little spiderhunter went, '*Which? Which?*' flying at top speed through the understorey. My eyes alighted only on empty spots, in mid-air, from where each call was emitted, by which time the bird had already zipped past, eluding me. From the skies high above, an unseen crested serpent eagle laughed shrilly, '*Heeeee heee hee*.'

I began to get worried, impatient. I had limited time on this trip, less than two weeks in the field, and only one morning to explore this trail, which snaked along the forested slope above Teirei lui. For the bird survey to be comprehensive, I needed to explore different trails and habitats and identify all species accurately. By preparing a complete checklist, documenting changes in bird communities across habitats, the survey could add to the knowledge on biological diversity in Dampa and potentially contribute to the conservation plans for birds here. I knew that cold, misty mornings were not ideal for birding, as bird activity tended to be low. Notorious skulkers like tesias and wren-babblers were hard enough to spot on brighter days, leave alone finding, identifying, and counting them on murky mornings. Such times are best avoided for a systematic census of birds, on point counts—stationary counts made at different points in the forest—or transect counts made along straight lines or trails. Right now, though, I was not constrained by rigid survey methods; I was willing to wait for bird activity to pick up as the sun rose higher, and watch quietly till the birds showed themselves. For this to work well, the birds needed to co-operate, too; it wasn't just a matter of my skill. Or was it?

'*Qu-ick*,' said a bird from the shrubs. '*Qu-ick!*'

An unknown call, yet strangely familiar, like the voice of a long-forgotten friend who calls you out of the blue, 'Do you recognize who this is?' I scanned the undergrowth even as I racked my brains trying to recall if this was a bird I'd once known. I'd studied birds for many months in Dampa earlier, but all that was nearly two decades ago. Then, I had learnt to identify by sight and sound over 200 species in a matter of weeks, while trying to strictly adhere to a policy that 'no record at all is better than an erroneous one.' When I finished that study in the summer of 1995, my list held over 210 bird species, and I prided myself in knowing the calls of virtually all the birds I encountered on my birding walks. 'Here sings a black-naped monarch,' or 'There calls a red-headed trogon!' I would note, or merely pause to listen to the soft, subtle notes of a snowy-browed flycatcher in deep rainforest. But now, I felt like I was back at square one, more neophyte than past-master. I felt compelled to reaffirm my acquaintance with these birds again, as if I were coaxing faded friendships into comfortable familiarity once more. But perhaps, in the return, there was opportunity, too: like when rotational jhum farmers returned to cultivate a fallow, after it had regenerated a full twenty years, finding soil rested and replenished by age.

'Come on ... come on! I can see the leaves shaking there near the ground where you are flitting around. Come out where I can see you.'

A five-minute wait and he refuses to show. I imagine him saying, 'Sorry! Can't come out now.' More likely, he didn't care a whit for my plight. His voice had triggered a cloudy memory, and a name had been forming in my head: buff-breasted babbler. But before I could confirm, he just left. Vanished.

'Bastard!'

This is when things started to get hairy. Here I was, having travelled thousands of kilometres to do this work, ostensibly to benefit the birds of Dampa, and they were simply failing me. Or—and the thought came close on its heels—I was here on my own work and failing myself. Had my field skills declined in these two decades? Did I need to toil that much more, strain my ears a little harder, watch much more keenly, to do the same things that I had managed without much effort? Had I forgotten the habits of the birds, their individual quirks and mannerisms that had earlier guided my eyes and ears? I had foolishly walked into a rainforest without the tape recorders and playback equipment that others use to lure birds out; without the mist-nets to snare them and securely identify them in-hand; without cameras and long lenses that captured them in a trice and allowed me to comfortably identify them later on a computer screen. Was I a stubborn, old-fashioned geezer? An outmoded snob who believed, as I still did, that all you need for a good bout of birding was a choice pair of binoculars, field notebook, and pen?

I had no time to reflect on the answers—it was easier to deflect self-doubt and self-loathing onto the birds. And that's how, instead of naming the birds I found, I began calling them names.

After failing to find birds the previous evening, I had returned to my room at the Teirei Forest rest house to find solace, as I often do, in reading. As darkness fell and the temperature dropped below 10°C, I tucked into my sleeping bag and opened the book I'd carried along—a book on writing by Anne Lamott which, oddly enough, was titled *Bird by Bird: Some Instructions on Writing and Life.*

In the second chapter, titled 'Short Assignments,' Lamott explains the book's title. She recalls how her elder brother, when he was ten years old, was struggling to finish a report on birds, which he'd had three months to write, and which was due the next day. Close to tears, he sat at the table, ringed by binder paper and pencils and bird books that lay unopened, frustrated by the huge task ahead of him. Then, his father sat beside him, placed a reassuring arm on his shoulder, and said:

'Bird by bird, buddy. Just take it bird by bird.'

Well, there I was the next morning, on my own short assignment, ready to put aside my failures and take it bird by bird. But the damn birds refused to show. I don't know what you would have done in this situation. I screamed, mutely, at every mysterious bird call. I let fly, motionless, at every fleeting glimpse. In complete silence, I cursed.

With that, my luck turned. The tesia was just an early victim. My patience exhausted in fifteen minutes, I pinned him to the twig with one cutting comment.

A little later, briskly turning a bend, I spooked a bird that exploded from virtually at my feet.

'*Freeze, asshole!*' I said, behind gritted teeth. The instruction applied to both of us. I stood, binoculars glued to my eyes. The bird alighted on a slanting bamboo culm 30 metres away and glared back. Feather for feather, he was one of the most beautiful emerald dove males I had ever seen: coral-red beak and silver-capped head on wine-lilac neck, zebra-patterned rump immodestly flaunted under emerald wings. I could have stood there for ever. Only, the feeling was clearly not mutual and the bird hustled away in a clapping flutter.

'And what are you fussing and churring and whistling about? Yes, you with the nervous tic, with your bunch of buddies on the branches. Show yourself clearly or shut your frigging mouth!'

It seemed rather extreme, even to me, to thus lambast what turned out to be a coterie of shy brown-cheeked fulvettas winding its way away through the bamboo. They were nondescript and dull birds, brown with a touch of grey on their heads, foraging in the shadow of bigger and more colourful peers.

The fulvettas made a dignified exit after their brief showing post my rant. Only, after they left, other words—far greater than mine—came to mind and refused to leave.

> The fault must partly have been in me.
> The bird was not to blame for his key.
> And of course there must be something wrong
> In wanting to silence any song.
>
> — 'A Minor Bird,' Robert Frost

The fault was partly in me. A many-layered fault, of finding excuses when I failed at finding birds, of being stubborn, snobbish, or merely impatient. It was like blaming friends—who I had forgotten for years, not seen, nor cast a thought in their direction—for failing to show up when I wanted them to. Like them, the birds lived neither for my convenience nor my disposal; they had lives of their own, and were free to roam at will. It was I who needed to make an effort to see them, to understand, once again if need be, who they were or weren't. As Anne Lamott writes:

> ... if you want to get to know your characters, you have to hang out with them long enough to see beyond all the things they aren't. You may try to get them to do something because it would be convenient plotwise, or you might want to pigeonhole them so you can maintain the illusion of control. But with luck ... you will finally have to admit that who they are isn't who you thought they were.

And what if, like the birds, the knowledge that I sought was not something to be chased or coerced into revealing itself? If the best I could hope for was to remain receptive and observant, and let the story show in its own time as a reward for attentive and repeated effort?

Finding the bird, identifying the species, knowing their calls and habits: these were just the first but crucial steps of a long chain of things I needed to do to translate a confirmed sighting to something of larger substance. I still needed to systematically cover various habitats, from streams and rivers to fallows and forests, resurvey transects I had walked two decades ago, measure vegetation attributes such as tree density and canopy cover to quantify habitat change; enter, verify, and analyse the collected data; then interpret and write my findings, all in the hope that it would lead to greater scientific understanding and better conservation and management efforts on the ground.

Years ago, this had formed the bedrock of my work on the effects of shifting cultivation on rainforest birds in Dampa. Now, while surveying the same areas again, I had the opportunity to take my earlier work ahead, deepen my understanding of recovery of rainforest vegetation and bird communities. After this short trip in December, I would come again in February for several weeks, but I was beginning to wonder if even that much time was enough.

Even during this short visit, I was already concerned about the changes in land use around Dampa. Monoculture teak and rubber and oil palm plantations were replacing diverse secondary forests, and traditional livelihoods based on shifting cultivation on community lands were being beaten back by government and corporate interests to bring in economies based on cash and private ownership. In such a context, Dampa's birds appeared inconsequential and irrelevant, but they, too, helped appreciate the changes: the presence and kinds of birds in various sites serve as revealing titres of transformation in land use when habitat alteration reaches its threshold and that little extra drop of disturbance irretrievably changes the colour of the landscape. But, I realized, the birds were not the primary instrument of the assay—they were living measures of change in landscape. *I* was the blunt instrument making the measurements, scrawling notes and observations into my fraying field notebook. What if I was not up to the task? If all I could achieve was a mismeasure of a pertinent conservation issue, a partial diagnosis stemming from my own limited capacity? Would I be able to describe my results clearly—after recording the right birds, find the right words, too? Again, I took encouragement from Anne Lamott.

> If you don't believe in what you are saying, there is no point in your saying it.... However, if you do care deeply about something—if, for instance, you are conservative in the great

sense of the word, if you are someone who is trying to conserve the landscape and the natural world—then this belief will keep you going as you struggle to get your work done.

To be a good writer, you not only have to write a great deal but you have to care.

The trail that led to the river was overgrown. I could hear the rush of the river over rocks a hundred feet away, but could not see it through the tangle of vegetation. The plants were wet with morning dew, sparkling in the sun that now rose over the trees. As mist steamed off the plants, I waded through grass and fern and sedge, wet to my thighs, squelching along. I dodged swinging banana leaves and shoved bamboo culms and branches out of my way. With all the noise and disturbance of my passage, there was no question of finding birds. The trail almost petered out, and so did the morning. I decided to turn back before I started cursing the plants. I would come back, later, begin afresh.

Through a small break in the vegetation, I saw a small segment of the Teirei. On a rock near the middle of the river sat a small brown bird, who flicked out into the air, and returned to her perch. A female plumbeous redstart. She kept sallying out, to catch flying insects perhaps, returning each time to the same spot. Out and back, out and back. To me, she looked loopy with life.

# Abode of Rainforest Rarities

From deep in the rainforest a shrill, plaintive whistle rose to sunder the stillness of the night. Then there was silence, but for the chirping of insects and the faraway wail of a flying squirrel. Then the whistle sounded again, strange and mystifying. In over a year's stay in the rainforest camp at Sengaltheri, I had heard nothing like it. Picking up a powerful torch and a pair of binoculars, I slipped out into the forest. The whistle grew louder as I approached and resolved into three distinct notes, rising and falling. With slow, silent footsteps I crept up the narrow trail, watching out for pit vipers and wondering. Was that the call of a nocturnal mammal or a bird—was it the ... yes, there it was! The two large eyes reflected a deep red in the beam of the torch. I was staring into the eyes of one of the rarest creatures inhabiting the rainforest—the Sri Lanka bay-owl.

For a quarter of an hour, I watched the bird. The owl flew about soundlessly through the dense vegetation, adept at avoiding the tangles of branches and lianas. From farther away, the bird's mate responded to the calls. A flying squirrel launched into the air from the canopy of a tall tree nearby and glided away into the darkness. Returning after the exciting nocturnal foray to my base camp in the Kalakad-Mundanthurai Tiger Reserve, I pondered over how, even after months of observation, the rainforest had an endless repository of surprises.

Few parts of the world hold greater fascination for the naturalist than the tropical rainforest. Copiously supplied with rainfall, often exceeding 300 centimetres annually, the rainforest is resplendent in lofty, luxurious, evergreen vegetation, because of which it is also called the tropical wet evergreen forest. Globally, biologists have identified 35 regions as 'hotspots' of plant and animal diversity, including rainforests. It is in this ecosystem that life blooms in virtually unsurpassed and exuberant variety.

Within India, a unique and magnificent tract of rainforest occurs in the hills of the Kalakad-Mundanthurai Tiger Reserve in southern Tamil Nadu and the adjoining hill tracts in the state of Kerala. Spread over 895 square kilometres in the Agasthyamalai-Ashambu Hills, the Reserve boasts of nearly 400 square kilometres of mature tropical rainforest. Today, few areas in the Western Ghats, of which these hills form the southern extremity, can boast of such a vast, continuous tract of rainforest from the foothills to the summits of high mountains.

Three key words define the biological importance of the Kalakad-Mundanthurai Tiger Reserve: *diversity*—of habitats and species; *endemism*—the occurrence of species unique to the Western Ghats; and *rarity*—the existence of globally rare and threatened species. On each count, the Reserve is critically important for conservation in the Western Ghats.

A measure of the amazing biological diversity in the Kalakad-Mundanthurai Tiger Reserve is evident from its plants. About

2,000 of the 4,500 species of plants known from the Western Ghats are believed to occur in the Reserve, and botanists are still working on the growing list. Two hectares of rainforest may have over two hundred species of plants. Among the rainforest animals, spectacular ones such as great hornbills and lion-tailed macaques are well known, but it is the diversity of lesser creatures that is truly breathtaking. Butterfly and moth of every hue, and all manner of invertebrates from pinhead springtails to monstrous giant wood spiders and stick insects occur. There is no doubt that innumerable species of insects are waiting to be discovered. The 1998 discovery by field biologists from the Wildlife Institute of India of species such as the orange-lipped forest lizard is only the tip of the biological iceberg of wealth sheltered in these forests.

A fundamental factor explaining the diversity of plants and animals is the variety of natural habitats contained within the Reserve—from dry thorn scrub, deciduous, and semi-evergreen forests to rainforests, high-altitude grasslands, hill streams, and rivers. Each habitat has its own associated community of species and level of diversity, with the rainforest tending to pack in more species per unit area than the other habitats. Man-made environments such as tea and cardamom plantations, the inevitable *Eucalyptus*, and reservoirs add to the habitat diversity. However, as these habitats tend to be colonized by common, widespread, and non-forest species, their significance for conservation of plants and animals typical of the region is moot.

Biologists use the term 'endemic' to describe species restricted to a region. The Western Ghats is an important centre of plant and animal endemism. The conservation of endemic species is important since their local extinction is equivalent to global extinction. The Kalakad–Mundanthurai Tiger Reserve is a haven for a number of Western Ghats endemics. Besides the famous lion-tailed macaque, recent research has shown that mammals such as the Nilgiri marten, the brown palm civet, and Malabar spiny dormouse, thirteen species of birds, and a large number of reptiles, amphibians, and plants that are endemic to the Western Ghats occur within the Reserve.

Rarity is a factor more significant for conservation than diversity or endemism. Rare species are often, though not always, the ones closest to the threat of extinction. Where the rare species are also endemic, the situation is compounded. Rare species can be roughly categorized into two types—those that are common in some small, scattered patches of particular habitats and are hence rare when viewed from a broad-scale perspective, and those that are rare throughout in a given area or habitat. The broad-tailed grassbird and Nilgiri tahr, restricted to the high-altitude grasslands in the Reserve, are examples of the former, whereas species such as the bay-owl, the orange-lipped forest lizard, and many large birds and mammals belong to the latter category.

A critical issue at stake in dealing with rare species is the area of habitat. A single pair of hornbills or owls or a single leopard may require an area of several square kilometres to meet their year-round needs for food, shelter, and breeding. This is what makes large protected areas such as the Kalakad–Mundanthurai Tiger Reserve so important: first, that many rare species have substantial habitat to maintain their populations, and second, that there are very few such areas left elsewhere. The lion-tailed macaque is a case in point. The 300–400-strong population of lion-tailed macaques in the Reserve is perhaps the largest left, and represents more than 10% of the macaques remaining in the wild.

Maintaining the integrity of the rainforest is critical not only to conserve plant and animal species, but also to safeguard the benefits it provides millions of people in the form of a single natural resource: water. Several perennial streams originating in the higher reaches of the Reserve supply rivers, such as the

Tambiraparani, Servalar, Ramanadhi, and Manimuthar, that help feed millions of people in the hot, dusty plains of Tirunelveli District. Besides the ecosystem benefits from the forests, their biological wealth represents an invaluable reservoir of ecological, genetic, and economic potential. As an accessible haven in a sea of humanity, the Reserve has immense potential for environmental education and biological research, and it has only begun to be tapped.

To achieve their goals of protection and conservation, the Reserve's managers have to tackle a number of problems. Maintaining vigil against poachers of wildlife, sandalwood, and the three Cs—cane, cinnamon, and *Canarium* (black dammar)—is a major task. An equally tricky problem is the control of tourists, especially the deplorably large numbers who misbehave and litter, as well as the crowds of pilgrims who visit the temples within the sanctuary. The management is also trying to solve problems posed and faced by the people living on the periphery of the sanctuary through eco-development initiatives involving village committees. Given all this, there is an urgent need to create awareness of the immense values of the Reserve, and build support for its continued conservation.

# Shadowing Civets

Deep in the rainforest, the monotonous *plip-plip* of the radio receiver kept strange rhythm with the *chill-chill of the stream frogs*. Dawn's gentle fingers, probing through the dense canopy, barely lit the leaf-strewn forest floor. Through the lifting mist we walked, the rich earthy scent of the forest in our nostrils, the morning chill fresh on our faces, dew condensing on our eyelashes. For miles on all sides stretched unbroken rainforests within the Kalakad–Mundanthurai Tiger Reserve. These exquisite forests swathed the slopes of the southern end of India's Western Ghats, renowned for its biological diversity and endemic species.

Ears glued to her headphones and directional Yagi antenna held aloft, biologist Divya Mudappa worked her way through the undergrowth. We were on the trail of the first radio-collared brown palm civet, one of Asia's most elusive and poorly known small carnivores. This civet species is endemic to the tropical rainforests of the Western Ghats mountain range running along India's west coast, a global biodiversity hotspot. In a swift operation the previous night, after weeks of preparatory baiting, Divya and her colleagues from the Wildlife Institute of India had caught and radio-tagged a sub-adult male labelled K70, then released him at the site of capture. The activity-sensitive collar had signalled a night of lively activity leading to a lull as dawn broke. Our task was to locate K70 within his home range and identify the day-bed he had chosen.

Civets are small cat-sized mammalian carnivores, primarily active by night and known to use tree hollows and notches to rest during daytime. Yet, when we finally located K70, we were in for a surprise. This was no hollow or tree-notch; K70 was curled up in a leafy nest of the Indian giant squirrel placed on a stout branch around 24 metres up a *Syzygium* tree. Giant squirrels are sprightly arboreal creatures that break twigs and leaves to build nests, called dreys, in the canopy of Asia's tropical moist forests. Being rather timid and careful in disposition, their dreys are chosen with care on tall, large-boled trees. A single giant squirrel may build four to eight dreys within its home range of about a hectare, not using all at the same time. Within a civet's larger home range, there may be dozens of such nests, most not being used by squirrels. The civets capitalize on the availability of these giant squirrel dreys, using some as day-beds. Although no direct interaction between squirrels and civets has been seen at these nests, perhaps they occasionally operate a civet-by-day and squirrel-by-night time-share with an amicable switch performed at dusk. As Divya's research continued and more civets were radio-tagged, she found that over a third of the daybeds used by brown palm civets were giant squirrel dreys, the remainder in tree hollows, vine tangles, and forks of branches.

In retrospect, it is almost puzzling that even this basic aspect of the civet's ecology was not known when Divya began her studies in 1998 with a Wildlife Conservation Society (WCS) Research Fellowship. Yet, that was the case—and it holds true for a great diversity of civet species in south and south-east Asian rainforests. Being nocturnal and elusive, often restricted to dense forests, civets have been difficult to observe and document in the wild. Even today, for most species, we know little about their distributional range, food habits, habitat requirements, social behaviour, or breeding in the wild: almost all aspects we need to know to understand civet ecology.

Yet, the tide is changing. Using a slew of methods from traditional spoor surveys and nocturnal spot-lighting with binoculars to modern techniques such as radio-telemetry, motion-sensitive infra-red camera traps, and DNA analysis, backed up by hundreds of miles of footwork, biologists are throwing light on the darkness that has shrouded these elusive creatures. Studies from south and south-east Asian forests, particularly over the last two decades, are helping unravel the civets' distribution and ecology. These are timely studies, as large parts of this region have high rates of forest loss and conversion, coupled with high human populations. The handful of research surveys carried out thus far has yielded a mixed message. Civets appear resilient to habitat loss and alteration in that they continue to occur in altered tropical habitats such as logged forests or shade-coffee plantations. Yet, their populations are reduced and their long-term survival in these rapidly transforming landscapes remains questionable. Whether all civet species can continue to survive in plantations if there are no remnant forest tracts in the vicinity is an aspect yet to be clearly documented.

Much useful information on the brown palm civet was also generated in Divya's study in the Western Ghats. Meticulous collection and observation of the scats of these rainforest civets over three years showed that these animals, despite being taxonomically categorized as carnivores, consumed a great diversity of fruit throughout the year. Around 60 species of rainforest trees and lianas provided fruits—mainly pulpy drupes and berries—that civets eat. The predilection of palm civets (belonging to the family Paradoxurinae) for fruit has been anecdotally recorded for long, although the extent of their fruit consumption and their year-round dependence on it had not been clearly evidenced.

A little further on the trail from our base camp in Sengaltheri that morning, we encountered a small log fallen across the path. On it was a palm civet scat with eight *Holigarna* seeds, dark from the black, blister-inducing sap of this tree belonging to the mango family (Anacardiaceae). The scat also had numerous small fig seeds and the remains of a beetle exoskeleton. A significant aspect of the civet's fruity diet is that they mostly consume only the pulp, passing the seeds virtually undamaged through their scats. The population of civets at Sengaltheri was thus carrying thousands of mature seeds from ripe fruit picked from the branches of rainforest trees or lianas and depositing them at various distances in the surrounding forest—an invaluable service of seed dispersal. Through germination experiments, Divya found that some species appeared to benefit from such civet dispersal; seeds from scats germinated better or faster than seeds from ripe fruit that had fallen under the trees. In some trees, such as the elaeocarps that fruited copiously—with perhaps too many fruits for them all to be consumed by the civets—clusters of rotting ripe fruit lay on the forest floor beneath the trees, their doomed seeds providing a bonanza of food for seed-boring beetles. Recent research on many tropical rainforest plant species around the world has highlighted how animals such as hornbills and civets provide a survival benefit to seeds they disperse away from the vicinity of the parent tree. Nevertheless, palm civets tend to deposit their scats on fallen or rotting logs, tree trunks, and rocks, and the fates of these seeds in terms of survival or regeneration remain to be discovered.

The verdant and diverse tropical rainforests of Kalakad-Mundanthurai Tiger Reserve appeared to be an excellent habitat for the brown palm civet. Encounter and photo-capture rates were higher here than in other disturbed forests or plantation areas. Divya's work also showed that individual civet home ranges here were less than 60 hectares. This was considerably smaller than the home ranges larger than 300 hectares reported for other civets by WCS Scientist Alan Rabinowitz from studies in seasonal dry tropical forests of Thailand and also smaller than the 100 hectare home ranges reported for Malay civets in a south-east Asian rainforest of Borneo—the only comparable studies available. Divya attributed this to an adequate, year-round supply of fruit and the dense, relatively undisturbed wet evergreen forests all around. On the flip side, she surmised that in degraded habitats, owing to lower plant diversity, fruit availability, and fewer suitable day-bed sites, civets would occur at lower densities and require larger home ranges to meet their needs. This was partly confirmed when she surveyed rainforests fragmented by tea and coffee plantations in the Anamalai Hills, another region in the southern Western Ghats. The endemic brown palm civet appeared to have declined in abundance, particularly relative to other more common and widely distributed small carnivores such as the small Indian civet and mongooses.

At first, these results did not appear to bode well for the brown palm civet. With annual deforestation rates still over 1%, the Western Ghats forests continue to be altered and threatened by various forces of development, urbanization, and forest conversion. In the previous decade, brown palm civets had been reported only in a few localities in the southern half of the mountain chain and their conservation status appeared precarious. Yet, Divya's initial survey and later work with Nandini Rajamani, another field biologist, were encouraging. In nocturnal surveys, they noted the species in dozens of tropical rainforest localities, and even in some shade-coffee and cardamom plantations, from the southern tip of the Western Ghats north to the state of Goa.

Concurrently, an offshoot of the seed germination work Divya had carried out at Sengaltheri began to yield results that seemed to point the way forward in fragmented landscapes. At the end of the germination study, it seemed only natural that the numerous rainforest tree and liana species raised could be planted in degraded rainforests. So, in 2000, we launched a reforestation project in the Anamalai Hills, a former rainforest landscape now covered by tea and coffee plantations. Within these plantations survive over 40 rainforest fragments ranging in size from 1 to 200 hectares. Many of these fragments are degraded because of logging or invasive weeds. Research indicated that these parcels retained high conservation value because they helped remnant populations of endemic and threatened 'species' such as the lion-tailed macaques and great and Malabar grey hornbills, survive. They also act as corridors for large wildlife such as Asian elephants and wild dogs or dhole that move between adjoining protected areas. The viability of these wildlife populations was, however, in question as forest loss and degradation continued.

As an initial step for the restoration programme, the Nature Conservation Foundation, based in Mysore, established a nursery near Valparai in the Anamalai Hills with support from local plantation companies. Since 2001, more than 150 tree and liana species native to the mid-elevation tropical rainforests of this region have been raised from seed and nurtured in this nursery. After lobbying and earning the support of plantation companies including Tea Estates India, Parry Agro, and Tata Coffee, more than 55,000 seedlings were planted in seventeen degraded rainforest fragments. Two-thirds of the seedlings have survived beyond two years. Coupled with natural regeneration in these protected sites—they had only an open canopy and weedy understorey a few years earlier—they are beginning to look more like recovering rainforests. If this hopeful trajectory continues, one can visualize the re-emergence of the tall rainforest canopy with macaques, hornbills, and brown palm civets eating fruits, dispersing seeds, and reviving natural cycles.

On that day in Sengaltheri, these events were still in the unknown future. Later that night, we went radio-tracking K70 again. As darkness descended over the slopes, he emerged from the day-bed and moved easily, swiftly through the high canopy. If not for the radio-collar, we would have lost him in a minute. He walked through the canopy, the branches barely shaking, and headed towards a fruiting tree. We shadowed him to the tree but could not see him despite receiving the signals from directly overhead. The mewing call of an Indian giant flying squirrel sounded from above and we sensed more than saw its swooping glide from a high branch to land with a soft thud against a low tree trunk nearby. We saw a mouse deer—rather the glint of his eyes in the undergrowth—before he merged into the forest. The tree canopy almost closed the sky over our heads, and only a slight glimmering of starlight showed through; the moon was yet to rise. Oblivious to us, K70 moved briskly but silently away, a mere shadow in the darkness, in the rainforest of the night.

# Kalakad

## *Three Years in the Rainforest*

A place that is marked by the presence of people is not unusual, but a place that leaves an indelible mark on people is extraordinary. In the ancient mountains at the southern tip of the great Western Ghats ranges, sheltering among rocky peaks and rugged slopes draped with tall evergreen forest, lies one such place. A place of beauty and challenge and diversity, which if you have really experienced, you will declare has no real equal. And if you have lived and worked there, wherever you go, the place will go with you. It will remain a benchmark, a touchstone, a reference point in felt memory and field experience, against which you will forever measure other places, newer knowledge.

* This essay is written with Divya Mudappa.

A place that does all this, slowly, gently, but inevitably, is Kalakad-Mundathurai Tiger Reserve.

Near the southern tip of the Indian peninsula, the Kalakad-Mundathurai Tiger Reserve sprawls over an expansive forest landscape within the Western Ghats of Tamil Nadu state. Occupying 895 square kilometres, it adjoins other wildlife sanctuaries (Neyyar, Peppara, and Shendurney) and reserved forests lying across the administrative boundary in Kerala state, forming a forest tract nearly twice as large over the Agasthyamalai-Ashambu hill ranges. Biologists consider this landscape one of the most significant areas for conservation of biological diversity in the Western Ghats. It retains one of the largest and last remaining unbroken tracts of over 400 square kilometres of tropical rainforest, much of which has not been logged or converted to plantations, ripped by roads, or ravaged by mining like many other parts of the Western Ghats have been. Partly for these reasons, Kalakad-Mundanthurai Tiger Reserve offers an unparalleled opportunity to understand the ecology of rainforest plants and animals in a relatively undisturbed setting: an understanding that is a vital step to help conserve such a place for posterity.

From the wide sweep of the Tirunelveli plains, the Kalakad mountains rise abruptly in looming grandeur. South of Tirunelveli, on the national highway that runs down to Kanyakumari at the southern tip of the Indian peninsula, the road turns sharply west towards the mountains. It passes through a rich countryside where paddy, banana, and other crops are grown in flatlands amid scattered lakes, old village ponds, and rocky outcrops. Past villages at the foothills, the road ascends the mountains to a Forest Department camp at the edge of the rainforest.

A narrow foot trail about a mile long snaked uphill along a torrential water channel, then across a river, and through the rainforest. Up this trail, on which everything from rice and gas cylinders and pipes and field supplies had to be carried, in the shadow of Kulirattimottai mountain, we had established a base camp that became our home for three years. It was an abandoned house with a cardamom drying room, the remnant of an expired plantation lease. There was no electricity or any other modern facilities, but as a camp literally a step away from the rainforests, it was perfect. People said we were cut off from the world. Yet, there in the rainforest, we felt more immersed in the world than ever before.

We had come there to study small mammals and birds, posing fundamental questions of ecology: on the distribution and abundance of species in relation to their environment. What were the small mammal and carnivore species—from rodents and shrews to civet and marten—that lived in the rainforest? And what was the community of birds? How did the distribution and abundance of all these species change from lower to higher elevations or from abandoned plantation and previously logged forest to undisturbed mature tropical rainforest? How did endemic species, such as the nocturnal brown palm civet, thrive in the rainforest: how much area did the civets need; what did they feed on; where did they roost by day before they set out to feed by night?

With a bunch of such questions tucked into our belts, we set out to answer them through field research and observations. We laid quadrats to measure vegetation and grids and catch-and-release traps for studying rodent populations. We surveyed transects and point counts for birds and walked trails with tagged trees to document monthly patterns of leaf-flush, flowering, and fruiting of rainforest trees and lianas. We radio-collared brown palm civets to track and study this elusive and enigmatic species. With eyes and ears on the mountains and feet on the earth, we tried to discern the pulse and flow of the rainforest.

Immersed in the rainforest our work slowly brought us to appreciate the enduring rhythms of nature and its cycles of renewal. From early morning counts of birds, daytime surveys of plots and trails and transects, through nocturnal tracking of civets onto the next day: this was our diurnal rhythm. Every morning, the spot-bellied eagle-owls tucked into their tree hollows, and as the sun crested the mountains, the black eagles came skimming over the treetops. At the end of the day, as the giant squirrels went to roost in their tree nests, the flying squirrels and civets emerged to roam by night.

Then came the pulse of the seasons. The year opened cool and dry, or laced with the moist departure of the northeast monsoon, and *Canarium* trees flared red amid a sea of rainforest green. After the elephants passed by in March, peeling tree bark and snacking on *Ochlandra* reed bamboos, came two hot and tempestuous months with pre-monsoon thunderstorms that revived the wilting shrubs and replenished rainforest streams. Then, from June to September, the southwest monsoon reigned, with short sunny mornings and rain-lashed afternoons under dark, gloomy skies. The forest turned damp, as did our clothes and books and everything else in the camp, and fruits of *Palaquium* trees littered the forest floor and little seedlings sprung up on the moist leaf litter.

Then, as one monsoon withdrew, depressions in the Bay of Bengal brewed another. The northeast monsoon brought persistent, torrential rains and thick mists that swallowed the rainforests barely twenty metres away from our doorstep and poured in through the windows into our home. The swelling rivers, which sometimes flowed over the trail cutting off our base camp, thundered down the valley, carrying revivifying waters to the people in the plains. Even during a deluge it was remarkable how there was so little erosion—the slopes were swathed in dense forests—and the waters remained clear and pure to drink. Finally, as the year wound down, the winds and clouds and rains withdrew, cool, clear skies would open over the forests again, and the crimson flush of *Canarium* would flag the beginning of another year.

The rainforests were a place of eternal surprise. Even as we went in search of answers, looking for our study species, other creatures, puzzles, and wonders confronted us. We could take nothing for granted: all our senses had to be on alert all the time.

The trail cameras had been set, the civets collared, but dense vegetation kept much hidden. In the night, our spotlight would reveal little more than the shining eyes of flying squirrel or civet in the canopy, a shy mouse deer nibbling on fruits fallen on the forest floor. Even by day, we noted birds more by their songs than by sight, although a glimpse of an elusive Malabar trogon or the sweet songster, the endemic white-bellied blue flycatcher, was an almost daily joy.

Such sights and sounds hinted at what the forest held, and yet they constantly surprised us. That loud honk was not the alarm bell of a distant sambar, but the courtship call of a nearby frog; that black blur on the branches was not a scampering giant squirrel, but a Nilgiri marten on his hunt; that repetitive pulse was not the beep of a receiver left on by mistake, it was a tiny cricket cheeping in the undergrowth; that flash of yellow streaking from tree trunk to trunk was no darting woodpecker or butterfly, but a *Draco*, the flying lizard; that whistle emerging from the dark rainforest by night was no forlorn cry of mystery mammal, it was the haunting call of the rare Sri Lanka bay-owl. In the rainforest, even a sudden silence or a carpet of fallen *Mesua* leaves revealed something—the hushing of an unseen cicada on tree bark under the scanning eye of a treepie, or the passing of a sated troop of langur in the trees.

Watching animals, we learnt more about plants. The civets, although carnivores, ate more fruits than animal prey, and so we tried to document and identify the fruits and the plants they consumed. And fruits were always there: every month, through the year, some species provided sustenance to civets and macaques and birds such as hornbills and mountain imperial-pigeons. Seeing seedlings sprouting from civet scat or trail side, we grasped how so many native rainforest plants could be regenerated from seed, into seedlings that could be planted to bring back rainforest in abandoned plantations and other degraded sites.

We had come to the rainforests for our research, but when we left three years later, we left with so much more. Working by day and night, more than what we came to study, we learned about natural history and ecology of the rainforest. And what we gathered informs and guides us to this day. As we completed our doctoral research, wrote our thesis and papers and reports, we began a project to ecologically restore degraded rainforest fragments in the Anamalai Hills, a range about 200 kilometres to the north. The work was inspired by field experiences in the Kalakad rainforest. It was this place that taught us to not just take away new knowledge, but try to give back something through informed conservation actions. It taught us how we could assist the civets in their task of forest regeneration, how we, too, could contribute to renewal as farmers of the forest.

Fourteen years later, the saplings planted in the Anamalai Hills now reach towards the sky having become young trees over twenty feet tall. In the restoration site, the young *Canarium* flames upward year after year, alongside quick *Elaeocarpus* and slow *Palaquium* and many other species, and on the leaf litter below, a passing civet has deposited a fresh batch of seeds.

The plants evoke a recollection of a distant rainforest, a home by the river running below the rocky dome of Kulirattimottai, where we would like to be again—to be reinvigorated, to learn, to be surprised anew. Yet, in this moment, the forest does not seem to be outside of us at all: seeing seed and scat and surging sapling before our eyes, we perceive the rainforests of Kalakad.

# Feathered Foresters

In the din of rushing water, we did not hear the hornbill in flight. Neither the loud whoosh of her wings nor her clanging calls. It was the large shadow that darted over a gap of sun-drenched forest floor that made us look up. In smart black and white plumage and characteristic heavy beak, a female Malabar pied-hornbill was flying over the forest and across the river. Even in flight, the horny protuberance, called casque, was visible on her beak: a curving black-topped wedge, like a beak upon beak.

She flew to a tall *Bombax* (wild silk cotton) tree on the opposite bank of the rushing river, and was instantly busy, hopping up the branches and on to the tree trunk, exploring. She flapped her wings and perched by the side of a hollow in a tree. Thrusting her head and her long curved beak into it, she probed and pulled

* This essay is written with Divya Mudappa.

out bits of debris from the tree hollow and tossed them into the air. Little twiggy pieces of bark and rotting leaves danced their way to the ground, or caught the air and sailed into the river. In February, with the year still young, she appeared to be examining the hollow as a potential nest for the season.

We then noticed another hornbill, perched on a branch nearby. This was a male, identified by his slightly larger black casque, and a dark, rather than pale, eye-ring. The birds were large, but were dwarfed by the large, emergent *Bombax* tree on which they were perched. The pair we watched was in the forests along the River Chalakudi, at Athirapilly in the Anamalai Hills of the southern Western Ghats. Here, the riparian forests, which are the preferred habitat of the Malabar pied-hornbill, are but a remnant of the past. A string of dams has submerged or transformed the riparian forests upstream, and even here, there is a looming threat of a proposed hydroelectric project. Further downstream, the forests give way to oil palm and rubber plantations, stripped of character and biological diversity by intensive monocultures, and of habitat and hornbills. At Athirapilly, that day, we felt lucky to have sighted the hornbills: a rare opportunity to observe the behaviour of a rare bird in a rare and threatened habitat.

We were in the midst of a quest along the Western Ghats, surveying for places where this and other hornbill species still occur, and where they have a greater chance of surviving in the future. These mostly comprised some of the most extensive tracts of forests, the sources of life-giving rivers. The link between the hornbills and healthy ecosystems and rivers was pervasive all along our survey that took us from the Sahyadris in Maharashtra to the forests of the Agasthyamalai region in Kerala and Tamil Nadu.

The search for hornbills, particularly the Malabar pied-hornbills, led us to the lowland riverine forests in Dandeli in the northern part of the central Western Ghats—along River Kali. We walked many miles in the forests, in and around Dandeli town, waited on the bridges across the river and in the timber depot of the Forest Department, watching and counting these birds. In the larger landscape around, the forests were vast and grand, made grander by the hornbills. For it was not just the Malabar pied-hornbills that had made it their home, three other species had as well: the great hornbill, the Malabar grey hornbill, and the Indian grey hornbill.

The most prominent hornbill in this region was the Malabar pied-hornbill, although the smaller, endemic Malabar grey hornbill was more abundant. We watched them, enthralled by their antics, assiduously picking the most ripened fruits from the tips of tree branches that bent with the weight of these birds. At dawn, they would disperse across the forest, in pairs or small flocks, in search of their favourite food: fruits. Like other species of hornbills found in India, the Malabar pied-hornbills consume fruits of a wide diversity of tree and liana species. Wild figs (*Ficus*) are an all-round favourite, but so are some of the sugary fruits such as *Bischofia javanica* or lipid-rich fruits of the Lauraceae family, such as *Litsea*, and fruits of lianas including *Strychnos*.

The hornbills were a delight to watch, their careful balance on the trees, their expertise in picking ripe fruit with the tips of their long, curved beaks and tossing them into their mouths. Sometimes, we also observed a male hornbill offering fruits to the female as a courtship gift. Though they ingested most fruits, the seeds were eventually excreted or regurgitated, leading to their dispersal away from the parent tree. This served a vital purpose, as the seeds of some species suffered heavy predation by rodents or beetles if they fell in concentrated clusters directly below the parent tree. By moving seeds by the thousands every day, hornbills perform a vital role in scattering seeds in areas where they have a higher chance of germination. It is for this role of regeneration that hornbills have been called 'farmers of the forest' or 'feathered foresters.'

It is not just the forests and trees that are checked out by the hornbills on their daily rounds. In Dandeli, some of the Malabar pied-hornbills regularly visited a factory nearby, where, in the pile of dust and ash left behind, they enjoyed a thorough bout of dust-bathing. This behaviour, fascinating to watch, probably serves an important role in plumage maintenance and care. Another interesting behaviour, often seen around the breeding season is the spectacular beak-to-beak contact in mid-air, called aerial jousting or bill-grappling. A pair of males jousting in air, or occasionally, a gentler contact between male and female, has been seen close to roosting sites in the evenings near Dandeli. Larger hornbills, such as the great hornbill, are also known to clash their beaks in mid-air, a behaviour called 'casque-butting'. Biologists believe that such behaviour may be related to play-chasing or agonistic—aggressive—interactions, at sites of hornbill congregation such as roosting trees or fruiting fig trees.

With the onset of the breeding season, the females occupy their chosen nest hollows and remain sealed-in for nearly three months, incubating the eggs. The males scour the forest for fruit and prey to bring to them and feed them these choice bits through a vertical slit in the nest entrance. The hornbills nest in large trees that provide hollows of the right size and character. They often return in subsequent years to nest in the same tree hollows. For those observing nests, this is a familiar and reassuring aspect of the annual cycle: the return of the hornbills, year after year, to their chosen trees, the great birds whooshing over the forest canopy.

At the nest, the female and young are careful to expel the seeds outside, and soon enough, a mini fruit-tree forest emerges around the nest tree's base. The nest, thus, is the focal point not only of the hornbills' regeneration but also that of their habitat. Once the breeding season is over, the hornbills could once more be seen in flocks, and flocks of up to eighty birds would settle along favoured roosting sites in Dandeli in the evenings.

One such evening at Dandeli, when the grey dusk descended over the Kali, still remains etched in memory: At the river bank, the trees and bamboo were in tumult. A dozen Malabar pied-hornbills landed with clamouring calls and the whoosh of wings after a day spent wandering widely in search of fruits. The bamboo and the branches, leaning out over the river for light, swayed and shook as the birds hopped to favoured perches to settle for the night. Even as the hornbills settled, other large creatures took wing from a nearby tall tree. With a flash of red-brown fur, the bats opened their skin-tight wings wide and flapped away. The flying foxes went off looking for fruit, to do by night what the hornbills did by the day.

What does the future hold for the Malabar pied-hornbill in the Western Ghats? A crucial aspect is that these birds are found mostly in forests, particularly riparian tracts, below 600-metre elevation along the Western Ghats. Such areas are also under intense pressure for other land uses, that is, for agriculture and hydroelectric impoundments. Projects such as the proposed Athirapilly dam also signal danger for the remaining population of the hornbills. The loss of forests that these projects will lead to is often misleadingly projected as only a fraction of a large forest landscape. What is forgotten is that such projects may lead to a 100% loss of the unique low-elevation riparian forest tracts that remain. A troubling question then emerges: will the hornbills, these farmers of the forests, remain secure in their last remaining habitat—or will they vanish under uncontrolled expansion of concrete and machinery? And that brings up a bigger question: what will remain of the forest and river ecosystems that the hornbills help sustain?

# Namdapha

## *Deep Forest*

At the foot of snowcapped Dapha Bum (4,571 metres) a strip of land pokes its way along the spectacular valley of the Noa-Dihing like a finger into Myanmar. Spanning nearly 2,000 square kilometres of rugged terrain, swathed by tropical rainforests of incredible biological diversity, this land, the famed Namdapha Tiger Reserve in eastern Arunachal Pradesh, is a truly special landscape, a wellspring of nature. With more than 1,000 plant species, over 500 bird species, and a documented diversity of animal species that is still climbing steeply, Namdapha will entrance any wildlife enthusiast. It is the home of the leopard, clouded leopard, and tiger, and at the higher reaches, of possibly even the snow leopard. Over the next 12 days, Divya and I, along with a couple of naturalist friends, Sarath Champati and Kalyan Varma, are to trek through this landscape. We're accompanying our colleague at the Nature Conservation Foundation, Aparajita Datta, to learn about and help with her research and conservation work. We feel a wonderful anticipation, there is much to see and work to do, too.

To reach Namdapha, we enter Arunachal Pradesh by the road from Dibrugarh to Miao. We stop at the state border for the mandatory check of our Inner-Line Permits, essential for all non-Arunachalis. It is a moment to pause and reflect; the cultivated plains of Assam are behind us, and ahead, like a dream turned real, an undulating landscape stretching into unknown hills: Arunachal awaits!

Our journey into Namdapha begins at the rather undistinguished town of Miao, where we obtain our permits to enter the Tiger Reserve from the Forest Department headquarters here. We intend to proceed on foot on a long trek deep into Namdapha and beyond, into the villages of Gandhigram and Vijaynagar. The walk would take us through the spectacular forests in the tourism zone up to Firmbase, beyond which we would traverse the core area of the Tiger Reserve not open to tourists. Depending on logistics, weather, and landslides, the trek could take anywhere from 5 days (for those who are really fit and in a rush) to 15 days or more. We are in no rush and want to take in the sights, yet logistics dictate that we keep moving. The first night, however, we rest in the wood and bamboo home of Akhi Nathany, our Lisu field co-ordinator at the Lisu *basti* in Miao.

Awakening the next day, a cool November morning, in the basti dwarfed by a huge strangler fig tree, to the sight of a pair of wreathed hornbills flying high, we prepare for our trek. Supplies are purchased, porters arranged, leech socks tried on, and everything dumped into the back of a 4×4 truck that would take us a short distance along the 100-mile-long Miao–Vijaynagar road (M–V road). The M–V, as we were very soon to realize, was a road only in the notional sense. It was wide in parts, reduced to a narrow foot-trail in many others, slushy and overgrown, and punctuated by landslides that had erased its very trace off the slopes. A few kilometres on from Miao, at Mpen, we show our permits to the forest staff at the park entry gate and enter Namdapha. At the best of times the M–V road is motorable up to Deban in a four-wheel drive vehicle. This was not the best of times. In the aftermath of rains from the Super-cyclone Sidr that had swept past Bangladesh into northeast India, a huge landslide around the 12th mile had torn down the road and even many large trees. We were dropped off some distance away, and our team prepared to go the rest of the journey on foot.

Through a gap in the rainforest canopy, towering over the Namdapha landscape, the ethereal Dapha Bum gleams in afternoon light. Namdapha spans an altitude range from 200 metres to over 4,500 metres. The lower reaches are covered in extensive tropical wet evergreen forests or rainforests, characterized by giant dipterocarp and buttressed trees. As one goes higher, one traverses sub-tropical and temperate broadleaved forests, and pine forests, into alpine meadows, and permanent snow and ice. One goes from the land of the clouded leopard to the abode of the red panda, and perhaps, the snow leopard.

The flagship mammals—the ones most likely to be seen—are the primates and the Malayan giant squirrels. Hoolock gibbon families, with their clarion hoots carrying for miles in the clear morning air; troops of handsome, leaf-eating capped langurs, and itinerant Assamese and stump-tailed macaques may be encountered. At night, one may see slow loris and flying squirrels or, with luck, one of the elusive smaller carnivores, from the clouded leopard and marbled cat, to spotted linsang and civets. Several squirrels, deer, takin, bamboo rats, hog badger, wild dog, and other mammals add to the diversity. Yet, one has to work hard to see many species here; the rainforests are dense and the legacy of hunting casts its shadow on the landscape.

We stay at Deban Forest Rest House the first night of the trek. After a relaxed morning exploring the forests and watching gibbons, squirrels, and many birds, we cross the river in the afternoon by boat to camp on the windswept banks of Deban nullah. In other places, the rivers had swelled with water and we would form a human chain to cross at specific locations that our experienced Lisu guides knew well. Fed by rain gathered in the vast watershed, and by snow-melt, the streams and tributaries course into the sparkling waters of the Noa-Dihing. Even as landslides carve the shape of the hill slopes, the rivers shape the curves of the valleys, creating a rugged, pristine ambience where one feels nature at work.

Even with their heavier loads, our Lisu assistants and porters make their way through the terrain easy and fast, as if they were built for it. We squelch through the slush of landslides, sinking knee-deep at times, soldiering on along the forest tracks and river crossings. Carrying sturdy cane baskets and heavier loads, the Lisus walk and climb expertly, across landslides and rivers with a smile on their lips.

The rainforests beyond Deban, towards the hot springs of Bulbulia and around Hornbill camp must rank among the most spectacular and biologically rich forests in northeast India. Around Hornbill, our next campsite, the terrain is relatively easy; the forests are dense but yield their riches with a little effort. We watch hoolock gibbons, capped langurs, and Malayan giant squirrels in the canopy, as well as a thrilling array of birds, many found only in northeast India within the country. Even the names are enticing: mesias and tesias, malkohas and hornbills, peacock-pheasants and wren-babblers, cochoas and laughing thrushes, parrotbills and yuhinas.

Namdapha comes alive when hornbills are in flight. Spectacular flocks of great and wreathed hornbills grace the skies. Five species of hornbills are found here, including the rare rufous-necked hornbill, the majestic great and wreathed hornbills, and the smaller Oriental pied-hornbill and brown hornbills. In their itinerant wanderings across vast areas of forest in search of fruiting trees, large flocks of wreathed hornbills criss-cross the skies each day, the whooshing of their wings audible from far. Aparajita's work on hornbills in Arunachal's forests have established the hornbills' importance as seed dispersers and architects of the rainforest plant communities.

Among the many rare and threatened birds found only in northeast India's forests within the country are species such as rufous-necked hornbill, grey peacock-pheasant, red-headed and Ward's trogons, chevron-breasted babbler, snowy-throated babbler, hill partridges, and beautiful nuthatch, to name a few from a long list. A great diversity of insects, fish, amphibians, and reptiles also distinguishes Namdapha's forests and rivers.

Even as we walk slowly through the rainforests past Hornbill camp, our assistants have gone ahead to pitch camp. With cane, bamboo, and a handy *dao* (machete) the Lisu can work all manner of wonders from utensils and baskets to pipes and shelters in no time. After a long weary trek punctuated by exciting wildlife sightings, we welcome the sight of the pitched camp where hot cups of tea await. The vista of the Namdapha river valley close to its confluence with the Noa Dihing, bound by forest-clad hills and snow-capped peaks, feels like one of the last truly wild places in India. Next to the murmuring river, under silvery starlight, we huddle by the warm glow of the fire and tuck into a simple but tasty dinner of rice, dal, and boiled *lai patta*, leafy greens similar to mustard leaves, a staple in northeast India.

The wide river valleys strewn with rounded boulders are punctuated by stretches of tall grass, home to the hog deer and birds of the open grassland such as chats and shrikes. We learn it is as easy to lose our way in the tall grass as it is in the dense forest, and are glad to have our trusty guides to show the way. The Noa-Dihing river is alive with birdlife: black storks and cormorants, sprightly black-and-white forktails and wagtails, dippers and ibisbills, redstarts, and river lapwings.

Holding hands for support and wading waist-deep across the Noa-Dihing, we trek past Firmbase to our next camp at Ngwazakha, or 38th mile settlement, a reference to a notional milestone on a nearly non-existent road far from this small cluster of huts. The next morning, with misty vapours lifting from unbroken and unexplored forests in the distance, we trudge on to camp near 52 mile. Another day of a bitingly tough trek brings us to our putative campsite near 62 mile, only to find it wet and sloping. We are forced to move camp a mile or so farther; in our tiredness, this seems the longest mile yet.

Our walk takes us through rainforests and bamboo thickets, rivers and streams, over boulders and landslides, partly along the notional M–V road, but frequently off-road. The M–V road, built in 1972 and defunct since the 1990s, serves mostly as a frame of reference in the landscape. We do parts of the last stretch towards 77 mile and on to Gandhigram village on the M–V road, catching glimpses of enthralling forests and wildlife, but with little time to stop.

Finally, 8 days and nearly 140 kilometres from Miao, we emerge from the forest and behold the picturesque village of Gandhigram—Shidi to the Lisu—sprawling in the valley of the Noa-Dihing. With rice fields and bamboo houses, pigs and chicken, not to mention the children playing on the streets, the village seemed scarcely conscious of its isolation at India's eastern extremity. Yet, the forest-clad hills around and the flocks of hornbills overhead were a constant reminder of the rich wildlife in the landscape. After two days of rest and feasting on delicious pork, luscious pineapples, native popcorn, and rice cakes we reluctantly left Gandhigram for Vijaynagar. The final 18 km trek took us past forests and jhum fields, crossing cane bridges and streams.

Vijaynagar town, at Namdapha's doorstep and yet quite detached from it in spirit, marks its events in relation to the arrival and departure of the irregular AN-32 flights and the sundry occupations of the Assam Rifles, the village, and the Government offices. With the weather threatening to worsen and the temperature dipping with snowfall on the high peaks, we decide to wait for a sortie to head back. It is a tense wait as there's little assurance of a flight. Luckily, we get a flight on the second day; strapped aboard along with luggage, we are off. Below us, the landscape that took us 12 days to trek through flashes past in 20 minutes. Yet, the range after range of hills and unbroken forest entice; even as we leave, we feel we shall return.

# CONSERVATION

## a world of wounds

Now, I truly believe, that we in this generation, must come to terms with nature, and I think we're challenged as mankind has never been challenged before to prove our maturity and our mastery, not of nature, but of ourselves.

—Rachel Carson, writer and marine biologist (1907–1964)

As a conservation biologist, as a *practitioner* of conservation, you would actually have to build upon science, management, policy, advocacy, *implementation*—very important—and of course education for the larger grouping of people.... It is not a narrow field, you actually have to be multi-multi-disciplinary for working with conservation.

—Ravi Sankaran, ornithologist and conservationist (1963–2009)

# The Beleaguered Blackbuck

In the heart of Chennai city in southern India lies a small but significant wildlife refuge occupying about 270 hectares, the Guindy National Park. The city is fortunate to have such a Park, with its population of chital deer and blackbuck antelope, and a remarkable diversity of other plant and animal species. In some ways, the Park is fortunate to be located in the city, too: It lies just a few kilometres from the headquarters of the State Forest Department, the primary authorities tasked with its protection and management. It is accessible to a range of educational and research institutions, and evokes the interest and stewardship of a community of nature lovers and conservationists in the city. Managing a small nature reserve, one that's located nearby, and has no human settlements or other land uses within it, should be simpler than protecting and conserving a large tract of wilderness threatened by imminent destruction in a remote region. Yet, it is not so easy. There are problems in being small and perils in being surrounded by a city, too.

Guindy National Park owes its existence to a fortunate series of historical events. In the early seventeenth century, the area where the park is located now must have had a tract of the same sort of tropical dry evergreen forests and scrub jungles that exist as small remnants today along the Coromandel and Circar plains along the east coast of India. Historical information suggests that around 1675, William Langhorne, the then Governor of Madras, established a house in this forest and called it Guindy Lodge. Much later, in 1821, Sir Thomas Munro, another Governor of Madras, purchased the area and established it as a gubernatorial country residence and weekend resort. Guindy Lodge continued to house a succession of eminent people, and in 1946, became the official residence of the State Governor. Now called the Raj Bhavan, and located on the park periphery, it is a prominent landmark in the city.

Around the same time that Guindy Lodge became Raj Bhavan, one of the Park's present-day star attractions was introduced—the chital—according to naturalist and writer, M. Krishnan, writing in his pioneering studies on the mammals of peninsular India in 1972. Although some reports going back at least to 1900 also mention the occurrence of chital in Guindy, Krishnan notes that while chital were introduced, the native animal of the Park was the blackbuck—the striking, spiral-horned antelope of the open Indian plains. The blackbuck, once widespread and common across the grasslands and drier parts of India, is today an endangered species with scattered populations across the country. By 1972, the time of Krishnan's writing, herds of blackbuck in the scrub jungles and open lands around Chennai had already disappeared due to urbanization and development, leaving only a remnant population in Guindy still protected.

In the 1970s, many local naturalists drew attention to the value of Guindy as a nature reserve in Chennai, highlighting that the Park was home to an impressive diversity of species, including over one hundred and fifty bird species, reptiles, amphibians, invertebrates, and plants of the tropical dry evergreen and thorn forests. Due to their efforts and the support of Government authorities, Guindy was declared a National Park in 1978 under India's Wildlife Protection Act. As a result, even today, on the open grassland called the Polo Field in Guindy, a visitor can delight in the spectacle of blackbuck clashing horns in territorial fights, courting females, or simply grazing, oblivious to the bustle of the city right outside.

In the latter half of 1990, I began my fieldwork on chital and blackbuck in the Park, which would continue till the summer of 1993. Through systematic observations and application of field methods, I studied the population size and structure of the two mammals; how they used the woodland, scrubland, and grassland habitats in different seasons; and other aspects, including their grouping behaviour and breeding seasonality. After repeated visits several times every month over 34 months of fieldwork, I began to see the Park in a newer light. It was not merely a repository of plants and animals, set aside like a zoo, or a museum piece for the edification of the city's citizens. It was a layered crucible of life, set to its own rhythms, familiar yet ever-changing, inextricably linked to the landscape and community of the city. My field research raised several pertinent conservation concerns related to the Park. And I learnt several crucial lessons on what conservation means in the real world.

What I found in my very first year of field study raised an alarm. The population of blackbuck in the Guindy and Raj Bhavan areas had plummeted from at least 250 animals in the late 1970s and early 1980s to about 85 or fewer animals in 1991. This decline had presumably occurred over the years and had gone unnoticed. What helped in detecting this rather drastic decline in blackbuck population was comparable scientific information and wildlife census estimates made between 1977 and 1982 by earlier observers, particularly R.K.G. 'Cutlet' Menon and R. Selvakumar. Menon had carried out a three-year long study of blackbuck behaviour in Guindy and Point Calimere Wildlife Sanctuary, while Selvakumar carried out his Masters thesis project on blackbuck in Guindy. The first conservation lesson, then, that the decline in blackbuck population illuminated was the need for field research and systematic monitoring, for a dedicated science that can provide reliable and rational guidance for wildlife management.

The causes of the decline in blackbuck numbers were difficult to untangle as the Park had suffered many changes over the years, all of which could have affected the antelope. The most likely cause was the loss of open grasslands and territorial areas. The Polo Field, an open meadow of less than 4 hectares, a critical habitat and centre of blackbuck activity, had become slowly overgrown

by weeds and scattered woody vegetation, R.K.G. Menon had noted as early as 1986. This grassland, although it was a small patch that occupied less than 2% of the total area of the park, provided vital blackbuck foraging and territorial grounds. So, when another 2½ hectares of grassland was fenced off by the Forest Department to grow grass for the animals in the adjacent Children's Park zoo, it would have hit the population hard, too.

By 1991, we found that there had also been much increase in the woody vegetation in other areas, particularly of introduced invasive trees—earpod wattle (*Acacia auriculiformis*) and mesquite (*Prosopis juliflora*)—besides native woody plants such as *Clausena dentata* and other species. Thorny *Prosopis* bushes and weeds such as *Sida cordifolia* and *Croton bonplandianum* dotted the Polo Field and surrounding areas. During 1977–9, Menon and Selvakumar's studies reported five territorial males on Polo Field, and aggregations of 50 to 70 blackbuck were not uncommon. In the early 1990s, the sight of one or two territorial males and small herds of 15 to 25 blackbuck on Polo Field presented a sorry contrast. Making matters worse later, the Forest Department, in apparently well-intentioned but misguided efforts, planted trees in open areas in the scrub, further destroying the open habitats that the blackbuck need and prefer for their foraging, ranging, territories, and breeding. Fences were even erected on the Polo Field meadow to grow fodder for the zoo, disturbing the main territorial area of the blackbuck. Loss of habitat is one of the main threats to wildlife the world over, and can happen even within a National Park under the noses—and because of the actions—of a management authority; a second self-evident lesson from Guindy.

Other changes that occurred over the years may also have contributed to the decline of blackbuck. The overall area of contiguous habitat practically halved, from over 550 hectares to only 270 hectares today. Between 1961 and 1978, before

Guindy became a National Park, 170 hectares were sliced away for educational institutions and memorials of departed leaders. Gandhi Mandapam, Indian Institute of Technology, Guru Nanak College, and the Cancer Institute: tall granite walls and fences now separate the Park from these adjacent campuses, making it a small, closeted ecosystem. This fragmentation and isolation disrupted the distribution of blackbuck territories between Guindy National Park and Raj Bhavan, and the normal seasonal movements of the deer and antelope in search of forage.

In contrast to the blackbuck, the chital population in the park was very high in the 1990s: over 200 per square kilometre according to the results of my study. This high density was probably because the chital population had been artificially fed and shielded from natural mortality. Since the mid-1970s, throughout the dry season and summer—the 'pinch period'—the Forest Department provided tons of para grass, a species native to Africa and the Middle East, as supplemental fodder for chital. The Department distributed one ton of fodder every day at ten feeding spots. Over the years, this brought in huge biomass into the park ecosystem. Often 30–40 chital would aggregate at these feeding spots, but the blackbuck rarely, if ever, consumed this grass. Provision of fodder in large amounts year after year is likely to have curbed the natural dry-season mortality of chital, while augmenting the nutritional condition and fertility of does during the main breeding season. The Park also lacks predators that could keep chital populations in check, but for some feral dogs and jackals that occasionally hunt fawns.

Packed into a small habitat, the chital and blackbuck may have competed with each other to the detriment of the latter smaller species. In the 1990s, during the wet season, from June to November, I often saw herds of up to a 100 chital crowding into the small Polo Field meadow, physically disrupting the blackbuck herds and activities of territorial males. At such high densities, chital may have reduced the food available for blackbuck, particularly as several grass and herb species are consumed by both ungulates.

Conservationists are now concerned even over the chital population, which also appears to have plummeted. On 11 July 1992, in a single morning of fieldwork, I recorded in my field notes 67 chital in 22 herds in Guindy, besides 11 blackbuck. Visiting the park with a few friends exactly 24 years later, on 11 July 2016, we saw no blackbuck or chital inside the park, only a single chital feeding on garbage dumped behind the Forest Quarters at the periphery. In the absence of reliable research and monitoring, one cannot assess the chital population or the causes for its decline, although undue habitat alteration and mismanagement could well be a reason for it.

The final lesson, then, is that even in a small, accessible, and apparently well-conserved nature reserve like Guindy National Park, one cannot be too complacent. To regain lost values as a wildlife refuge the Park still needs serious and well-informed efforts that are neither *laissez-faire* nor unduly protectionist. A top priority is the maintenance of the grasslands and open areas, especially for blackbuck, including the removal of weeds and woody plants that have established or were planted. Further measures include control of invasive plants such as earpod wattle and mesquite, stopping tree planting particularly in open grasslands and scrub, removing fences, and discontinuing the artificial feeding of chital. Careful management coupled with research and monitoring of the effects of these measures can help ascertain whether conservation objectives are indeed being realized. Guindy's plight stands as a synecdoche for Indian nature reserves, both small, isolated ones as well as larger parks.

Today, Guindy National Park is significant not only because it still harbours a population of the endangered blackbuck, but also because it contains remnants of the endangered tropical dry evergreen forest vegetation that has shrunk to less than 10% of its potential area along the Coromandel Coast. Its rich diversity of plant and animal species will undoubtedly enlarge with further discoveries and documentation. The Park, considered a 'green lung' for the city, and other remnant scrub jungles around Chennai, also present excellent opportunities to educate biologists, students, forest managers, and urban citizens. For a number of prominent Indian field biologists, including R. Selvakumar, Raman Sukumar, Ragupathy Kannan, Kavita Isvaran, V. Santharam, and Abi Tamim Vanak, Guindy and these scrub jungles were an early training ground.

Guindy National Park will also have to come to terms with the surrounding populace. Unlike many other nature reserves in India, this Park has not witnessed intense conflicts between local people and the Forest Department. A minor conflict with people who used to enter the park to access water in one of the ponds from adjoining Velacheri—a place now swallowed up in Chennai's urban sprawl—was sorted out when the Forest Department began to operate a borewell pump to supply water outside the park boundary. The Department also stopped the indiscriminate collection of fuelwood and *Aristida setacea* grasses for brooms. Potentially, it could remove invasive trees and shrubs such as earpod wattle and mesquite as part of habitat management and distribute the wood to local people in need.

Even in the early 1990s, a recurrent problem in Guindy was tourism. During my field research, I sometimes observed tourists driving recklessly, disturbing the blackbuck on Polo Field, making noise, and leaving garbage behind. It is a shame that the Park's enormous potential for educating tourists about nature and conservation lies untapped. While it may not be able to sustain the huge crowds that throng the adjoining Children's Park and Snake Park, managers could envisage planned, guided tours for small groups on foot or in battery-powered, pollution-free vehicles to enable Chennai citizens and tourists to appreciate the free-ranging wildlife and natural wonders at their doorstep. Few cities can boast of a place such as Guindy National Park. Only time will tell whether the city of Chennai will finally wake up and be moved enough to conserve this beautiful remnant of the natural world.

# A Bounty of Deer

The loud, throaty bellow of the chital stag reverberated through the fresh morning air. Hidden behind a few trees and bushes, I watched silently as he stood on an open grassland near a small herd of chital does and their young fawns. Farther away, there were two other herds of deer. One herd had larger animals than the other and comprised about a dozen individuals. Their tawny, orange-tinted coat, and the highly branched antlers of the males, indicated that this was a herd of swamp deer or barasingha. The second herd was a group of three hog deer at the edge of a patch of tall grass. From that particular vantage point in Dudhwa Tiger Reserve, I could thus observe, at a glance, three species of deer. Yet, this was not all. Later that day, I saw two more species, the sambar and the Indian muntjac or barking deer, in the dense sal forests within the sanctuary.

Dudhwa, on the Indo-Nepal border in Uttar Pradesh, is one of the few places where one can see, even today, five of India's eight species of deer in the Family Cervidae. Deer are among the commonest, most visible, and attractive mammals in India's many forests and grasslands. They display a fascinating variety in their appearance, behaviour, ecology, and natural history, which is interesting in its own right, but it also offers insights into their evolution and adaptation, and into the ecology of their forest and grassland habitats across India. Unfortunately, several factors have brought some deer species to the brink of extinction.

The Indian subcontinent has an unusual bounty of deer species. The 68 extant deer species in the world are distributed in three families: the true deer or cervids (Family Cervidae), the musk deer (Moschidae), and the chevrotains and mouse deer (Tragulidae). Of the 13 deer species that occur in India, the true deer (Cervidae) include eight species, all in the Subfamily Cervinae: chital, sambar, swamp deer, hog deer, hangul, sangai, Indian muntjac, and the leaf muntjac or leaf deer. The latter was reported to occur in India for the first time in 2003 in Namdapha in the northeastern corner of the country. Among the musk deer species, India is lucky to have four species, distributed along the higher reaches of the Himalaya from Kashmir to eastern Arunachal Pradesh: Kashmir, alpine, Himalayan, and black musk deer. Finally, there is the Indian spotted chevrotain or mouse deer.

All these deer belong within the larger group of herbivorous, hoofed mammals or ungulates, including camels, pigs, antelope, cattle, and hippopotamuses—the artiodactyls or even-toed ungulates. Besides their even number of toes (two main toes, two reduced to dewclaws above and behind the hoof), deer are characterized by a four-chambered stomach and the presence—in the males of most species—of antlers, which are bony outgrowths of the frontal bones of the skull.

In a herd of deer, the striking pair of branching antlers is often what first attracts an observer's attention. The growth and development of antlers plays a major role in the life cycle and reproductive behaviour of deer. Antlers occur in 50 of the 68 extant species of deer in the world, being small and inconspicuous in species such as the muntjacs and the tufted deer. The musk deer in the Genus *Moschus* and the water deer of China and Korea lack antlers. Antlers are not horns. Horns, which are present in other ungulates such as buffaloes, antelopes, goats, and sheep, are composed of keratin and grow continuously through the life of the animal. Unlike horns, antlers are composed of bone and are shed and grown every year.

In the chital or spotted deer, the first pair of antlers appears from stub-like pedicels on the head when the male is almost a year old. These antlers grow to be simple and spike-like, about ten centimetres long. The following year, as the male grows in body size, these antlers are shed from the pedicels. A new pair of antlers begins to grow, attaining a final length of 25–30 centimetres in a few months. This process occurs annually, and the antler size increases roughly in proportion to the animal's age.

Growing antlers are covered by a layer of skin and hair, called 'velvet,' richly supplied by blood vessels. The velvet, when backlit, appears like a bright fringe around the antler, whose growing tip is still usually rounded rather than pointed. With the onset of the breeding season, the levels of testosterone increase in the blood, and the antler undergoes mineralization and calcifies. The velvet gradually peels off, or the male rubs it off on shrubs and branches. This exposes the underlying bony, hard antler.

The antler of an adult has a branch called the brow tine, emerging just above the pedicel and curving forwards and upwards. The main tine or beam curves upwards from the pedicel, branching again to form a branch called 'bez.' In other deer, such as the swamp deer, there may be a further 'trez' branch and many tines, giving it the highly branched appearance that also gives this species its local name of barasingha (from the Hindi, *barah*, twelve, and *singha*, tines). The antler of the sangai or Manipur brow-antlered deer is peculiar in that the brow tine and the main beam form a continuous sweeping arc over the head.

The time of year when most adult males are in hard antler forms the peak of the breeding season or rut. The peak rut differs according to the species. Over most of India, adult chital stags attain hard antlers between April and June, whereas sambar rut during the winter. Spurred by the testosterone, the neck and body muscles develop, making the animals appear larger than usual. Males begin to rove widely in search of oestrous females and avidly court them. In other deer, such as the swamp deer and the hangul, males defend territories during the breeding season. By displays such as roaring and herding, they attempt to attract females to their territory and jealously guard their harem from other males. The females may not be mere passive spectators of male prowess; anecdotal observations suggest that females actively choose the males they mate with.

In more primitive deer (on the evolutionary scale) such as Himalayan musk deer and the barking deer or muntjac, the male secondary sexual characteristics and breeding systems are different. Musk deer and muntjac are forest dwelling, relatively sedentary, and territorial. The musk deer lack antlers, but instead carry a pair of nasty-looking tusks that are merely the elongated upper canines of the males. The muntjac male also has tusks but, in addition, carries a pair of small, spike-like antlers on his head, often with only a small protuberance representing the brow tine. The muntjac are an intermediate form between primitive deer and the more advanced cervids, such as the chital or barasingha.

Males mainly use antlers in serious and playful fights, though antlers may also play a role as status symbols that indicate dominance. Antlers occur only in males, except in the reindeer.

Female reindeer attain hard antlers in winter, when males are in the vulnerable velvet-antler stage. They use their antlers to ward off males that compete with them for food in their cold tundra habitat in North America and Eurasia.

Males spar and fight using their antlers, the frequency and intensity of fights being higher around the time of the rut. Being hard, bony, and free of any soft tissue, they are less prone to undue damage when males interlock, shove, and clash with each other. Broken antlers do occur at times, but these can be shed at the end of the season, and new ones grown the following year. Biologists have proposed that antler casting may have evolved to enable such repair. This way, antler size can also increase in tandem with body size.

Actual combat is infrequent, though some form of aggressive—agonistic—interaction is quite common between males. This is because each species of deer has evolved certain characteristic assessment displays that lets males size each other up before combat. Roaring rates, hard antler size, and parallel walks are displays of this nature. After several such encounters with one another, males establish a dominance hierarchy. Subordinate individuals learn to avoid or merely spar playfully with dominants. Serious fights usually occur between males are that are almost evenly matched, and may lead to deaths or life-long injury.

One result of the competition and courtship during the breeding season is the annual crop of fawns. In most deer, females give birth to fawns during the season when resources are abundant. Thus, hog deer in wet grasslands of the terai fawn mostly during April and May when, after the annual burn and the pre-monsoon showers, there is a flush of new grass sprouts. In contrast, chital, close relatives of the hog deer, produce most fawns between December and March, during the onset of the dry season. It is not clear why they do this. Perhaps this helps the does to coincide their period of late lactation, when their energy demands are high, with the onset of rains and abundant food in May and June. It also entails pregnancy during the wet season, enabling does to meet the needs of the developing embryo while storing resources for the following dry season. When the fawns are born, they are usually kept in hiding for a few days or weeks, as they are vulnerable to predators. Soon after, the fawns begin to follow their mothers; gradually, they learn to forage on their own after weaning. During the fawning season, one can commonly see does and fawns forming small herds and foraging together. The tendency to form herds varies, however, with the social system and habitat of the species.

Herding or group-forming behaviour is a characteristic of many ungulates or hoofed mammals, including several species of deer. Some, such as the small, forest-dwelling musk deer and muntjac, occur solitarily or in pairs of a male and a female, or a female with her young. Other species, such as the chital and barasingha form large herds of a hundred individuals or more. Usually, forest species form smaller herds compared to species of the open grasslands or the forest–grassland interface. In chital, average herd sizes vary widely in different months of the year, from about two to over thirty or so individuals. During the wet season, groups of a hundred or more individuals sometimes occur. In scrub and grassland habitats, herd size and density of chital directly relates to the amount of rainfall and grass growth.

Besides the abundance, availability, and dispersion of food items, the need for safety in numbers when facing predators also influences herd size. Larger herds can detect predators earlier, as many animals are simultaneously watchful. On detecting predators, such as a pack of wild dogs, a large herd of chital will often bunch up into a compact group. They sound alarm calls and stamp their feet on the ground, the latter serving to deposit a scent, secreted from glands between the digits of the feet and serving as a warning. In such a situation, large males often muscle their way to a secure position in the centre of the herd, leaving females and young on the periphery. So much for chivalry in the species!

Predators do take a regular toll on individuals, but they are a part of the natural environment and dynamics of deer populations. In fact, the absence or removal of predators often leads to a cascade of ecological changes, including deer overabundance, overgrazing impacts on plant communities, and negative effects on other species that share the same habitat. This is particularly the case when deer are introduced into oceanic islands that also lack predators. The chital, introduced into islands as distant as Hawaii in the Pacific Ocean and, closer home, into the Andamans in the Bay of Bengal, has acquired the status of a pest or invasive species. Still, in their native distribution range, for many deer species including the chital, overabundance is rarely an issue. Instead, there are more potent threats to their survival that have led to population declines and brought a number of species to the brink of extinction.

Several aspects of deer biology have contributed to this situation. Species such as the swamp deer and the hog deer, being specialized to grassland habitats, have suffered from habitat loss to agriculture and development activities. Three subspecies of swamp deer exist today, all highly endangered: the hardground barasingha in Central India, and the two subspecies in northern India and eastern India that occur in wet grasslands. Several thousand hardground barasingha probably existed in the Central Highlands of India, in grassland habitats along the Vindhya and Satpura hill ranges. Hunting and loss of habitat brought down their numbers so drastically this century that by 1970, only 66 survived in Kanha Tiger Reserve in Madhya Pradesh. After ecologists attracted attention to the barasingha's plight, wildlife managers undertook several conservation measures, including relocation of villagers occupying and cultivating some grassland areas. The species appeared to respond positively to these measures and, within a decade, there were about 280 animals again at Kanha. The swamp subspecies of the barasingha occurs in the terai, a unique, marshy, tall grassland habitat along the Indo-Gangetic and Brahmaputra plains. This habitat occurs today in a few sanctuaries and protected areas in Uttar Pradesh, Nepal, Bhutan, West Bengal, and Assam. Similar threats as faced by the hardground barasingha and hog deer have contributed to the decline of these subspecies in north and east India.

Within India, two other deer species occur in single, isolated, protected areas, hovering at the brink of extinction in the wild: the hangul and the sangai. A combination of habitat loss and hunting for trophies and meat may have reduced them to this plight. The hangul is a subspecies of the European red deer. Another subspecies that occurred in the eastern Himalaya, the Sikkim stag or shou, is now perhaps extinct in India. From over 2,000 hangul that existed around 1947, probably less than 300 exist today, mostly within a single sanctuary, Dachigam. This sanctuary, established as a game reserve by Hari Singh, the last Maharaja of Kashmir, today occupies some 141 square kilometres in the Kashmir Himalaya. The hangul population thrives in subtropical forests, migrating between the higher slopes of the mountains and the autumn rutting grounds in the main Dachigam valley.

The sangai, also known as thamin, is perhaps the most endangered subspecies of deer in the world. Its population declined from about 100 individuals in 1959 to only 20 individuals or so in two decades. Today, the only existing wild population occurs in the Manipur valley of northeast India, occupying a very peculiar habitat in Keibul Lamjao Wildlife Sanctuary, a part of Loktak Lake. Many floating islands, called the *phum* or *phumdi*, occur in this marshy habitat. The phumdi is a mass of decaying organic matter about one to four feet in depth. It floats on water during the wet season and settles on hard ground when water levels drop. Sangai have modified, split hooves that enable them to move over the floating vegetation with a fluid gait that has earned them the name dancing deer. There are substantial numbers of this deer in captivity in various zoos today. Conservationists have therefore suggested establishing free-ranging populations in other areas using captive stock, although there are strict norms and guidelines for reintroduction projects that need to be followed. This may reduce the threat of extinction that can speedily dispose of a single, small population. Yet, conserving this species is not an easy proposition, as threats continue to affect its survival. In 1983, the construction of the Ithai Barrage with the commissioning of the Loktak Hydroelectric Project has changed the hydrological regime and ecology of the lake, further affecting the sangai's survival.

India's musk deer are also endangered. The species is famous for the aromatic musk, contained in a pouch under the male's abdomen. As musk is valuable in illegal wildlife markets, it attracts poachers. Even females and young are sometimes snared. Today, the law protects the musk deer, and programmes for captive breeding and reintroduction of the species into the wild are in the offing.

India is fortunate in its bounty of deer, from the chevrotain and musk deer to the huge sambar. For thousands of years, they have thrived in India as an essential part of natural ecosystems. Today, their future appears uncertain. Will the hangul, the hog deer, the sangai, and the barasingha vanish like the shou? Or will conservation efforts enable them to persist through the 21st century? The only certainty is that: if these species disappear, India's forests and grasslands will lose an irreplaceable element of their charm.

# Hornbills

## *Giants Among Forest Birds*

Among the surest contenders for the title of most spectacular—even bizarre—birds of Indian forests are hornbills. Contrasted against the familiar house sparrow, for instance, the great hornbill—full 4 feet in length with striking black, white, and yellow hues, and a beak as large as a crow—appears as different as a giraffe from a rat. The birds' natural history and their unique traits and habits makes for some of the most curious tales from the world of birds. See hornbills in their natural habitat, and you'll find yourself captivated by their charms.

Imagine yourself in a luxuriant tropical rainforest, walking silently along a trail. From the distance, over the towering tree

* This essay is written with Divya Mudappa.

canopy you suddenly hear a deep, persistent *woosh, woosh.* As it approaches, the sound grows louder, almost as if one of Gulliver's Brobdingnagian giants is bearing down upon you, heaving with the effort. It is a giant, in fact—a giant among birds and largest of the hornbills—the great hornbill. If you are lucky, you will get an awe-inspiring glimpse, like most others do, of a pair flying majestically over the rainforest.

Fascinating and grand, with curious beaks and breeding habits, hornbills are a naturalist's delight. One of the first things to strike you about their appearance is their huge, grotesquely caparisoned beaks. Atop their large, curved beak is a horny protuberance or ridge called the casque. A unique feature of hornbills, the casque's various eccentric or exaggerated shapes and colours gives several species their names: helmeted hornbill, knobbed hornbill, even rhinoceros hornbill. The casque is not as heavy as its size would suggest since it is hollow in all species except the helmeted hornbill. In this species, the casque is solid and the skull thus forms nearly 10% of the bird's body weight. The uses of the enlarged beak and casque are many. The beak helps procure various, often large, food items including fruits (figs, for instance, are a hornbill delicacy, and nutmegs), arthropods, lizards, snakes, even the occasional flying squirrel. During nesting, the female uses her beak to plaster the nest cavity opening with a 'cement' made from her droppings. In the helmeted hornbill and the great hornbill, the beaks are used in a rare and spectacular form of aggressive interaction called aerial casque-butting, where the males clash their beaks mid-flight. Males usually have a larger, differently coloured bill and casque than females and the colours may play a role in their displays.

Although most hornbills are strikingly coloured, by avian standards they are modest. Their feathers lack any pigment besides melanin—offering a circumscribed range of possibilities from white, greys and browns, to black. This is, however, offset by a couple of adaptations. A gland on the bird's rump, which normally secretes an oil useful in preening, has been modified to secrete a cosmetic wax, yellow to red in colour probably due to the presence of carotenoid pigments. During the breeding seasons, great hornbills, for instance, smear their bills, casques, and white feathers with this secretion, turning these parts a bright yellow or orange. Several species have also developed bare patches of coloured skin—the wreathed hornbills occurring in northeast India have a bright yellow or blue throat pouch. Differences in the colouration of the bill, casque, feathers, or eyes, besides size dimorphism, help distinguish males from females and adults from young in most species.

Hornbills are today considered taxonomically unique enough to be placed in two separate families, Bucerotidae (savanna and forest hornbills) and Bucorvidae (ground hornbills), under the order Bucerotiformes. Earlier they were placed along with birds such as kingfishers, bee-eaters, rollers, trogons, and hoopoes in the order Coraciiformes. They are most closely related to hoopoes, which also lack pigments besides melanin, in contrast to other colourful coracids. Of the 57 species of hornbills in the world, 25 occur in Africa. Only 14 of those occur in forests and the rest are savanna birds, because of the large spread of savanna habitats in the region. The Oriental region (South and Southeast Asia) has 31 species, and one species is found in the Australasian region (Papua New Guinea). These regions are dominated by forest habitats, which have left a biogeographic imprint on the hornbills' habitat preferences—only a single species occurs in wooded savannas in Asia, while the rest are moist forest birds. One can scarcely talk of hornbills of Asia and Africa without also mentioning the toucans of Central and South America. The toucans are remarkably similar to the hornbills, being large-billed fruit-eating birds found in the neotropical forests of the Americas. Evolved independently and placed in a different family, Ramphastidae—in the order Piciformes that includes woodpeckers and barbets of the world—toucans are an example of convergent evolution.

Besides their striking appearance, hornbills are also noteworthy for their peculiar breeding habits. Hornbills nest in cavities or hollows in trees or rotting branches, barring two species of African ground hornbills that sometimes nest in holes in rocks or earthen banks. The tree-nesting species are called secondary cavity nesters since they only use pre-existing cavities. These cavities may be natural ones or enlarged nests excavated and used earlier by primary cavity nesters such as woodpeckers. Hornbill nest cavities are large, commensurate with their size. Each species prefers, however, cavities that are just the right size for a female to squeeze into. The nest may be 10–30 metres high on trees of large girth, such as many rainforest trees in Asia and massive baobabs in Africa.

Ornithologists in the Western Ghats in southwest India have studied how hornbills—the smallest, the Malabar grey hornbill, and the largest, the great hornbill—select nest sites. The research on the former species by one of us (Divya) and the latter species by Ragupathy Kannan in the rainforests of the Anamalai Hills showed that the choice of nest trees is not random, and that hornbills select trees that are much larger than expected by chance. There are also clear differences between the species. The great hornbill prefers trees that are more than 1 metre in diameter at breast height (that is, about 1.3 metres from the ground) and that achieve an average height of 44 metres. The Malabar grey hornbill mostly selects smaller trees, of about 0.6–0.9 metres in diameter at breast height and with an average height of 28 metres. In drier forests in southern India, the similar-sized Indian grey hornbill uses trees of about 1 metre in diameter and 27 metres height, in particular the white cedar or mountain neem, *Melia dubia*. In the forests at the foothills of the Eastern Himalaya in northeast India, biologists Aparajita Datta and G.S. Rawat, in a study from the Wildlife Institute of India, found three species of larger hornbills (great and wreathed hornbills, and oriental pied-hornbills) using large nest trees, particularly *Tetrameles nudiflora,* a tall emergent deciduous tree. Studies on hornbills in southeast Asian rainforests also report that some species of dipterocarps and eugenias, which form large emergent trees, are preferred for nesting.

The most peculiar aspect of hornbill breeding begins after nest selection and preparation are done. The female occupies the nest to lay her clutch of eggs and then, intriguingly, she commences the apparently masochistic act of sealing herself in within the cavity. Using her beak as a trowel, she 'cements' the nest opening with her own droppings, which consist of a sticky paste with fig seeds and other waste. Finally, only a thin slit is left through which she can extrude her bill-tip. For weeks, she will remain incarcerated thus, fed regularly by the doting male who brings and regurgitates one to several dozen fruits and animal prey. Immobilized and cooped up, the female moults her flight feathers and sometimes her tail feathers, too. To maintain nest sanitation she twists around each time to forcibly eject, like a bazooka blast, her droppings through the narrow slit of the nest cavity. When the young hatch, they emulate the female in this eminently sensible practice of nest sanitation for the duration of their incarceration. In the larger hornbill species, only one chick is raised in a nest, whereas in some species there may be up to four.

In southwest India, the breeding season of the great and Malabar grey hornbills usually begins in February, but could start as early as December. The eggs of the former hatch after nearly two months, while those of the latter smaller species take about forty days. The female great hornbill stays with the chick for over a month. Then she emerges from the nest to help the male feed the young in the final phase of nesting. The chick reseals the nest in the same manner as the female and remains incarcerated for another fortnight. Four months after the female first occupied the nest, the chick breaks out fully fledged. In the Malabar grey hornbill, the female and fledged young break out of the nest together at the end of the twelve-week nesting period. The brown hornbill, found in northeast India and southeast Asia, has an unusual, co-operative breeding system. Among these birds, the male is assisted by one to several 'helpers', usually young male offspring of previous years, who help in feeding the incarcerated female and young.

Why do hornbills have such peculiar nesting and breeding habits? Since hornbills, like other large-bodied birds, have long incubation and fledging periods as well as low fecundity and breeding rates, they are particularly susceptible during the nesting phase of their life cycle. Adaptations, however strange they may seem, that enhanced their nesting success and survival would have evolved by natural selection. By nesting high up in trees and sealing the entrance, the chances of predation by carnivores such as the yellow-throated marten and king cobra, or of nest usurpation by other hornbills, are reduced. Hornbills often achieve a high nesting success of 75% or more under natural conditions.

While male hornbills do bring invertebrates and animal prey to feed the nest inmates, the bulk of their diet comprises fruit. The male perches at the edge of the nest or on a nearby branch and regurgitates up to several dozen fruits one by one—a process that takes a few minutes. Depending on species and locality, about one-fourth to three-fourths of their diet during the breeding season comprises figs. The hornbills' predilection for fruit is accentuated during the non-breeding season, when various other fruits are also consumed.

Figs are notoriously aseasonal in fruiting, unlike many other fruit-bearing trees. The super-abundant fruit crops on fig trees are a major resource for a variety of birds and mammals during annual periods of fruit scarcity. They have therefore been labelled 'keystone resources' in many tropical forests. With the onset of the rains, more seasonal fruits of trees in families such as Myristicaceae (nutmeg) and Lauraceae (laurels and cinnamons) become available. In contrast to figs that have numerous tiny seeds and are rich in sugars, these fruits have single seeds and are rich in lipids.

In return for the nutriment the fruits provide, the hornbills disperse their seeds far and wide. With home ranges spanning 4–30 square kilometres or more, the great and wreathed hornbills often fly several kilometres in a day, effortlessly transporting their gut-cargo of seeds. Dispersal is also effected through the seeds dropped and defecated in piles, called middens, which accumulate at the base of the nest trees. Some tree species with large fruits, such as *Myristica* (nutmeg) and *Beilschmiedia*, appear to depend exclusively on large frugivores such as hornbills and imperial pigeons for dispersal. Dispersal may benefit trees in different ways. For some, it would enable seeds to reach favourable germination sites away from the parent tree. For others with hard seed coats that are whittled down during their passage through the hornbill's gut, germination and viability may be improved. Such reciprocally beneficial relationships have been termed 'mutualisms' by biologists. Hornbills have been called 'mobile links' and 'keystone mutualists' because of their wide-ranging habits and role in seed dispersal as well as regeneration of rainforest trees. The functional role of hornbills in maintaining tree populations and regeneration makes them a critical component of tropical forest ecosystems, across Africa and Asia from open savanna to rainforest.

Despite their importance in tropical forest ecosystems, several hornbill species are highly threatened and endangered today because of human impacts on their populations and habitats.

India has nine species of hornbills, all of which are considered endangered and receive protection under India's Wildlife Protection Act of 1972. In fact, the Malabar grey and Narcondam hornbills are endemic and not found anywhere else in the world. Two other Indian species, Malabar pied-hornbill and the Indian grey, are endemic to South Asia, with the latter being relatively common and widespread over its range. The status of the Narcondam hornbill is most precarious, as there exists only a single population of about four hundred individuals. This population is restricted to the small island of Narcondam in the Andamans, which spans about 680 hectares. A population of goats introduced in the island around 1976 went feral and increased in numbers, affecting forest regeneration. Conservation scientists, particularly Ravi Sankaran of the Sálim Ali Centre for Ornithology and Natural History, Coimbatore, worked with the authorities to remove the goats from the island.

To understand the causes that have led to the endangerment of hornbills and to devise effective conservation measures, several aspects of hornbill natural history and ecology are relevant. The hornbills' preference for particular types of nest trees is of great significance for their survival. When nest trees are logged or trees with cavities removed as part of silvicultural operations, it can trigger declines in breeding populations. Another vital factor is nest-site fidelity—pairs use the same nest every year. In the Western Ghats, we have found hornbills using the same nests and nest trees even after 27 years. Identifying and protecting these trees will go a long way in maintaining hornbill populations.

Poaching poses another threat. Often, it is for the meat of adults or unfledged squabs in the nest. Since hornbills have low breeding rates, poaching at the nest hits the population where it hurts most. In many tribal societies of northeast India and southeast Asia, hornbill casques and feathers are valued as ornaments or symbols of hunting prowess. In the past, when forest cover was extensive and human populations low, traditional hunting for subsistence or ornaments may have had little impact. Today, elders and local hunters in many parts of southwest and northeast India themselves recount the local extinction of hornbills from hunting in otherwise suitable forest areas in the vicinity of their villages. In southeast Asia, commercial hunting threatens populations of helmeted hornbill, killed for illegal trade in the so-called 'hornbill ivory' of its casque.

The destruction, conversion, and fragmentation of mature tropical forests due to logging, plantations, or shifting agriculture, are other inexorable forces affecting hornbills. In the rainforests of Arunachal Pradesh, Datta's research in the mid-1990s showed that plantations and commercial logging for timber particularly affect large species such as the great and wreathed hornbills. Hornbills have become locally extinct or very rare in large areas of northeast India covered by dense bamboo forests that result from shifting agriculture with short fallow periods. Fragmentation of rainforests is proceeding apace in tropical forests of south and southeast Asia. When the size of forest patches falls below a threshold, it may become inadequate to meet the year-round habitat, nesting, and food requirements of the hornbills. Where the birds are left unmolested because of legal protection or local taboos, a pair may cling like vestiges to a small forest patch that contains their traditional nest site, as in Puthuthottam in the Anamalai Hills. Their survival here is, at best, tenuous.

Given these threats, it is ironical, but fortunate, that some of the modern conservation success stories involve the endangered hornbills. Using hornbills as flagships for their forest habitats, many countries including India have protected them and their habitats under wildlife laws. Appreciating the monogamous fidelity of breeding pairs, people in a part of Sumba island in Indonesia have stopped hunting the Sumba hornbill for meat and have instead made it a mascot. Two other hornbill species have also been adopted as state birds in Indonesia. Researchers in Mahidol University of Thailand have also successfully used hornbills as flagships for environmental awareness and conservation efforts. In northeast India, Aparajita Datta and her team from the Nature Conservation Foundation, Mysore, have established an innovative hornbill nest adoption and monitoring programme involving the local community, which has led to widespread awareness on the plight of hornbills and their habitat, while directly protecting nests and breeding hornbills and supporting livelihoods of local tribal people.

In the Anamalai Hills of southwest India, following the studies of researchers, several management measures were launched. The Forest Department in the states of Kerala and Tamil Nadu now employs local tribal people who, along with researchers, monitor several dozen hornbill nests every year during the breeding season. The lopping of fig trees and saplings to provide fodder for camp elephants was also stopped after their significance for hornbills and other frugivores was highlighted. In the rainforests on the Karumalai Gopuram hills of Parambikulam Wildlife Sanctuary, great hornbills had suffered a brief period (three years) of local extinction. Owing to the efforts of the Kerala Forest Department and the awareness and co-operation of local tribals, the species has staged a comeback and resumed nesting in the area. Recent field surveys have shown that the Anamalai Hills in the Western Ghats hill range have one of the best populations of hornbills, particularly the great hornbill and the endemic Malabar grey hornbill.

There is much that remains to be done to conserve hornbills and their forest habitats. Let us hope that the efforts of local people, researchers, forest staff, and conservationists alike will continue, like fig trees, to bear copious fruit. The sight of hornbills in majestic flight over spiring rainforests will then continue to inspire awe in the future.

# A Life of Courage and Conviction

On 17 January 2009, India's birds lost a great friend and leading champion of their cause when Ravi Sankaran—born 4 October 1963—one of India's most articulate, loved, respected, sincere, and endearingly forthright ornithologists, passed away from a cardiac arrest. His demise leaves a permanent void in the conservation community, and for all who knew him personally it is the man himself—far more than even his exceptional work—who will be missed. It is difficult to talk of him in the past tense. He was such a man.

Ravi's life and work were marked by a deep passion for the conservation of birds and natural areas, an unbridled exuberance and enthusiasm for his work, and the fire of a pioneer spirit that

* This essay is written with Divya Mudappa.

made him take on the most challenging, difficult, and remote assignments. His work spanned the deserts of western India and the swamps and grasslands across northern India to the snow-clad mountains of the Himalaya (Nanda Devi Expedition) and the dense forests of both northeast India and the Andaman and Nicobar Islands. His well-honed field skills and adventurous exploits—wading into mangroves and walking the beaches, exploring caves and remote forests, enduring heat and cold in the desert, rain and flood in the rainforest, and battling multiple bouts of malaria—made him a legend among field biologists in India.

In his work, Ravi took up issues that few dare to—working for conservation outside protected areas, with hunters in western and northeastern India, and with harvesters of swiftlet nests in the islands. He went out to study little-known species—lesser florican, Bengal florican, Narcondam hornbill, Nicobar scrubfowl or megapode, edible-nest swiftlets—and came away, not just with valuable scientific information but with a plan for their conservation. Being Ravi, this naturally meant working to implement those plans as well, and often using methods arrived at by complete out-of-the-box thinking.

In India, the conventional strategy to conserve a species has been to list it in a relevant Schedule of the Wildlife Protection Act and hope that this would eventually translate to conservation on the ground. While Ravi did believe in having inviolate areas, protection, and strong legal backing for conservation, he was an astute pragmatist and free thinker who would argue, and argue well, for unconventional methods if the situation demanded. In the case of the swiftlet project, this meant actually advocating the de-listing of the species from the Wildlife Act to enable the sustainable farming of nests and thereby reduce the pressure on wild cave-nesting birds, whose legal protection on paper, leave alone in practice, has hitherto amounted to nothing.

In India's northeast, where great diversity and remoteness go with few protected areas and considerable community forests, and where the diversity of people is matched by the diversity of the forest resources they harvest or hunt, Ravi took on a challenging project of conserving what remains with community support. Although he initiated this project only a couple of years before he died, Ravi managed to mobilize people in village after village in Nagaland, and build support and catalyse the creation of a crucial community forest reserve in an otherwise heavily altered and exploited landscape. His passing away at a crucial juncture is a blow to the project, which still holds so much potential for reviving community lands and people's livelihoods in the region.

Ravi had a knack of working with and among people; he had a deep appreciation, curiosity, and respect for diverse cultures and cultural traditions. He tried to make his work count not just for the species he studied, but for the people who shared space and resources with the species. Thus, he worked on-ground for the restoration and management of grasslands for people and the lesser florican, and for managing forests for sustainable use and saving species from local extinction in Nagaland.

Ravi's interest in nature and wildlife was apparent from his childhood. His friends recall how Ravi would often try to introduce them to snakes and insects. His schooling was at Rishi Valley, Madanapalle, Andhra Pradesh, an institution known for its holistic approach to education, involving, among other things, the appreciation of nature. Ravi then strode towards his professional career, obtaining a BSc degree in Zoology from Loyola College, Chennai, before joining the respected Bombay Natural History Society (BNHS) to work ultimately towards his doctoral degree, conferred in 1991. A firm believer in long-term work, Ravi continued monitoring florican populations and their grassland habitats every year, even as he took up other challenging projects on bird species in the Andaman and Nicobar islands. Although he retained links with the BNHS and his other colleagues and collaborators from his florican days, Ravi's latter work was taken up at the Sálim Ali Centre for Ornithology and Natural History, Coimbatore, which he joined as a scientist in 1992, becoming one of its most active and dedicated scientists.

Over the years, his scientific work has resulted in numerous substantive publications that will endure as key references on the species he worked on, including a dozen research papers in the *Journal of the Bombay Natural History Society*, five key publications in *Forktail* and *OBC Bulletin*, papers in *Biodiversity and Conservation* and *Biological Conservation*, and dozens of other book chapters, technical reports, popular articles, and research notes and observations. Ravi was an energetic and engaging speaker as well, and his voice reached thousands of people at seminar presentations as well as popular and informal talks to diverse audiences, from school and college students, conservationists, and administrators to graziers, hunters, and harvesters in the villages at his field sites.

In May 2008, 16 years after he had joined, Ravi was appointed Director of the Sálim Ali Centre, and hopes soared around the country for the revival of this key institution. Transformed almost overnight into a man charged with greater responsibility, he rolled up his sleeves and took to the task with gusto—much as he would have done to save an endangered species from extinction—working late hours and clearing the backlog of work and policy matters. As he tried to streamline and tame the institution, the gravity of the job also, albeit a tad, tamed and streamlined Ravi, who was now more likely to be seen in formal attire or issuing a more carefully worded statement. But his conviction, the causes he strived for, and his approach carried on in as unflagging and uncompromising a spirit as ever.

A day before he passed away, he was working on polishing, to his satisfaction, a proposal on the future direction and peopling of the institution, picking up the phone to solicit feedback from others around the country to ensure he had not missed anything. Ravi, in any case, missed little.

Ravi was neither an armchair scientist nor one to just mail reports off to someone else to get conservation action going. He prepared careful reports based on his research and surveys, tailor-made for different audiences who he recognized as key for conservation implementation, ranging from Lieutenant Governors and Police officials, Forest Department and Ministry executives in the State and Central Government, and the wider conservation community. But he was a forceful advocate of conservation action, and he would follow through with the implementation himself, year after dogged year. In a speech addressed to the next generation of bird researchers at a workshop in Haridwar just a month before he passed away, he had said:

> As a conservation biologist, as a *practitioner* of conservation, you would actually have to build upon science, management, policy, advocacy, *implementation*—very important—and of course education for the larger grouping of people.... It is not a narrow field, you actually have to be multi-multi-disciplinary for working with conservation.

In short, one has to be like Ravi to be able to make conservation work.

More than anything, Ravi was honest, frank, and forthright and an extremely lovable person, whose tongue was as sharp as his mind, and who could yet soften the blow of his words with an endearing and contagious guffaw or smile. He was truly inspiring in the field in India and someone you could discuss both scientific ideas and on-ground conservation realities and get the best direct advice and feedback you could hope for, with an unhesitant crushing rebuttal of any wishy-washy idea.

He leaves behind his wife, R. Rajyashri (Deepa) and five-year old daughter Yamini, who live in their farmhouse near Coimbatore, and whose loss cannot be measured by our words. He leaves behind trained students and inspired collaborators, who will hopefully shoulder the mantle to continue and extend his work. He leaves behind unforgettable memories and a great many friends filled with a deep sense of loss. As one of them explained, 'Lots of people like me are still grieving in silence, not for the loss to conservation, but for having lost a great friend; no words can express our sorrow ... time cannot heal this wound....'

We, too, have lost our wonderful friend, our critic, our role model, and like a kite that has lost its flier, are now cast adrift.

## Postscript

Ravi's wife Rajyashri and their daughter Yamini accompanied him on many of his trips. Rajyashri was a wonderful, fun-loving person, who welcomed friends and students into their home at all times even if they dropped in at the oddest hours. She was also keenly interested in conservation and frequently accompanied Ravi to help in research and conservation work. After Ravi's death, she started working in Rishi Valley school in Andhra Pradesh. On 14 September 2010, both Rajyashri and Yamini died in a tragic accident on the road to school.

# Cavities, Caves, and a Caveat

Scientific papers and field reports on wildlife conservation often carry a caveat. It goes something like this: conservation must be species- and site-specific. To save a threatened species or ecosystem, the research provides some conclusions of a general nature, but the action needed on the ground depends on the specific context of each species and site.

In the Andaman and Nicobar Islands, this caveat is well illustrated by two endangered bird species. The Narcondam hornbill is a large bird of the rainforests that eats fruits and small prey and disperses seeds of forest trees. The edible-nest swiftlet is a small, drab greyish bird that lives on the wing above the forest canopy, hawking insects from the air.

The nests of the hornbill and the swiftlet rank among the strangest nests in the world of birds. The hornbill nests in cavities

* This essay is written with Divya Mudappa.

in large, tall trees. The female stays sealed in for weeks incubating the eggs, with only a slit-like opening retained in the cavity wall for the male to feed her and the chicks. In contrast, the swiftlets live in colonies of dozens to hundreds of birds and make nests using their own saliva, plastered as little cups against the walls and ceilings of limestone caves.

The Narcondam hornbill survives as a single population of a few hundred only on Narcondam, an extinct volcanic island spanning just about 680 hectares. The swiftlet ranges over a huge swath of islands and coasts in southeast Asia. Partly because of this, the *IUCN [International Union for Conservation of Nature] Red List of Threatened Species* lists the Narcondam hornbill as 'Endangered' and the swiftlet as 'Least Concern.' Yet, both species are considered endangered in India and listed in Schedule I of the Wildlife Protection Act.

What makes them peculiar makes them endangered. The hornbill, endemic to Narcondam, was threatened by the alteration of its forest habitat accompanying the establishment of a police outpost and the introduction of goats on the island. The swiftlet was threatened by exploitation of its nests because of the bizarre taste for edible-nest soup in China and eastern Asia. With nests fetching lucrative profits in wildlife trade, over-exploitation of unprotected swiftlet populations in the islands ensued.

The research and conservation effort for these two species were spearheaded by stalwart field biologist Dr Ravi Sankaran, a doyen of Indian wildlife scientists who passed away in 2009. The field research and on-ground conservation programme he launched in the 1990s through the Sálim Ali Centre for Ornithology and Natural History (SACON), Coimbatore, has been developed and continued by the Centre's scientist Dr Shirish Manchi.

In Narcondam, the forests were under threat. With the police outpost established in 1968, ostensibly over a dispute with Myanmar, some forest was cleared for plantations and disturbances such as hunting and alien plants followed. Introduced as a food source for the outpost of less than twenty people, the population of goats boomed from a handful in 1976 to around four hundred in 2,000, over half being feral. Goat grazing stripped the forest regeneration and posed a major threat to the future of the hornbill's habitat. Translating field research to action, the Forest Department was persuaded to evacuate the goats. Virtually all goats were removed and strictures against hunting declared to the outpost. On-paper protection under the wildlife laws, of the hornbill under Schedule I and of the island as a wildlife sanctuary, was changing towards actual on-ground protection.

The swiftlets' situation was quite different. Research by Sálim Ali Centre's scientists showed that swiftlets had vanished from 60% of the unprotected nesting caves and declined by over 73% in numbers, suggesting initially a need for stricter protection and listing in Schedule I. Field efforts helped swiftlet populations increase by around 52%, as thousands of chicks fledged from caves protected by erstwhile nest collectors, their incentive being that they could harvest the used nests after the breeding season. In the harsh and remote landscapes of the islands, protection without incentive was virtually impossible, as caves left unprotected even briefly were plundered overnight, killing thousands of hatchlings.

Simultaneously, another solution emerged: the edible-nest swiftlets could be cross-fostered in the nests of their more common cousin, the glossy swiftlet that nests even in houses. Hatchlings from foster nests return to their natal houses to breed. Nests may then be harvested from these 'house ranches', a technique successfully implemented in southeast Asian countries, where swiftlet populations are thriving. Nests may be harvested after chicks fledge without destroying broods or domesticating free-ranging swiftlets, thereby providing lucrative livelihoods for local people from a relatively ecologically benign activity. Ongoing field research also indicates that this is a feasible solution.

Yet, for hornbill and swiftlet, problems fester like the rumblings of a dormant volcano. On Narcondam, a few feral goats are still reported, and disturbances emanating from the police outpost and their plantations continue. In 2011, a proposal was tabled before the Standing Committee of the National Board for Wildlife to clear additional area on Narcondam to install radar surveillance and diesel power generation stations by the Indian Coast Guard, with little heed to ecological consequences. The proposal entailed road construction and other disturbances that would have irrevocably altered the habitat and the already precarious situation of the species. Fortunately, widespread protests and a campaign by ornithologists and conservationists outlining the threats were successful. On 31 August 2012, the Ministry of Environment and Forests rejected the proposal so as to safeguard the island and ensure the conservation of the endemic hornbill. Owing to its extraordinary evolutionary and conservation importance, the fragile island and its endemic hornbill deserve highest levels of protection, including strictures against plant and animal introductions, rather than being bandied about for insensitive uses.

In contrast, higher protection of the edible-nest swiftlet—listed in Schedule I in 2003—may paradoxically compromise the species' survival. While lending paper protection, the listing stymied effective action for house ranching by mandating a bureaucratic quagmire of Central and State permits. Well over a decade ago, Ravi Sankaran, stressing immediate protection of swiftlets and nesting caves, suggested house ranching enterprises by local people and the eventual downlisting of the swiftlet as its populations recovered. A new lease of life for the swiftlet may yet come if the authorities recognize this and act on findings of field research and experience. In the Andamans, today, swiftlet conservation efforts continue, including nest cave protection and experimental house ranching, and preliminary results seem to indicate an increase in the populations of this species.

The endemic hornbill, threatened by various disturbances on Narcondam Island, and the swiftlet, whose nests are a commodity in wildlife trade, provide lessons for conservation. In their own way, these two bird species show how conservation may require very different approaches if it is to achieve its fundamental goals. Saving threatened or endemic species in islands often requires highest protection, but not always. Thus, protection efforts should be sensitive to the ecology of the species, the site, and the people involved.

# Life in the Garbage Heap

From a boat on Assam's Deepor Beel—the freshwater lake lying southwest of Guwahati, the largest city in the Northeast—you can look east past thousands of waterbirds and a carpet of floating leaves to see the city's seething, smoking garbage dump. Under spotless blue skies, a thin brown haze blankets the lake from fringing forest to quarried hillock, from skirting township to the Boragaon dump yard. As another dump truck lurches to a halt and tips its load of filth over, an unruly mob of black kites and a cloud of dark

mynas explode from the murky earth, flapping like pieces of tattered cloth caught in a gust. The truck deposits another mound of unsegregated waste—a fraction of the more than 600 tons generated daily from this city of nearly a million people—all plastic and putrefaction, chicken heads and pigs entrails, street dirt and kitchen waste, broken glass and soiled cloth, bulbs and batteries and wires and electronics, metal, paper, and much more. Beside the truck waits a line of people: women, adolescents, and children. And behind them, a phalanx of greater adjutant storks—tall, ungainly birds with bayonet-like beaks and naked yellow and pink necks—is next in line.

In company with cattle and dogs, the people will scavenge first. Driven by poverty, with little or no land or possessions of their own, these people from poor families living around the lake have turned to the dump yard, despite its appalling prospects for their health, to scour a livelihood from the residues of urbanization. In the waste thrown out of home and market, hospital and motel, collected and dumped again by the trucks, in that twice-discarded garbage, they will rummage to gather things to sell, to use, to survive. Without even a cloth draped over their noses against smoke and stench, they will sift valuable scraps from the offal with metal hooks and bare hands.

Then, it will be the turn of the greater adjutants. The birds will parade over the dump, pick up and swallow rotting meat, skins, and bones, fish tails and goat's ears, eyeballs and hooves, and some will carry it back to their nests on tall trees in villages many kilometres away, and regurgitate it to feed their hungry chicks. The stork, whose world population is estimated at around 1,200 to 1,800 mature individuals, about 650 to 800 of which lives in Assam, is considered endangered, its population trending downhill. Yet, there are days when nearly half the world population may be seen in the city of Guwahati, congregating at the garbage dump. In its decline, the stork has learnt to survive off the thrice-discarded filth of humanity.

With the reek of burning, decaying garbage, the air carries the droning of flies and keening of mosquitoes, the chatter of women and children, the clatter of stork beaks. The air lies thick with humid vapours that burn the nostrils, clog the windpipe, catch at the throat as if to stifle the breath of life. And above it all, a twister of storks turns slowly. The lanky birds rise and rise in a thermal spiral, wings held wide, yellow pouches hanging at their throats, like penitent beings weighed by remorse seeking the heavens.

Seated on low thwarts, the three of us—Jaydev Mandal, a young research scholar from Gauhati University studying greater adjutants, Divya, and I—are rowed out by the boatman into Deepor Beel. The floodplain lake, a now-festering fragment of the great River Brahmaputra that has swung its course to the north, is a Ramsar site, a wetland of international importance. Every year, thousands of waterbirds gather at the lake, from resident swamphens, lapwings, and herons, to wintering migrant shorebirds, ducks, and geese. Rowing in deeper water or punting through shallows with his oar, the boatman guides the flat-bottomed boat, freshly waterproofed with sticky black tar, over water clouded with sediment and plankton, towards the distant flocks of waterbirds. The boat skims the canopy of a swaying forest of soft underwater plants, topped by floating waterlily leaves that look like plates, like expanded hearts.

Fed by river and rain, the lake receives water and storm runoff along feeder streams and drains. It mixes city wastewater with the purer flows from the hills of the Rani and Garbhanga Reserve Forests to the south. Polluted by pesticide and fertilizer runoff, contaminated with faecal bacteria, muddied by erosion and sewage, Deepor Beel is slowly turning into ditch water. Hemmed in between stolid hills and restive city, pincered by highways to the east and west, cleaved from the southern forests by road and railway line, the lake is shrinking, too. Over two decades beginning in 1991, Deepor Beel lost 41% of its open water and shrank from 712 hectares to 421 hectares, becoming more fragmented. With the loss of wetland area, the birds, too, appear to be in decline. Only the city and its impressive garbage dump are growing and growing.

On the lake, the water parts for our boat, closes in our wake, the parted plants mark our passage on the surface. In distant boats, fishermen fling their nets onto open water, or pull at them, stooping to pick the day's catch from the tangle. Our boat slips over a long, taut fishing net stretched wide across the lake. The water is calm, it is the horizon that is now swaying. To the west, vehicles ply on the busy highway, to the north, the city burgeons, to the east, the unsavoury Boragaon dump yard moulders, and to the south, a train thunders along.

The tracks shrill and clatter under metal wheels, as they will almost every hour: the trains will not stop. There is blood on these tracks, the blood of elephants—herds, calves, tuskers—who tried to cross the tracks from the forest seeking water and forage in the lake. The elephants were slammed, were dragged, were extinguished: a slaughter wrought in passivity, for who in their right mind will attribute active intent to trains? Only the journeys of the elephants shall come to a stop.

The clamour of hundreds of whistling-ducks accompanies our passage, their pulsed whistles and squeals lance over the water. Suddenly, the air reverberates. The beat and rush of wings roils the air overhead as hundreds of pintail ducks and greylag geese take wing from rippling lake to splayed-out sky.

The storks still soar in the distance. From high, they must see the vast braid of the Brahmaputra winding through the landscape. They must see, far, far to the north, the grandeur of Himalayan peaks, dusted with snow and weighed by glacier. They must see that the lake below is but a drop of water on land. The ducks and geese wheel and quarter, they sweep and swerve in the air. They begin to descend, as the storks do, too, out of seamless, unmarked skies.

As dusk settles over the landscape and the milling flocks of birds settle for their roost, the fishermen return to their villages in their boats. The trucks and trains have passed, and perhaps the elephants, too.

By night, as the city flickers to life before us, the garbage dump, its people, its birds, all become invisible. Tomorrow, the fishers will return to the lake, and the women and children to the dump. And, from the skies above Guwahati, it is again to the lake and garbage dump that the birds must drop.

# Death on the Highway

Crunch! Splat! Thud! A daily massacre is occurring under the wheels of our vehicles. Thousands of lives are snuffed out tragically, instantly, and yet, we hardly notice.

Around India, as in other parts of the world, millions of animals risk daily encounter with increasingly fast vehicles plying on an expanding meshwork of roads and highways. Roads through our countryside and forests and the people who drive vehicles on these routes cause the highest toll. This is a toll of actual lives—a headcount of animals crushed to death or else grievously injured and mutilated. Even leaving aside domestic dogs and cats, an indiscriminate diversity of wild species—from butterflies, squirrels, lizards, and partridges to more threatened species such as leopard cats to tigers and lions, chevrotains to sambar and elephants, lorises to langurs and lion-tailed macaques, and sheildtail snakes to king cobras—comes to a sticky end.

The scale of the problem is imposing. India boasts of having the second largest road network in the world, second only to the United States. India's road network increased from 3.4 million kilometres in 2001 to 5.6 million kilometres in 2016. Although 63% of this is surfaced and less than 2% of this comprises National Highways, the latter alone account for 40% of our total traffic. Like many things in India, the 'total' in that expression is a very large number indeed. From 2001 to 2015, the number of registered motor vehicles in India increased from 55 million to 210 million. Over the same period, the total passenger movement by road in India soared from 2,143 to 15,415 billion passenger kilometres per year and is still growing.

With such traffic, it would be scarcely surprising if animal kill rates were high, too. Roads passing through forest and other natural areas such as grasslands and wetlands are of greater concern from a conservation point of view. The few studies that are available from Indian forests indicate a grave situation already. Studies have documented kills ranging from dragonflies and butterflies to many larger mammals and birds, including carnivores. Around noon in Nagarahole–Bandipur in southern India, as 50–100 vehicles zip past every hour, a study patiently documented around 40 kills of insects such as butterflies and dragonflies for every 10 km every day, doubling over the weekends with increased traffic. A rough calculation indicates that vehicles here kill around 15,000 animals every year in just that 10-kilometre stretch.

In the Anamalai Hills of southern India, a 2001 study of road kills of reptiles and amphibians found that around 6 were killed per 10 kilometre of road each day during the monsoons. Conservative extrapolation would suggest that a 100-kilometre stretch of road through forests here witnesses an annual slaughter of around 10,000 amphibians and reptiles. Even this estimation is based on a study carried out 10 years ago, when traffic volumes were much lower. The widening of roads and unregulated, ill-planned tourist influx has, if anything, made things worse. Using slightly different methods, a 2017 study in the same landscape found about 12.5 reptile and amphibian kills per 10 km—double the earlier estimate.

Such patterns of death on the highways are a common feature wherever roads traverse our forests, grasslands, and wetlands. Along the Western Ghats, a hill range much touted, ironically enough, as a centre of amphibian and reptilian diversity, with so many new species even now being discovered and described—hundreds of thousands are probably killed every year. These numbers should not make us complacent that there are enough animals out there that we can overlook such slaughter—that would be a dangerous assumption, a form of denial, or unwarranted optimism. Neither can we take heart from areas where few deaths are now seen along roads; this could also be because populations have already been pushed over the brink.

Planners and managers somehow don't take the problem seriously. Even when they are aware of the issue, they feel nothing needs to be done because they believe that while many are killed on roads, many others escape and the species can survive. What they fail to understand is that the additional mortality on roads can tilt the demographic scale against a population that already grapples with various natural factors and human-caused disturbances for survival. Studies from elsewhere have revealed that the negative effects of high traffic density can be as serious as direct loss of forest cover for amphibians, and that traffic needs to be avoided or maintained at low density for up to two kilometres around breeding ponds if frog diversity is to be conserved in the landscape. Another study estimates that even if 10% or more of the adults annually risk being killed by vehicles along roads near breeding areas, the population will eventually perish.

In most cases, all that the animal is trying to do is get to the other side, like the proverbial chicken. The road surface and corridor itself is of little use to most animals. Perhaps a dove or myna would find some fallen scraps of food worth eating; a lizard or snake may be attracted to bask on the hot surface, as to a rock on a sunny day. Dragonflies and mayflies may be attracted to the polarized light emanating from the asphalt, a form of light pollution that fools them into believing that they are over the surface of a water body. As they fly around to feed or defend territories or even try to lay eggs on the water-road, they imperil their own survival. And then the road becomes an ecological death-trap, where the very adaptations evolved over millennia to enable these species to locate their food and thrive in their environment now nudge them to their death.

The roadkill threat is not something only ground-dwelling animals face. The threat of roadkills is particularly acute for many tree-dwelling species that do not normally cross on the ground. With roads mercilessly slicing through our forests, and government departments and road contractors recklessly widening roads and slashing all vegetation, including regenerating trees and saplings on either side, the tree cover breaks over the road. Besides loss of natural vegetation and native species typical to each area, this causes increased soil erosion and landslides. Which, of course, leads to further expenditure in road maintenance—providing further opportunity for ecological damage. All of this adds to wastage of public money, while also wrecking the tree cover that would have allowed many species to safely cross the road overhead.

Unable to cross overhead using the overlapping branches of intact forest canopies, the animals now face a permanent problem—a serious, life-threatening challenge—of a gap caused by the break in tree cover over the road. That crossing, even if takes only a few seconds or minutes, can be an agonizingly long and threatening one for an animal trying to get across even a moderately busy road. In the absence of tree cover, arboreal animals are sometimes forced to use electric wires of power lines to cross, leading to the double jeopardy of electrocution deaths for species such as lorises and lion-tailed macaques. The roads and power lines through our forests are increasingly turning into graveyards of tree-dwelling species such as monkeys, lorises, civets, squirrels, and tree shrews.

Animals may also be seriously stressed or change their behaviour in the vicinity of roads. Studies from Africa on elephants and chimpanzees have shown how they tend to avoid roads and change their behaviour due to the associated risks, as one would expect from such highly intelligent species.

Other factors may compound the road problem. The building of culverts, fence rails, barricades, chain-link and barbed-wire fences, and other concrete and metal structures along roads makes the crossing even more difficult. Parapet-like walls running without a break for hundreds of metres or kilometres along roads, especially on hill roads, become insurmountable obstacles for species such as porcupines, pangolins, turtles, young birds and mammals, to name just a few. On hill slopes disfigured by such roads, even large animals such as sambar and elephants have to negotiate the upper slope, cross the road, and try to somehow step or jump over roadside walls and culverts to step or land safely on the steep lower slope. Another compounding factor is the attraction of scavenging animals to road-killed carcasses, which may lead to further deaths from speeding vehicles until the carcass is safely disposed away from the road.

As roads become wider and busier, the number of animals crossing and the rate of roadkill usually increases, but beyond a point it may actually begin to decrease. This usually happens when roads become four-lane highways or expressways catering to tens of thousands of vehicles every day. The reduction may be due to the decimation of wildlife populations along the road as well as a 'barrier' effect, where many animals actively avoid the road and avoid crossing it. A road like this passing through a forest or key natural habitat essentially cleaves it into two pieces. For many species, this is an added fragmentation of an already fragmented habitat.

In addition, roads are now well known to cause various ecological changes, leading to a wide range of impacts including many, often unnoticed, detrimental effects on wildlife. The disturbance associated with roads and the opening created by the road corridor does favour some species; unfortunately, these are mostly undesirable ones. Alien weeds spread along roads using them as invasion highways into ecosystems. The exposure along the road desiccates and dries vegetation, making it more prone to fires. Trees are more exposed, too, and may fall because of high wind speeds along the road or suffer from stress related to altered ecology. All of these contribute to permanent and chronic changes in the environment and habitat, thereby affecting wildlife and ecosystem health.

Yet, this is only a small part of the story. No study has yet comprehensively addressed all animal taxa from invertebrates such as snails and ants to large creatures such as peafowl and elephants. Even the studies carried out so far may underestimate the true damage. Many animals are struck and badly wounded by vehicles along roads but manage to flee or drag themselves away from the road corridor to die unseen and unrecorded by researchers some distance away. It is not unusual for road-killed animals to be removed off the road or consumed by scavengers, including people; these kills go unrecorded. Even when dead animals on the road are noticed, other pervasive problems related to the road within forest areas are overlooked. This includes animals killed during road construction, earthwork and annual maintenance operations, particularly slow-moving and burrowing species such as turtles, snakes, and soil fauna.

No study has yet even catalogued the extent of roads through natural areas, especially forests, across India or the loss of forest cover due to roads. A notable exception, from Garo Hills in Meghalaya, showed that just in this region 456 hectares of biodiversity-rich forest were lost to roads between 1971 and 1991. A 2018 study in the Western Ghats found the deforestation decreased by 88% if forests were protected and away from roads. Another long-term aspect is the issue of increased access: people moving in and settling or polluting otherwise remote areas.

While more studies on road ecology are required in India, there is also an urgent need to use existing information and experiences from other countries to reduce and avoid this carnage. This requires the immediate attention and close co-ordination of ministries and departments related to roads and forests (or other natural ecosystems). Most important, it requires the attention of the citizen, the casual driver, the tourist—particularly the vehicle-based 'eco-tourist'—whose individual initiative, sensitivity, and care could save thousands of animal lives.

A range of measures could help remedy the situation. Some are merely engineered quick-fixes that can help in certain locations or in the short-term, such as artificial 'canopy bridges' for movement of arboreal mammals. Other measures include proper deployment of speed breakers in roads through forests, creation of underpasses and overpasses that are well-designed keeping in mind the ecology and behaviour of the species whose mortality rate is sought to be mitigated. Signboards informing people to look out for and allow wildlife to cross and measures to check overspeeding may also be implemented. Such short-term measures, if implemented based on research that has identified roadkill 'hotspots' can have very positive effects. For example, the installation of just four speed-bumps along 1.5 kilometres of highway passing through a forest in Zanzibar helped reduced the mortality of threatened Zanzibar red colobus monkeys by as much as 85% right in the first nine months. Prior to this, every year, vehicles used to kill 15% of the colobus monkey population living near the road. Slowing down vehicles at key locations is a very crucial aspect that reduces the likelihood of roadkill while providing greater reaction time for drivers and animals to evade a collision.

Longer-term and more sustained measures require a deeper understanding of the landscape through which roads pass and a greater sensitivity to the species we share this world with. The number, extent, and width of roads passing through forests and wetlands should be strictly regulated. Improvements to the quality of the road surface and adequate signage should be the emphasis for driver comfort and safety, not increasing the number of lanes or width of the road, or the speed with which vehicles can traverse these crucial stretches. As there is virtually no understanding of these issues among planners, land managers, and the wider public, despairing conservationists today regard narrow, bad roads as a great boon, one that is surpassed only by the complete absence of roads.

A key long-term measure is to encourage natural vegetation on either side of the road. Currently, vast amounts of public money is wasted in slashing all vegetation on either side of thousands of kilometres of road, with the spurious claim that this improves visibility or makes the road safer. In fact, dense weed growth rapidly chokes up the opened spaces on roadsides, replacing more pleasing and open, natural, native vegetation. In forest areas where tree cover would have naturally shaded out weed growth—performing a public service at no cost and with considerable aesthetic benefits—the opened spaces with obnoxious weed growth now represent a wasteful annual cost of repeated slashing in the guise of road maintenance. The lack of any understanding that good, stable, and safe roads really need consideration of ecological aspects as well is one of the glaring failings of the government and road construction companies.

The design and adoption of regulations is urgently needed. Forest roads should mandatorily retain and maintain tree canopy connectivity over the road. Where such connectivity has been lost, at a minimum, for every 200 metres of road, a 50-metre-wide stretch needs to be marked off with signs and speed breakers and the tree canopy with overlapping branches re-established overhead. Efforts to establish and maintain such stretches should begin as a top priority along all roads through our wildlife sanctuaries, national parks, tiger reserves, reserved forests, and their buffer zones.

Guidelines need to be evolved keeping specific species and landscape considerations in mind. For instance, in tropical forests of equatorial Africa, the home of the highly endangered great apes (gorillas and chimpanzees), the International Union for Conservation of Nature has prepared best-practice guidelines on a range of issues, including road planning. This includes recommendations to plan roads at least 5 kilometres away from protected area boundaries, reduce road width of primary roads to less than 7.5 metres (less than 12.5 metres including graded portion and shoulders) and that of secondary roads to less than 4.5 metres (8.5 metres including shoulders), avoiding road construction in closed-canopy forests, minimizing the number of secondary roads, and re-using old roads rather than build new roads. There has been some effort to develop such guidelines in

India, including manuals prepared by the Wildlife Institute of India, but there is much more to be done.

Forest areas around the world, including in India, are transected by a large number of old, unused, and unnecessary roads (for example, old logging coupe roads, roads built during dam construction, or as 'game' roads for hunting). It is time to undo the damage wrought by these roads by actively removing them and ecologically restoring natural vegetation. Although the methods available for road removal may cause some short-term disturbance, research has clearly established the conservation benefits in the medium- and long-term.

An overarching need, although perhaps the most difficult one, is the sensitization and involvement of individual drivers. A vast majority of drivers probably have no deliberate will to kill animals. They presumably have no wish to cause lasting harm to the environment or to the public exchequer by insisting on roads made and managed by ecologically illiterate and insensitive agencies. When individuals become aware and begin to care, it can have two useful effects. As drivers, they can adopt more responsible driving practices, watch out for and respect animal crossings, and avoid other unsavoury practices such as feeding animals by roadsides. This, as a direct contribution, can help save hundreds to thousands of animal lives over an average driver's lifetime. Second, by example, by persuasion, or ultimately by their vote in a ballot box, they can indirectly influence others to save thousands of lives, minimize ecological damage, help improve roads, and make the driving experience through natural areas infinitely more pleasant. When the paths of people and animals cross, each can go their own way, leaving behind not a flattened carcass but the memory of a pleasant encounter.

# Natural Engineering

## *India's Green Infrastructure*

Every year as parliament prepares for its budget session, there is talk, among other things, of India's infrastructure. The *Economic Outlook for 2009–10* prepared by the Economic Advisory Council recently noted constraints to India's economy from a shortage of physical infrastructure. From economists, development gurus, and politicians, this is a common refrain: infrastructure development needs due emphasis and allocation to achieve growth and abolish poverty.

Yet, a simple question comes to mind: what is India's infrastructure, really? Take a look at a brochure or website of any company that peddles infrastructure. High on the list will be dams and hydropower, roads and bridges, and energy. In the national Economic Survey, railways, ports, and communication are also included. These are meant to represent infrastructure, the backbone that industry ostensibly needs to enable India to stand proud in the developing world. But pause for a moment and humour this tangential interpretation.

Assume for a start, not unreasonably, that sectors such as agriculture and fisheries, forests and public health are at least as important as industry and services. What kind of infrastructure enables and supports the flow of goods and services in these sectors?

Imagine an infrastructural edifice built thousands of feet high, able to snatch the moisture out of the atmosphere and funnel it down for our use, clear as crystal. Stretch this for hundreds of kilometres to benefit millions—think of the Western Ghats, and the Himalaya.

Imagine marine infrastructure with components hard as concrete and pliable as reed, structured and self-renewing, guarding beaches and fringing islands—our densely populated and productive areas—from the incessant battering of the sea. Ring it around islands, fish-filled lagoons, and bays to fuel the prosperity of coastal communities. Think coral reef and mangrove, coastal dunes and seagrasses.

Think of a giant reservoir, built without dams and concrete, holding not water but dynamic snow and ice, safeguarding irrigation and drinking needs for centuries. Freeze it in the cold weather, melt it down for the hot summer. Think of thousands of glaciers on the Himalayan peaks.

There is other ingenious infrastructure—a substance spread over vast areas, porous and friable, inches to feet in depth, with the right ingredients of nutriment and water retention to support productive crops. Soil. A self-repairing giant sponge capturing water over thousands of square kilometres and releasing it, measured and pure. Forests. An immense water-network, linking peoples, quenching thirst, and sharing its fertility through flood-carried silt and aquatic life. Rivers.

If by infrastructure is meant that which builds the structural foundation of our country, then this is the infrastructure that we must revive and guard—our mountains and rivers, forests and grasslands, reefs and mangroves, soil and air. These are the real bedrock of the nation, its fibre and substance.

What happens, then, to our roads and railways, ships and power lines? One could argue for an attitudinal frame-shift, where we see these not as the foundations of India's development, but as superstructures built over our natural infrastructure, which temporarily moderate and regulate our passage through these lands and waters. We live presently in the blink of a geological eye; alone among all creatures we can perceive this and look far ahead, but do we?

And yet, the superstructures and the activities that go with them are not gentle on our lands and waters; they whittle away at the natural foundation at alarming rates. Logging, industry, and mining tear away at our last forests, leaving only sorry vestiges and artificial 'compensatory' replacements. Roads, rails, power lines, and pipelines thoughtlessly sunder habitats and cleave forests and wetlands, bringing fire and fragmentation, alien weeds and desiccation. Roads in hilly terrain are known to increase rates of soil loss through erosion and landslides over ten-fold compared to retaining undisturbed forests. In the Himalaya, glaciers continue to melt and retreat due to climate change, endangering lives, livelihoods, agriculture, and ecosystems. Droughts bite into human lives and GDP, and we believe water security will arrive through motors and metal pipes. Dams continue to submerge thousands of square kilometres of forests, degrade more in the surrounding watersheds, destabilize hill slopes and eat sediments in their life of brief, concrete bloom and extended, muddy decline. Unrepentant, unconcerned even, we build more.

The very surface and stability of our natural infrastructure erodes daily. It must be said: the superstructures that penetrate them and burgeon like cankers are responsible. Our built capital runs roughshod over our natural capital.

Yet, conservation, as Aldo Leopold put it, is not just about restraint; there is need for skill. Can the skill of engineering, which has taken us to outer space, be combined with that of ecology, which has helped us understand the dynamics of ecosystems? Will engineers now, through field training and textbooks, learn ecology, while ecologists strive to grasp the contours of engineering?

It is possible that ecologists and engineers, working together, could develop a more sensitive and salutary approach. One that replaces detrimental intervention emphasizing heavy machinery and concrete with more skilfully designed and deployed invention, placed in ways that respects, and sits gently, on our lands and waters. If such a collaborative skill can be nurtured, tempered by a humility learnt from history and geology and the marvel of evolution, perhaps our mark on the landscape may be permanent, and yet propitious.

# The Long Road to Growth

India's forests and natural areas now face a significant new threat. Unlike the external forces—submergence by dams; destruction through mining; clearing for agriculture, timber, and plantations—that have long dictated the plight of our forests and other natural ecosystems, this recent threat is internal. It arises from long intrusions that penetrate natural areas and whittle away at them from within. The intrusions—which cleave landscapes and carve out fragments, degrade habitat quality and integrity, and stretch ribbon-like over thousands of kilometres as a growing network—are linear infrastructure projects: particularly roads and power lines.

The central Ministry of Environment, Forest and Climate Change (MoEFCC) has been diluting norms for linear projects through orders and communications, including recently permitting central agencies executing linear projects in forests to cut trees after 'in-principle' or first-stage approval under the Forest Conservation Act of 1980. The orders allow projects to start cutting trees with approval from a Divisional Forest Officer, before second-stage clearance, which concerns compensatory afforestation and related procedures.

The issues at stake are wider and even more serious than the number of trees cut or how they are compensated. Linear projects such as roads, power lines, and railways support economic growth and other needs of a developing country, such as mobility and delivery of services. But linear intrusions also bring a litany of associated problems to natural ecosystems and rural and tribal communities, and present new challenges for conservation and sustainable development. It is worth examining these wider impacts and considering potential solutions to forestall or mitigate them.

Linear projects cause direct loss of forests and other ecosystems. In just two meetings in 2014 and 2015, the National Board for Wildlife considered proposals that involved over 2,300 hectares in and around wildlife sanctuaries and national parks for linear projects, including roads, power lines, canals, and fencing. Similarly, the Forest Advisory Committee, in four meetings in 2014, considered the diversion of over 3,300 hectares of forests for 28 linear projects, most of which are likely to be cleared.

Linear intrusions also cause habitat fragmentation: the break-up of continuous tracts into smaller pieces that are inadequate to meet the survival needs of many species, from rare plants and trees to hornbills and tigers. Many wildlife species avoid roads as they become wider and busier, effectively separating forest areas. Expansion projects and increasing the number of lanes in highways can negatively affect wildlife corridors, as in the case of National Highway (NH7) slicing crucial corridor forests between Pench and Kanha Tiger Reserves in Central India.

In mountains, roads may lead to severe forest destruction, landslides, and erosion, as evidenced daily in destructive road construction in many parts of the Himalaya and Western Ghats. A 2006 study noted that, on steep hill slopes, roads may increase landslides and surface erosion fluxes by 10–100 times or more as compared to undisturbed forests. The natural vegetation along hill roads in forests often helps to stabilize slopes and mitigate landslides. Road construction, dumping of debris, and slashing of roadside native plants, when carried out in a manner insensitive to terrain and local ecology, destroys natural cover, increases erosion and weed proliferation, thereby forcing additional maintenance work and avoidable expenditure.

Millions of animals, too, die along roads due to collisions with vehicles. Indian field research studies have documented that the spectrum of wildlife killed or injured ranges from small invertebrate, frog, and reptile species—many found nowhere else in the world—to birds and large mammals such as deer, leopard, tiger, and elephant. Estimates from a few studies range up to around 10 animals killed per kilometre per day, but the numbers could be higher, as injured animals are overlooked, and many kills go undocumented. With a length of over 5.6 million kilometres, India's road network is second largest in the world after the US. Studies estimate that vehicles along US roads kill around 1–2 million large animals, and 89–340 million birds every year. India may be a close second in roadkills, too, with species spanning insects to elephants.

Power lines, too, kill unknown numbers of wildlife. Poachers draw live wires to kill animals such as rhino and deer, while accidental electrocution kills many species, from birds like sarus cranes and flamingos to elephants. Railways, too, take their toll, gaining attention only when large animals such as elephants are killed along the tracks. The daily death of wildlife—including endangered species and those imbued with cultural values such as cranes and elephants—shows linear projects pay scant attention to conservation needs.

Linear intrusions affect areas larger than the area diverted for the project itself, because of negative 'edge effects' that diffuse to varying distances on either side. Each kilometre of road may affect at least 10 hectares of adjoining habitat. In Bandipur Tiger Reserve, a 2009 study found that tree death is two-and-a-half times higher along roads than in forest interiors. Linear intrusions also serve as conduits for the spread of invasive alien weeds such as *Lantana* and *Parthenium*. These may exacerbate ecological problems, degradation, and fire incidence, besides increasing annual costs of control efforts and maintenance.

Wildlife populations and behaviour in the adjoining landscape up to a kilometre or more are also affected. Even for species attracted to the vicinity of roads—reptiles that come to bask or deer and monkeys that come to feed—roads may act as 'ecological traps' literally drawing animals to their deaths. In 2009, a comprehensive scientific review of the effects of roads and traffic on animals found that roads had five times as many negative effects as positive effects. The scientists noted that evidence is already strong enough to mandate all road projects to routinely consider these effects and their mitigation.

When multiple linear intrusions—roads, canals, power lines, and railway lines—together shred the landscape into many pieces, their cumulative impact is higher. With trees and forests cut, tree-living animals like lorises, squirrels, and lion-tailed macaques are forced to cross roads on the ground or use electric wires to cross gaps in the canopy, leading to the double jeopardy of electrocution and roadkill. Local communities, too, may be affected by loss of land and traditional livelihoods, natural resources including useful trees along roads, increased risk of landslides, urban incursion, and other undesirable changes as exemplified by the notorious Andaman Trunk Road.

Linear infrastructure projects are needed for the economy but also impinge on ecology and communities. In a knowledge society, their development requires a fresh approach that uses the best technical resources and digital tools. It needs to consider wider issues and a larger landscape perspective than falls within the purview of a Divisional Forest Officer as in the Ministry's recent orders. Forests need to be recognized not as mere fungible assets to be compensated by artificial plantations, but as unique living systems of plants, animals, and dependent human communities.

Besides espousing economic benefits, linear projects must measure and mitigate long-term costs and ecological effects in a credible and transparent manner. The pursuit of mega-projects, often associated with lucrative contracts and corruption, spurs an undue emphasis on quantity and size (such as road width), which detracts from other priorities including quality, efficiency, and safety. Worldwide, a growing body of applied research in the field of road ecology and interdisciplinary efforts involving engineers, ecologists, and economists, is documenting such effects and needs, pinpointing realignments, better designs, and sustainable alternatives that can be adopted.

In India, the Standing Committee of the National Board for Wildlife prepared a detailed background paper and draft guidelines on linear intrusions in 2011, partly incorporated in the December 2014 sub-committee guidelines for roads in protected areas. The guidelines accord primacy to the 'Principle of Avoidance,' whereby wildlife protected areas and valuable natural ecosystems are not further disrupted by linear intrusions. They call for formal consideration of alternative alignments, routed around such areas and wildlife corridors to provide or enhance connectivity to peripheral villages and towns.

National and international guidelines also indicate appropriate restoration and mitigation measures for linear intrusions.

With site-specific inputs from wildlife scientists, overpasses, culverts, and underpasses can be designed to facilitate animal crossings, while speed and traffic regulation can reduce animal-vehicle collisions and enhance human safety and comfort. Infrared animal detection systems coupled to mobile messaging technology can alert train drivers and help prevent deaths on rail tracks. Structural modification of power line height and visibility, or using insulated or underground cables, in areas of risk will save species like elephants and flying birds from electrocution. Measures to retain overhead tree canopy continuity and roadside native vegetation help conservation while enhancing aesthetic values and driver experience along roads. Such scientifically informed, ecologically sensitive, and well-designed measures can guide linear infrastructure projects along better paths into the future.

# Watering Down Forest Protection

*October 2015.* India's Environment Ministry couldn't have chosen a worse moment to dilute forest regulations. At a time when the country is suffering from a deficient monsoon and water shortage for the second year running, it should have been strengthening forest protection, recognizing the key role forests play in hydrology and water supply. Instead, as news reports suggest, it's doing the opposite.

While identifying which forests shouldn't be converted to other uses such as mining, the Environment Ministry has decided to ignore parameters related to water. It rationalized this by saying there's no relevant hydrological data—but this is both inaccurate and inexcusable. As a result of its disingenuous claim, vital forest areas are now likely to be cleared for mining or turned over to other uses.

The ministry's action reflects India's flawed forest governance at large, which allows ecologically and culturally irreplaceable forests to be squandered for destructive development projects.

The attempt to establish objective parameters to demarcate inviolate forests—forests not to be converted to other uses—began as a reaction to the 'Go, No-go' policy of the previous United Progressive Alliance government.

In 2009, the Environment Ministry, in a joint effort with the Ministry of Coal, had placed India's forested areas under two categories—Go and No-go—and banned mining in the No-go zones on environmental grounds. An initial assessment demarcated a third of around 600 coal blocks as no-go, but pressure from the Prime Minister's Office and the Ministry of Coal ensured this was reduced to less than one-fourth of the blocks. Unwilling to accept even this, the Cabinet Secretariat in 2011 constituted a Group of Ministers that scrapped this classification and decided to assign the task to an expert committee.

The Environment Ministry, in turn, constituted two committees, one to develop objective parameters to identify inviolate forests nationwide 'where any non-forestry activity would lead to irreversible damage', and the other to subsequently prepare geo-referenced maps of inviolate forests as well as areas available for forest diversion.

The first committee, headed by former Environment Secretary T. Chatterjee, identified six parameters to be evaluated and scored (each to a maximum of 100) at the level of 1 km × 1 km grid cells across the country.

- **Forest type**, based on Forest Survey of India data, would be scored based on extent, range, and uniqueness of natural vegetation types (giving high scores to very valuable or highly restricted forest types).

- **Biological richness** was to be scored based on the countrywide biodiversity characterization carried out by the Indian Institute of Remote Sensing.
- **Wildlife values** were to be scored on the presence of wildlife protected areas, such as national parks and wildlife sanctuaries and identified animal corridors.
- **Forest cover** was to be rated using Forest Survey of India data, as an average of gross forest cover and weighted forest cover, the latter giving greater weightage to areas with higher canopy density.
- **Landscape integrity** was to be assessed from satellite images and maps based on the degree of forest fragmentation around the grid cell. Larger, more intact forest blocks would score higher than areas that had been fragmented by other intervening land uses.
- **Hydrological value** of forests, finally, would be assessed based on whether they served as catchments for perennial streams, hydropower or irrigation projects, or water supply schemes, and in relation to their proximity to streams and rivers.

The T. Chatterjee committee proposed to keep inviolate all grids that fell within wildlife protected areas, contained 1 square kilometre of very dense forests, or held the last remnants of rare forest types that cover less than 50 square kilometres in India. Emphasizing hydrological value, the panel identified the need to keep inviolate the forests that were catchments of perennial first-order streams (that is, those which consist of small tributaries and flow into larger streams), or were within 250 metres of perennial rivers, wetlands larger than 10 hectares, and storage reservoirs.

Besides these, the panel recommended protection for other areas: all remaining grids that scored over 70 when averaged across the six parameters were to be kept inviolate, it said. Further, a mining block would be inviolate if a majority of grid cells within it were inviolate.

Though the panel's six parameters were an attempt at being comprehensive, its scoring system and arbitrary cut-off of 70 meant that areas with valuable forests or other natural ecosystems could still slip through the cracks. For instance, a coal block that contained inviolate grid cells with the last remnant patch of a rare forest type or a critical wildlife corridor could still be cleared if the majority of its grid cells were open to 'violate.'

Despite these limitations, the central government could have implemented the six-parameter approach, while remedying the loopholes through finer assessments, carefully redrawing boundaries of coal blocks, and taking on board stakeholder inputs to address issues such as livelihood needs of communities. Instead, the Environment Ministry, under the new National Democratic Alliance government and facing continuing pressure from the mining sector, has further diluted the committee's recommendations.

First, the ministry clubbed biological richness and wildlife values as a single parameter based on data of the Indian Institute of Remote Sensing. This overlooks the fact that biological richness mostly considers parameters based on remotely sensed vegetation maps on plant species, ignoring vital conservation concerns such as distribution of endangered animal species and wetlands. It also devalues biologically significant non-forest areas such as grasslands, scrub, and deserts—habitat for critically endangered wildlife species such as great Indian bustard and Jerdon's courser—along with wildlife protected areas and animal corridors.

Another serious dilution was the decision to ignore hydrological parameters. The ministry's claim that the required data don't exist contradicts the committee's statement that 'country-wide geo-spatial data on all the six parameters are already available.'

The Water Resources Information System of India, developed by the central government along with research institutions, already hosts relevant public data on catchments, rivers and streams, surface wetlands, and reservoirs, required for the hydrological assessment. Hydropower and irrigation projects, too, record data on catchments and feeder streams across the country.

'Fairly good spatial data on basins and streams, existing and proposed water resource utilization, secondary data for some sites, and instrumentation and techniques for rapid assessment of stream and river flows are already available,' said Dr Jagdish Krishnaswamy, a hydrologist at the Ashoka Trust for Research in Ecology and the Environment, Bengaluru. 'The claim that such basic hydrological data on perennial streams and surface waterbodies are not available today in India is unfortunate and unacceptable.'

In discarding vital scientific considerations and assessments, the central government's intent is clear: opening up more forests for mining. Instead of 206 inviolate coal blocks as per the initial assessment, less than 35 blocks will be off bounds to mining due to the dilution of parameters.

By ignoring hydrology, the government is clearing the way for the destruction of forested watersheds and perennial rivers. Mining often results in heavy pollution of streams and rivers, making water unsafe for humans living within and far beyond grid cells or blocks. A 2002 study showed that iron ore mines within Kudremukh National Park had a profound impact on the River Bhadra. As compared to upstream stretches, the average sedimentation load downstream was 25 times higher, with about 68,000 tonnes of sediment transported downriver in just 67 days during the 2002 monsoon. While this mine was shut down by the Supreme Court, other mines continue to pollute with impunity. A 2015 study from Meghalaya found that acid drainage from coal mines has wiped out fish populations in sections of River Simsang, affecting downstream ecology and fishing communities.

India's forests urgently need better protection, assessment and monitoring, to safeguard water, wildlife, and livelihoods. Forest governance must pay attention to the multiple ways in which forests are valuable. Handing forests on a platter to private companies and industry will only create wider landscapes of desiccation and degradation.

# Protecting the Wildlife Protection Act

*9 September 2014.* As it completes 42 years, the Wildlife Protection Act, a landmark legislation that came into effect on 9 September 1972, is poised at a crucial juncture. Its future is on the minds of wildlife scientists, conservationists, and concerned citizens after the Narendra Modi-led government recently constituted a truncated National Board for Wildlife (NBWL) and hastened project clearances through its Standing Committee (SC-NBWL).

As the NBWL is the nation's apex body for wildlife conservation under the Act, conservationists are concerned that the government's moves signify an attempt to undercut the Act's objectives. The government's actions and subsequent developments also tellingly illustrate how governance failures lead to wildlife protection being misconstrued as anti-development, even as conservation needs become increasingly urgent in India and worldwide, and forests and innumerable wildlife species continue to be in peril.

In late August, the Supreme Court, acting on a public interest litigation, put on hold decisions taken by the Standing Committee. The committee had met on 12 and 13 August to accord clearances to 140 projects, such as dams, power lines, roads, and canals, slated to come up in and around the country's wildlife protected areas (including wildlife sanctuaries, national parks, and tiger reserves).

Earlier, on 22 July, the government had constituted the NBWL without including as many independent non-official members (five non-governmental organizations plus ten eminent conservationists) as required under the Wildlife Protection Act. With a retired Gujarat forest officer and a foundation under the Gujarat government as two of only three non-governmental representatives, the NBWL's credibility and effectiveness was also called into question, provoking litigation and judicial intervention. As a result of this muddle, the careful scrutiny of these projects, which is both imperative and mandated by law and earlier orders of the Supreme Court, was compromised. Also, as the SC–NBWL had not met since September 2013, in part because of elections, project appraisals and clearances that had been pending for close to a year were again delayed. This illustrated the persisting inefficiency in governance systems and the rather ham-handed attempt by the central government to steamroller project clearances, which also occurs with distressing regularity in the case of forest and environmental clearances (and a majority of projects are in fact cleared).

In instances of environmental clearance, the problem is often compounded by shoddy environmental impact assessment reports commissioned by the project proponents themselves.

Efficient and independent scientific scrutiny of all projects proposed in and around wildlife protected areas is crucial. The Wildlife Protection Act, by mandating the inclusion of multiple independent experts and non-governmental organizations in the NBWL, helps bring complex issues related to India's diverse ecosystems and local community contexts to the table for prior consideration before project approval. Comprehensive scrutiny opens up avenues to identify alternatives, reduce negative impacts on wildlife species, habitats, and local communities, and to enable project developers to strengthen protection efforts by co-operating with the Forest Department implementing local wildlife conservation plans.

Instead of asserting the value of such scrutiny-based clearance to ensure sustainable development, the Minister for Environment, Prakash Javadekar, has by his own remarks only bolstered the negative impression that these clearances are roadblocks to development that need to be removed. If real roadblocks exist, his ministry's manner of clearing them—by triggering judicial interventions or long confrontations between civil society and developers—risks running into other dead ends. This scenario hardly helps the cause of either wildlife or development.

To be fair, problems related to project appraisals and unwarranted delays had already come to the fore during the tenure of the previous government. At that time, members of the NBWL, Forest Advisory Committee, and other conservationists had provided constructive suggestions for improving the system. These included involving empanelled wildlife experts, forest ecologists, and social scientists in project appraisals and site inspections, getting project proponents to submit better maps and project documents, and implementing measures for rigorous, yet timely and transparent functioning of appraisal committees. Another key need identified was follow-up to oversee whether projects being implemented were meeting the conditions imposed when clearance was given. Sadly, none of these vital measures has yet been instituted.

The consequences of such inaction are serious given other trends. As norms for infrastructure intrusions such as roads and river-interlinking canals have been diluted, wildlife areas including Pench and Panna Tiger Reserves in central India will suffer additional forest loss, internal fragmentation, and disruption of wildlife corridors. The outmoded approach of encouraging more large dams in the Himalaya will seriously affect places like the Teesta basin in Sikkim, where an earlier site inspection report by the SC–NBWL had already revealed multiple violations of law and Supreme Court orders.

In India's grassland ecosystems, with land-use changes and continuing neglect of local community involvement in conservation, critically endangered species such as the great Indian bustard and Jerdon's courser now face a real risk of global extinction. If they occur, these will be the first documented bird extinctions in India since independence.

While key concerns over grasslands, rivers, forests, and mountains are ignored, the environment ministry continues to woo business and industry with promises of quick and easy clearances. The rush to project an industry-friendly environment risks destroying opportunities for environment-friendly industry.

In the last week of August, even as procedures to improve scrutiny of projects languished for attention, the Central government placed five of India's environmental laws under review. The Wildlife Protection Act, Environmental Protection Act, Forest Conservation Act, and the Air and Water Pollution Acts are being examined by a four-member 'high-level' committee to identify amendments that will bring the Acts 'in line with current requirements to meet objectives.' The committee has been given all of two months' time, and the ministry's website allows citizens to post comments till 28 September, with comments restricted to a limit of 1,000 characters (roughly 160 words). For a review exercise targeting five major laws and related Supreme Court orders, this is grossly inadequate. The committee should extend the time and channels for receiving public feedback and submissions in either hard or soft copy, hold public meetings in different regions of the country, and co-opt domain experts in environmental and social sciences. Their draft report, too, should be made available for citizen feedback.

The Wildlife Protection Act created the framework that protects the nation's 660-plus protected areas, which have come to occupy less than 5% of India's geographical area. It was intended to provide for 'the protection of wild animals, birds and plants ... with a view to ensuring the ecological and environmental security of the country.' With the country's apex wildlife body caught in a political quagmire, the case pending before the Supreme Court, and the law itself under review by a government with evidently different priorities, it is time for concerned citizens and conservationists to once again speak up: on the need to protect the Act.

## Postscript

After the Supreme Court rapped the government on its knuckles, the SC-NBWL was reconstituted including the requisite non-official members. It began functioning almost like a clearing house for projects in wildlife areas, approving dozens of projects in meetings each of a few hours' duration. This worsened a situation that was already bad, as Prerna Bindra recounted in *The Vanishing*; as wildlife habitat and conservation policy got little attention, detractors quipped that the body had become the Notional Board for Wildlife. Between 2009 and 2013, under the previous government, the SC-NBWL had considered 571 proposals in 17 meetings, recommending 45.5%, rejecting about 12%, and deferring others. Under the Modi government, the reconstituted SC-NBWL looked at 794 projects in 18 meetings during 2014–18, recommending 65.4% while rejecting a mere 1% of proposals brought to the table. Forest clearances issued by the environment ministry also shot up. For linear infrastructure projects, the annual average increased from 355 clearances and 8,339 hectares during 2009–14 to 720 clearances and 14,500 hectares during 2014–18. In 2018, an online portal was unveiled for forest, wildlife, and environmental clearances called PARIVESH, an awkward—even absurd—acronym for Pro Active and Responsive facilitation by Interactive, Virtuous and Environmental Single-window Hub. It made the application process easier for project proponents and increased transparency, but still fails to address the substantive concerns related to *how* projects are appraised, cleared, and subsequently monitored, and whether credible biodiversity and environmental impact assessments were carried out. As for the high-level committee constituted to review environmental legislation, its report failed to gain much traction and was widely criticized and panned by conservationists, civil society organizations, and the Parliamentary Standing Committee on Science and Technology, Environment and Forests. Today, the Wildlife Protection Act still stands, buffeted by the winds and whims of policy. As for critically endangered wildlife species, great Indian bustards appear to be on their last legs, with less than 200 individuals left—the last few threatened by the loss of their grassland habitats and the risk of mortality from power lines. Recent surveys for the Jerdon's courser have proved futile—the bird could well be already extinct in the wild.

# Living with Leopards in Countryside and City

The leopards in India have started to slowly change their spots, at least in the public imagination. Studies and surveys from 2006 onwards indicate that leopards are not merely a species limited to India's forests and wildlife sanctuaries. From crop fields in the plains to tea plantations in the mountains, from villages in the countryside to the suburbs of Mumbai, leopards live alongside people in remarkably different landscapes.

It is worth reflecting on what this proximity of the carnivore and field research findings from these landscapes portend for its conservation and coexistence with humans.

Naturalists have long known that leopards occupy a wide range of habitats, from farms to forests. Still, without closer study, biologists and managers have earlier found it difficult to ascertain whether the leopards in an area are residents or merely transient visitors 'straying' outside forests and wildlife protected areas, as frequently reported in news media. A slew of recent scientific studies and surveys, using trail cameras, radio collars, and other field techniques, is now providing a better understanding of leopards and gradually transforming conservation actions as applied to the cat. The research also suggests that dealing with leopards in human-dominated landscapes by capture and removal or translocation to new areas is not as effective a solution as it was once thought to be.

Between December 2014 and April 2015, wildlife scientists carried out a leopard study and census in Sanjay Gandhi National Park (SGNP) in Mumbai, India's most populous city. The 104-square kilometre SGNP, one of the largest protected forests in any big city worldwide, has had leopards living amid high human densities for long. In 2012, photographs obtained from camera traps during a preliminary survey indicated that the park was home to a minimum of 21 leopards. Still, no accurate estimate had been made using scientific field methods of the number of leopards in the park. Although estimating leopard numbers does not directly address the incidence of conflicts with people, the results will set a baseline for long-term monitoring, help understand conflict contexts, and inform management decisions.

The 2014–15 study and census was carried out by Nikit Surve, a young wildlife biologist and masters student at the Wildlife Institute of India (WII), with S. Sathyakumar and K. Sankar, scientists at WII; Vidya Athreya, of the Wildlife Conservation Society, India; and the Maharashtra Forest Department and field staff. Surve set up trail cameras in a sampling grid covering an area of about 120 square kilometres, encompassing the Park and adjoining areas such as the Aarey milk colony, to capture photographs of leopards walking past the cameras and estimate their numbers and density in SGNP. He walked 120 kilometres of transects to estimate the density of wild and domestic prey (such as deer, dogs, and pigs) and analysed leopard scats or droppings to understand the cat's food habits. In the field, the project helped train forest staff in wildlife techniques, use of field equipment, and monitoring. The results of the study, which Surve presented in his 2015 Masters dissertation, indicated at least 31 leopards occurred in SGNP at a density of about 22 leopards per 100 square kilometres. The leopards' diet (biomass consumed) was 57% wild prey such as chital, rodents, and langur, the remainder being dogs and livestock. One-fourth of their diet comprised domestic dogs that occurred at a density of 17 dogs per square kilometre in the Park.

Consuming prey like dogs, rodents, chital, and wild pig, the leopards have survived in the SGNP landscape, often with little or no conflict, even as people in tribal colonies in the Park report seeing leopards nearly every week. Still, conflicts do occur when leopards attack people or livestock and cause injuries or deaths. This sometimes leads to a public outcry and media attention that in turns brings pressure on the Maharashtra Forest Department to provide solutions. Conflicts involving leopards have been recorded in SGNP at least since the 1980s. An early study in SGNP by Advait Edgaonkar and Ravi Chellam in 1998 documented that between 1986 and 1996, 14 people, mostly children, were killed and 15 others, mostly adults, injured because of leopards. Most incidents occurred along the Park periphery, where further urbanization and encroachment has continued over the years.

Earlier, the standard management response to conflict incidents was to capture the so-called 'problem' leopards and take them into captivity or translocate them into the Park interior or other forest areas. In what was common practice in Maharashtra as in other states, the Forest Department captured leopards not only after attacks had occurred, but also when leopards had been found in human habitation, possibly considering the latter to be 'straying' individuals. In the absence of careful scientific study or monitoring, it was not certain whether the 'right' leopard involved in an incident had been caught, whether the same individual was being repeatedly caught, or what happened to the animals after release. As in other states, the operations ran the risk of targeting the wrong individuals.

The capture of leopards in metal cages and their translocation also caused stress to the animals and resulted in leopard injuries or deaths. Between 1986 and 1996, of the 52 leopards captured in SGNP, Edgaonkar and Chellam noted that 9 died during capture and, doubtless, many sustained injuries before release.

As later events showed, this practice of capture and translocation of leopards has not provided a lasting solution to conflict incidence. Rather counter-intuitively, it may have even led to an increase in conflict, as most dramatically illustrated by the surge in leopard attacks in SGNP between 2002 and 2004. As leopard biologist Vidya Athreya and others note in a 2007 publication, 'The most common strategy of dealing with the leopard "problem" in SGNP has been their capture in baited traps and subsequent translocation into certain areas of the Park and adjacent forests ...'. Between March 2002 and March 2004, 24 attacks on people occurred, including 6 within the park. The Forest Department trapped 26 leopards between July 2002 and December 2003, translocating 21 back inside the forest. In June 2004, there were 13 attacks resulting in 10 human deaths, which spotlighted the Mumbai leopards as never before. More than 30 leopards were then trapped, indicating a significant population of the cat existed in the midst of Mumbai, although exact numbers remained unknown.

The leopard capture and translocation programme that was carried out in many parts of Maharashtra may have had the unintended effect of spiking conflict locally and regionally. As studies by Athreya and others in 2011 on leopards in the

Junnar region of Maharashtra showed, leopard attacks on people increased by 325% while attacks on livestock increased by 56% after the initiation of a translocation programme in the area. The number of attacks on people increased from an average of about 4 per year to 44 in the 3 years of the programme. Released leopards marked with microchips were implicated in fresh conflicts around release sites, places where such incidents had not occurred earlier.

Multiple factors, including facets of leopard ecology and behaviour, may account for increased conflict because of the translocation according to the research. Besides stress and injuries, translocated animals may face a highly altered landscape context in the release site, including unfamiliar habitat, prey, and human settlements, besides having to confront or avoid other resident territorial leopards. As the leopards navigate their new surroundings, they may encounter people or resort to consuming domestic prey more often. This results in a translocation of the problem to a new area. At the same time, conflicts may not decline in the capture site if other leopards occupy territories vacated by captured individuals. Finally, translocated leopards may not settle in release sites but try to navigate their way back to their original home range, which might result in additional conflict incidents in new areas that they are forced to pass through.

A 2014 study in Maharashtra by Morten Odden, Vidya Athreya, and others that monitored released leopards using radio telemetry showed that three leopards released less than 10 kilometres away from the capture site returned immediately to the original home range. Two other individuals, an adult female named Sita and an adult male named Ajoba, released over 50 kilometres away showed long-range movements of 45 kilometres and 89 kilometres, respectively, passing through industrial, agricultural, and densely populated areas. While Sita returned to her original range, Ajoba did not, and he appeared to have settled in the Mumbai suburbs, at least until the time when his radio collar stopped functioning. In true Bollywood fashion, Ajoba's story was made into a movie by the same name and was released in 2014. Still, the study of movements of translocated leopards raised questions over the programme's effectiveness. As Odden and the other researchers state:

> It appears that relocations of so called problem individuals may either have only short-term local effects, may simply move the conflict to another area, or in the worst case scenario, increase the level of conflict.

As findings from research and monitoring became apparent, the Maharashtra Forest Department stopped using capture and translocation as an official policy (but not without slipping back to the method at times). From 2014 through 2016, there were no leopard attacks on people in SGNP. A spate of 6 incidents in 2017 (5 on infants, one of whom died while the other 4 were injured, and one attack on two women resulting in injuries) ended after the leopard was trapped, taken into captivity in the rescue centre, and not released again. In 2018, there were again no leopard attacks on people. Although conflict incidents have declined, occasional leopard attacks do occur and continue to spark media attention and sensationalist narratives of 'killer leopards.' A growing understanding, mediated by active civil society groups like Mumbaikars for SGNP, appears to be taking hold suggesting that the leopards are there to stay and what is most needed are measures to foster and sustain coexistence. There is also a crucial role for news media to inform and sustain public engagement, foster a balanced perspective, and facilitate better management. Research studies on leopard numbers, ecology, and behaviour also add critical pieces of information to assist and better prepare people to live in a landscape with leopards.

Even as people are still learning to live with leopards in the neighbourhood, it is apparent that leopards, on their part, have adapted quite well. Recent studies show that leopards occur frequently in densely populated areas as well as those of intensive human use, including landscapes dominated by agriculture in Maharashtra, tea and coffee plantations with scattered rainforest fragments in the Western Ghats mountains, and large multiple-use landscapes in Karnataka. Leopards survive in these landscapes, subsisting on any available prey species: wild prey such as deer and Indian crested porcupine where available or domestic animals and dogs in other areas. By avoiding human habitations by day and visiting by night, leopards have also learnt to navigate human-dominated terrain.

Across India, in possibly hundreds of places, leopards may pass human habitations every night without incident. It is this nearly invisible domain of neutral interaction that perhaps deserves better recognition and enlargement in the public space. Instead, apparent increases in leopard sightings, which may only be a result of increased reporting, still trigger alarmist media reports. In India, it is still too early to claim that the public perception of leopard has shifted from that of an unwanted intruder to that of an elusive, if occasionally troublesome, neighbour. Scientific field studies and on-ground conservation solutions involving landscapes with leopards are certainly pointing in that direction. Where states like Maharashtra have adopted measures based on a more informed understanding of leopards, there are positive signs of change. At the national level, the Ministry of Environment and Forests in 2011 issued clear guidelines on human–leopard conflict management, but these are yet to be widely adopted and implemented by various states.

Changing the focus from 'problem animals,' which has proved ineffective, to 'problem locations' also changes the issues that need to be addressed and measures that need to be in place. It brings into consideration site-specific measures to reduce leopard visitations that may lead to negative interactions with people. This could include control of the domestic dog population, appropriate disposal of garbage and kitchen and medical waste, and better protection for livestock in safe corrals. It draws attention to providing better amenities for tribal and under-privileged people living in the area, including lighting, housing, sanitation (to reduce open defecation in forests, especially at night), and other public safety measures. Changing the focus from leopard to location also implies a change from reactive measures (such as capture after conflicts occur) to proactive efforts (such as creating safer surroundings to pre-empt attacks) that reduce negative interactions while enabling people and leopards to share the landscape.

Taking a global perspective, building coexistence with carnivores where they occur in landscapes with people is not a concern restricted to leopards in India. It is relevant to populations of several carnivore species such as lions, bears, cougars, coyotes that live in or are expanding their ranges into human-use areas in many parts of Africa, Europe, and North America. As with these species, India's leopards impel a recasting of the conservation landscape, one that extends beyond the boundaries of protected areas and nature reserves, spanning the city, the country, and the wild, over a wider realm where people and wildlife can coexist.

# The Culling Fields

In 2016, the difference of views on the killing of wild animals between a former and sitting Environment Minister of the ruling party—the former against and the latter in favour—hit the front pages. This was when the Ministry of Environment, Forest and Climate Change permitted the states of Uttarakhand, Bihar, and Himachal Pradesh to declare earlier protected wild animal species as 'vermin' under the Wildlife Protection Act of 1972, thereby allowing private shooters and others to kill these species with few safeguards and no risk of prosecution. Soon after, Maharashtra and Telangana followed suit, issuing similar orders.

The species—nilgai antelope in Bihar and Maharashtra, the rhesus macaque in Himachal Pradesh, and wild pig in all states except Himachal Pradesh—were listed for culling because the animals, whose populations are allegedly increasing, damage crops. This decision raises questions about which minister is right and whether it is right to kill wildlife that damage crops. More pertinent is whether the problem has been framed and assessed correctly, and culling the appropriate solution in the first place.

In parts of India, wildlife species such as wild pig, elephants, macaques, and nilgai occasionally damage crops or property. No reliable estimates of economic loss nationwide are available, but a number—almost certainly an underestimate of real and opportunity costs to farmers and property owners—of Rs 200–400 crore has been quoted in media reports. This appears minuscule in relation to the government's willingness to spend on, say, a statue for Sardar Patel (Rs 3,000 crore) or on grandiose river-linking projects, but economic losses can be serious and crippling for individual poor farmers and deserve urgent attention.

Field research by wildlife scientists in diverse landscape contexts, on different wildlife species and kinds of human-wildlife interactions, including 'conflicts,' suggests multiple solutions. Culling or removal of 'conflict' wildlife, often labelled 'problem animals,' is one among a suite of possible interventions recommended by conservation scientists and managers. Unfortunately, removal through capture or killing may not prevent recurrence of conflicts and may even exacerbate them.

Himachal Pradesh, for instance, has translocated hundreds of rhesus macaques since 2004 (shifting conflicts to new locations while failing to stop conflict incidence), killed hundreds of rhesus macaques in 2007 (with conflicts recurring within two years), sterilized over 150,000 macaques between 2007 and 2018 (even as conflicts kept increasing), yet now continues more of the same. This is despite the state's own data showing a decrease in macaque population from around 317,000 in 2004 to 226,000 in 2013 and the 2015 estimates of about 207,000 macaques by scientists from Mysore University and the Sálim Ali Centre for Ornithology and Natural History indicating that macaque populations are in overall decline and only 8 forest divisions (out of 44) record sharp increase. In 2014, Karnataka carried out a brutal and costly exercise, capturing 22 elephants in Hassan Forest Division, translocating 5 to another forest, and consigning 17 to captivity. But conflicts did not stop: during 2015–18, there were 9 human deaths, even as researchers recorded in the division hundreds of crop damage incidents. As risks to people and crop damage continue, the local communities remain disempowered and ill prepared to avoid or reduce conflicts.

A better approach to conflict management requires integration of scientific evidence, ecology, and behaviour of particular species, and landscape and socioeconomic context. Without this, the response of state authorities, often based on political compulsions and public perception, even if legitimate, may end up being inappropriate and confused in relation to the problem.

On 9 June 2016, when the option of culling was opened up, the Inspector General of Wildlife issued a worse-than-muddled clarification. In a note trying to explain why wildlife species causing crop damage were listed for culling, the only factoid presented is that over 500 people were killed by animals across India the previous year. Human injuries and deaths due to wildlife is a serious issue, but recent studies show that a large proportion result from accidental encounters with species such as elephants and bears. Government figures report that around 400 human deaths a year result from encounters with elephants. Conflating such human deaths with crop damage by very different wild animal species implies connecting an extreme response such as killing nilgai in Bihar through an unjustified comparison with human deaths resulting from other wild animals nationwide.

If human safety was the chief concern—as it should be where potentially dangerous animals pose a threat to human life—it is more appropriate to first adopt measures to reduce human injuries and fatalities due to wildlife. Effective measures for this include deploying animal early warning systems that provide timely information to local people on the presence and movements of species such as elephants to facilitate precautionary measures, and attending to health and safety needs that reduce the risk of wildlife encounters. Housing improvements and provision of amenities such as lighting, indoor toilets, and rural public bus services help reduce accidental human deaths. Improving livestock corrals can reduce livestock losses and carnivore incursion into villages, while better garbage disposal and avoiding deliberate or accidental feeding of animals reduces risks associated with wild animals like monkeys.

Crop damage by wildlife may occur when animals enter crop fields because of habitat alteration and fragmentation (by mining or infrastructure projects, for example), because crops are edible, or because the fields lie along movement routes to forest patches or water sources. Research reveals that a small proportion of villages in the landscape may be conflict 'hotspots' and, additionally, peripheral fields may be more vulnerable than central ones. Such site-specific scientific information helps design targeted mitigation with participation of affected people. This includes supporting local communities to install—and, more important, maintain on a sustained basis—bio-fencing and power fencing around vulnerable areas.

Crop insurance for wildlife damage, which the Environment Ministry recently recommended be included in the National Crop/Agricultural Insurance Programme, also deserves trial. An insurance approach recognizes wildlife as a part of the shared countryside and as a risk to be offset rather than as antagonists belonging to the state that one wishes away. Conservationists today also use modern technology such as mobile phones for SMS alerts, customized apps, automated wildlife detection and warning systems, and participatory measures for wildlife tracking and rapid response to monitor and reduce conflicts, save crops, property, and human lives.

Broadly, such interventions are proactive measures to reduce negative human-wildlife interactions, and are in stark contrast to traditional reactive measures such as killing, removal, or compensation—actions taken after conflicts occur. Identifying appropriate proactive measures, including where, when, and how they should be deployed, requires prior scientific research on conflict patterns in specific landscapes and locations. Without this, culling, or permitting culling by farmers and private citizens, becomes a mere public relations exercise meant to assuage farmers who have lost faith in the authorities' ability to forestall or recompense losses. In Himachal Pradesh, the exercise ultimately turned out to be a non-starter as only five monkeys were culled as of December 2018—possibly because farmers and local people were both untrained to use firearms and unwilling to adopt a solution that violated their traditional ethos of not killing primates.

Blaming a 'problem animal' may be easier than carrying out concerted efforts to deal with what are actually 'problem locations'. Focussing efforts on removing individual animals detracts from needed investments in location and amenities. When shooters from other states kill wildlife with high-powered rifles and leave, they also leave local people and forest staff no better prepared, trained, or empowered to deal with likely future wildlife intrusions. Needless to say, this also fails to enhance people's standard of living or their ability to respond to interactions with wildlife in the future.

Merely removing 'problem animals' will not make 'problem locations' disappear. Servicing human needs, enhancing local amenities, and adopting science-based and sustained interventions will provide more lasting solutions. A moratorium on culling will thus help redirect attention to where it is really needed and be in the best long-term interests of people and wildlife.

# Bamboozled by Land-Use Policy

## *Jhum and Oil Palm in Mizoram*

Two spectacular bamboo dances, one celebrated, the other reviled, enliven the mountains of Mizoram, the small northeastern Indian state wedged between Bangladesh and Myanmar. In the first, the colourful Cheraw, Mizo girls dance as boys clap bamboo culms at their feet during the annual Chapchar Kut festival. The festival itself is linked to the other dance: the dance of the bamboos on Mizoram's mountains brought about by the practice of jhum. In jhum, bamboo forests are cut, burnt, cultivated, and then rested and regenerated for several years until the next round of cultivation, in a cyclic ecological dance of field and fallow, of farmer and forest. While Cheraw is cherished by all, jhum is actively discouraged by the state and the agri-horticulture bureaucracy. Although jhum is a regenerative system of organic farming, Mizoram state, the first in India to enact legislation to promote organic farming, has pushed hard to eradicate jhum under its New Land Use Policy (NLUP), a policy initiated in 1984, implemented in fits and starts through the 1990s, and resurrected as a decade-long push again from 2008 onwards.

Labelling jhum as unproductive and destructive of forest cover, policymakers and industry now promote 'settled' cultivation and plantations, such as pineapple and oil palm, claiming they are a better use of land than jhum. However, oil palm, rubber, and horticultural plantations are monocultures that cause permanent deforestation, a fact that the *India State of Forest Report 2011* notes while explaining recent declines in Mizoram's forest cover. In contrast, jhum is a diversified cropping system that causes only temporary loss of small forest patches followed by forest recovery. Understanding this is crucial to formulating a land-use policy that is economically, ecologically, and culturally appropriate for Mizoram and other northeastern hill states and their tribal communities who live amid extraordinarily rich forests.

Jhum uses natural cycles of forest regeneration to grow diverse crops without using chemical pesticides or fertilizers. Early in the year, farmers cut carefully demarcated patches of bamboo forests and sun-dry the vegetation for weeks. They then burn the slash in spectacular—but contained—fires in March to clear the fields, nourish soils with ashes, and cultivate through the monsoon. In small fields of 1–3 hectares in area, each farmer plants and sequentially harvests between 15 and 25 crops—indigenous rice varieties, maize, vegetables and herbs, chillies, bananas, tubers, and other species—besides obtaining edible mushrooms, fruits, and bamboo shoots. After cultivation, these fields are rested, and the farmers shift to new areas each year. The rested fields rapidly regenerate into forests, including over 10,000 bamboo culms per hectare in five years.

After dense forests reappear on the original site, farmers return for cultivation, usually after six to ten years, which forms the jhum cycle.

Regenerating fields and forests in the jhum landscape provides resources for many years. The farmer obtains firewood, charcoal, wild vegetables and fruits, wood and bamboo for house construction and other home needs. The diversity of food and cash crops cultivated and ancillary resources provided by current and rested jhum fields complicate comparisons with terrace or monocrop agricultural systems. One-dimensional comparisons—such as of rice yield per hectare or annual monetary return—can be misleading, because one needs to assess the full range of resources from jhum field, fallow, and forest over a full cultivation cycle, besides food security implications.

Comparing monocrops like pineapple or wet rice paddies cultivated using chemical inputs with organic jhum is not an apples-to-oranges comparison. It's like comparing a pile of pineapples with a cornucopia of rice, vegetables, cash crops, firewood, bamboo, and more. Interdisciplinary, holistic studies, notably those led by P.S. Ramakrishnan of Jawaharlal Nehru University, New Delhi, indicate that at cycles of 10 years or more jhum is 'economically productive and ecologically sustainable'.

In Mizoram, bamboos coexist with jhum in a dependent cycle that is often overlooked: while we see jhum fires burning forests, we remain blind to forests and bamboo regenerating rapidly after a season of cultivation. The *India State of Forest Report 2011* estimated that bamboo-bearing areas occupy 9,245 square kilometres, 44% of Mizoram. For every hectare of forest cleared for jhum, farmers retain 5–10 hectares as regenerating fallow and forest in the landscape. (Also, forests along ridges, ravines, and other areas, left uncut by jhum farmers, contain bamboo species. Besides *mautak* (*Melocanna baccifera*) that dominates in regenerating forests, over two dozen native bamboo species occur naturally in Mizoram's forests and jhum landscapes.)

Yet, government policy tilts firmly against jhum. The State's NLUP deploys over Rs 2,800 crores over a 5-year period 'to put an end to wasteful shifting cultivation' and replace it with 'permanent and stable trades.' Under this policy, the State provides Rs 100,000 in a year directly to households, aiming to shift beneficiaries into alternative occupations like horticulture, livestock-rearing, or settled cultivation. The policy has created opportunities where little existed earlier for families seeking to diversify or enhance income, for farmers whose harvest is insufficient to meet year-round needs, and for skilled and urbanizing workers seeking other jobs and trades. Still, NLUP's primary objective to eradicate 'wasteful' shifting cultivation appears misdirected.

Maintaining or regenerating forests was another NLUP objective. Even before NLUP was implemented, despite decades of extensive shifting cultivation, over 90% of Mizoram's land area was under forest cover, much of it bamboo forests resulting from jhum. Over the NLUP implementation period, declines in forest cover have occurred even as area under jhum cultivation is actually declining and area under settled cultivation is increasing, suggesting that the land-use policy has been counterproductive for forests.

Certainly, some areas may need protection from jhum. My earlier research indicated that remnant mature evergreen tree forests, as in the core of Dampa Tiger Reserve in western Mizoram, need to be protected for specialized and endemic rainforest species. But as forests regenerating after jhum support diverse plant and animal species, I had suggested fostering jhum in areas such as the buffer zone landscape surrounding the Reserve. From perspectives of agroecology, biodiversity conservation, and human-wildlife coexistence, jhum is far preferable to monoculture plantations such as teak and oil palm that now increasingly abut the Reserve.

Oil palm, notorious for extensive deforestation in southeast Asia, is cultivated as monoculture plantations devoid of tree or bamboo cover, and drastically reduces rainforest plant and animal diversity. In Mizoram, 101,000 hectares have been identified for oil palm cultivation. Following the entry of three corporate oil palm companies (Godrej Oil Palm Limited, Ruchi Soya Industries Limited, and Food, Fats and Fertilizers Limited), over 17,500 hectares have already been permanently deforested between 2005 and 2014. Promoting and subsidizing such plantations and corporate business interests undermines both premise and purpose of present land-use policies. As forest cover and bamboo decline, people in some villages now buy bamboo, once abundant and freely available in the jhum landscape.

If these trends continue, Mizoram is likely to be 'bamboo-zled' out of its forest cover.

Detractors of jhum often concede that the practice was viable in the past, but claim population growth has forced jhum cycles to under five years, allowing insufficient time for forest regrowth, thereby making it unsustainable. The shortening of jhum cycles is serious, but evidence linking it to population pressure is scarce. As Daman Singh notes in her book, *The Last Frontier: People and Forests in Mizoram*, villagers actively choose to cultivate at 5–10 year cycles even when longer periods are possible. In reality, jhum cycles often decline because of external pressures, relocation and grouping of villages, or reduced land availability. Village lands once open to local people for jhum are now fenced off under government or private plantations and horticulture crops often belonging to people who are wealthier or live in distant urban centres. But none of this implies jhum itself is unsustainable.

Attempting to eradicate and replace shifting cultivation, as NLUP does, is inappropriate. Instead, a better use of public money and resources would be to work with cultivators and agroecologists to refine jhum where needed. The state can involve and incentivize communities to foster practices that lengthen cropping and fallow periods, develop village infrastructure and access paths to distant fields, and provide market and price support, and other benefits, including organic labelling to jhum cultivators. Today, the state only supports industry and alternative occupations, leaving both bamboo forests and farmers who wish to continue with jhum in the lurch. Unless a more enlightened government reforms future policies in favour of shifting agriculture, Mizoram's natural bounty of bamboos is at risk of being frittered away.

The *Economic Survey Mizoram 2012–13* made a bold claim. After quoting the Forest Survey of India's (FSI) *State of Forest Report 2011* that 90.68% of Mizoram is under forest cover, the *Economic Survey* claimed, literally in bold letters in a box, that the State's forests 'have suffered serious depletion and degradation due to traditional practice of shifting cultivation, uncontrolled fire, unregulated fellings etc.' The claim is a frequent one made by the state government and the agri-horticulture bureaucracy. Actually, what the 2011 FSI Report said was this:

> Due to change in customary cultivation practices, focus has now shifted to raising horticultural crops ... thus preventing secondary growth on old shifting cultivation patches. This has also led to the decline in forest cover assessed in the state.

Thus, Mizoram's forest cover may be taking a turn for the worse not because of shifting cultivation but because of the state's push to establish permanent cultivation, notably horticulture crops such as oil palm.

Permanent plantations and settled agriculture also result in permanent loss of forest cover, unlike the temporary loss that is characteristic of jhum. Unfortunately, the FSI reports, those prior to 2011 and subsequent ones till 2017, do not distinguish areas under plantations, nor do they carefully record patterns of regeneration, so on-ground change in land use and forest cover remains difficult to assess accurately. Most often, a decrease in net forest cover in this region continues to be blamed on jhum without specific or convincing evidence. As a result, *jhum*, Mizoram's remarkable organic farming system, remains much misunderstood and maligned.

The expansion and impacts of oil palm (*Elaeis guineensis* and *Elaeis oleifera*) cultivation in tropical regions, especially in southeast Asian countries, is now a global problem from social, conservation, and climate change perspectives. Palm oil now accounts for a third of vegetable oil use worldwide. The area under oil palm cultivation has skyrocketed, from around 3.6 million hectares in 1961 to over 16.4 million hectares in 2011 and 21.4 million hectares by 2017, much of it by the felling of mature, secondary, and peat-swamp tropical forests. The deforestation and burning of forests in Southeast Asia for oil palm has caused species extinctions, water shortages, and widespread pollution, besides contributing to climate change.

In India, oil palm plantations are actively promoted by government and private companies, including in Mizoram where large areas have already been permanently deforested for oil palm cultivation. Near Aizawl's Lengpui airport, a large hoarding now advertises the benefits of oil palm cultivation. But the photographs of irrigated oil palm trees on flat lands appear incongruous amid the surrounding steep slopes withering dry in the sun during summer. Farther ahead, in areas newly cleared of bamboo and forest cover, small oil palm plantations appear and bare slopes are studded with rubber saplings. Intended as permanent crops, these plantations are often touted as superior to shifting cultivation by government authorities and private companies.

Under multiple schemes besides the state's flagship New Land Use Policy, both centre and state have subsidized seedlings, fertilizer, and the construction of water tanks, besides the construction of oil palm mills to benefit private companies. Furthermore, in an unusual arrangement, the state has apportioned captive districts to the three private companies—Godrej Oil Palm Limited, Ruchi Soya Industries Limited, and Food, Fats and Fertilizers Limited—for palm oil business, thereby making these farms 'corporate plantations in effect,' as a 2014 news report put it.

In Mizoram, plantations of oil palm, rubber, and teak have burgeoned. Teak, a deciduous tree not naturally occurring in Mizoram, is planted extensively by the State Forest Department even in evergreen forest zones. Viewed through an ecological lens, all these plantations are worse than jhum. Shifting cultivation is preferable to industrial and monoculture plantations because it creates and maintains a dense mosaic of bamboo, secondary, and mature forests in the jhum landscape. In other parts of northeast India, diverse bamboo forests and jhum areas are being replaced by tea plantations, mining, and timber monocultures. Scientific research from rainforests of India's Western Ghats and of Southeast Asia attests that industrial monocultures, such as teak, tea, oil palm, and rubber, provide habitat for fewer wildlife species than natural, mature, and secondary forests. A study from Thailand revealed that rubber and oil palm plantations have 60% fewer bird species than lowland rainforest.

Oil palm is conventionally grown as monocultures after clear-felling forest, retaining little or no natural tree cover. Numerous studies have documented that oil palm plantations support very few rainforest plant and animal species. Oil palm plantations may shelter less than 15% of the forest biodiversity, besides reducing water availability and quality in hill streams. My own research in the Dampa landscape along with Jaydev Mandal, research scholar at Gauhati University, indicates that monoculture oil palm is much worse for wildlife than the jhum mosaic of regenerating forests and fallows in the landscape.

The reduction in biological diversity also signifies a loss of natural resources for local people. Rural people in Mizoram derive bamboo, fuel wood, forest foods and produce, and timber from forests and landscapes under *lo* agriculture. All these are likely to become increasingly scarce as forest and traditional village community lands are converted into oil palm plantations under private ownership. Without paying heed to the drastic repercussions, oil palm plantations have also been established in areas of very high biodiversity value. In Mamit District, oil palm plantations have been established right along the boundary of Dampa Tiger Reserve, Mizoram's largest wildlife reserve. A few planters have started grumbling about wildlife entering oil palm plantations. In the long run, as plantations expand, animal corridors may be broken and wildlife may come into greater conflict with people. The government will then have to face more compensation claims and park management issues. If better land-use zoning is followed and plantations are not established in important conservation areas, such problems can be avoided in the future.

Another major concern regarding oil palm expansion in Mizoram is over water. Oil palm is a highly water-demanding crop, each plant requiring around 300 litres of water per day. Mizoram already faces serious water shortages for several months each year in both rural areas and urban centres. Establishing plantations of oil palm that require 40,000–50,000 litres of water per hectare each day is likely to exacerbate regional water scarcity. By this calculation, 2 hectares of oil palm uses up more water every day than a citizen of Mizoram needs for an entire year. During the dry months, hundreds of water tankers ply up and down heavily used roads taking water from streams and rivers to Mizoram's villages and cities. As oil palm expands, direct diversion of stream waters by planters to irrigate oil palm plantations will reduce availability downstream.

To make matters worse, the conversion of forests to oil palm plantations will also have negative effects on the watershed. As a 2012 review paper in the journal *Advances in Agronomy* points out, scientific research in Southeast Asia has revealed that conversion of forests to oil palm plantations has multiple hydrological effects. Oil palm plantations are known to reduce water infiltration, increase soil erosion and surface runoff, and increase sediments in the now-warmer stream waters. Crucially, oil palm plantations also reduce dry season water flows. All these effects are likely to be exacerbated in Mizoram because of the steep and mountainous terrain.

All this merits a serious appraisal of the perils of monoculture plantations; however, the state is promoting oil palm without any detailed study of the impact this will have. In Mamit District, oil palm is even planted after clearing forests on slopes and catchments under the Integrated Watershed Management Programme. In a brazen travesty of the concept of watershed management, water is diverted from natural streams to tanks and taken through pipes to feed this water-demanding crop in newly deforested areas.

As many of these plantations have been established on community lands, and land titles are provided only to those who convert to plantations such as oil palm, areas for village communities to practise traditional agriculture are shrinking along with the loss of community land tenure. There are social consequences, too. A 2019 study by Purabi Bose found that after households shifted to oil palm cultivation, gender equity changes to the detriment of women. Land titles mostly go to the men of the household, while women, who earlier played a more prominent role in jhum cultivation, are forced to work inside houses to manually process oil palm fruit crop. Some oil palm plantations have also been established in valley areas, earlier under productive rice cultivation, for example, in parts of Mamit District. What will be the consequences of this for rural food security? Traditional *lo* agriculture reliably provided a diversity of food items, intermixed with cash crops, in the cultivated fields and forest resources from the wider *lo* landscape. Such benefits are compromised by the establishment of monoculture plantations such as rubber and oil palm.

In contrast to such support and subsidies for oil palm plantations, the state government provides no support for farmers practising *lo* shifting agriculture. Shifting cultivation is often considered an unsustainable practice. In reality, the major challenges today for sustainable agriculture and agroecology actually concern industrial agriculture and plantations: how to reduce dependence on agrochemicals and move to organic farming; diversify from single to multiple crops; integrate fallows and hedgerows and unplanted areas in plantation landscapes; retain native plant species and vegetated buffers along rivers, ravines, and ridges. Oil palm promoters and planters have not made any effort to retain valuable bamboo forest patches, valley agriculture fields, strips of forest vegetation along streams and rivers to prevent erosion and pollution, or implement other essential safeguards. All these aspects of sustainable agriculture, missing in oil palm plantations, are often already practised in shifting cultivation in Mizoram.

A 2012 review of the science and sustainability of jhum in Mizoram in the journal *Agroforestry Systems* by Paul Grogan of Queen's University, Canada, and F. Lalnunmawia and S.K. Tripathi of University of Mizoram, Aizawl, notes:

> ... in contrast to many policy-makers, shifting cultivation is now considered a highly ecologically and economically efficient agricultural practice *provided that* the fallow period is sufficiently long.

Grogan, Lalnunmawia, and Tripathi list options to enhance shifting cultivation, such as nutrient and water supplementation, optimization of crop choice to extend the site-use period, along with measures to further retain soil and fertility, including judicious use of commercial fertilizer coupled with organic inputs. Jhum farming, with or without refinements, and modified multi-cropping systems such as the Changkham model and

Sloping Area Agricultural Technology (SALT) are all preferable to monoculture plantations such as oil palm.

Similarly, a better informed, planned, and sustainable approach to oil palm cultivation is urgently needed in Mizoram. This would include avoiding its cultivation in crucial watersheds and areas of high conservation value. It will also require changing land-use practices in oil palm plantations to follow international standards, including the protection of strips of forests along rivers, the retention of native trees and bamboo, and the conservation of community forests and reserves in the wider landscape. Until such an approach is adopted, one will have to view oil palm expansion with great trepidation.

# The March of the Triffids

> In temperate countries, where man had succeeded in putting most forms of nature save his own under a reasonable degree of restraint, the status of the triffid was thus made quite clear. But in the tropics, particularly in the dense forest areas, they quickly became a scourge.
>
> —John Wyndham, *The Day of the Triffids*

In his dystopian novel, *The Day of the Triffids*, British writer John Wyndham describes a world overrun by mobile, carnivorous plants—the triffids. The deadly plants, biologically engineered to produce rich vegetable oil, were economically valuable when farmed and controlled. But when a strange meteor shower drives much of the world's human population blind, the triffids begin to gain ground, moving about, attacking people, threatening human survival itself.

The triffids could serve as a metaphor today for the tropical oil palm. Oil palm is remarkable for its high palm oil yield, its profitability, and its mobility—read expanding cultivation in vast monoculture plantations in tropical forests worldwide. Many studies attest that biological diversity declines when oil palm monocultures replace forests (thus 'eating up wild spaces and species'). According to figures from the Food and Agriculture Organization of the United Nations, in 2017 oil palm occupied around 21.4 million hectares worldwide. Consumer pressure in the developed world has triggered calls for sustainable palm oil production, but the reality in the tropics is that oil palm cultivation continues to expand rapidly, particularly in Southeast Asia. India, too, is expanding oil palm cultivation to meet domestic demand and reduce palm oil imports from Malaysia and Indonesia.

Oil palm expansion in India is occurring in an area that falls within portions of two global biological diversity hotspots, the Himalaya and Indo-Burma—the Northeast. Other monoculture plantations established or expanding in this region include timber plantations, notably teak, and other plantation crops such as tea and rubber.

Given the extraordinary diversity in northeast India, conservationists are justifiably concerned at plantation expansion. But forest loss is not the only issue. Also at stake is the loss of traditional multi-crop agricultural lands and practices.

The forested mountains of India's northeast are home to over a hundred indigenous tribes whose agricultural mainstay remains the practice of shifting cultivation.

Yet, since colonial times, foresters have decried jhum as destructive. In recent years, government authorities in states like Mizoram are deploying policy-based incentives to eradicate jhum and replace it with other livelihoods and land uses, including oil palm.

Our research was motivated by the question: what are the consequences of such forest conversion to monoculture plantations on tropical forest birds? Is jhum really destructive as claimed, or better than monocultures for bird conservation? In a 2001 publication from the same landscape, I had noted the need to assess whether 'monoculture plantations ... are superior in conservation value to successional habitats arising after jhum, especially bamboo and secondary forests, which harbor many forest bird species'. But data was lacking.

To find answers, Jaydev Mandal, a research scholar at Gauhati University, and I surveyed sites in the core and buffer zones of Dampa Tiger Reserve, a 500-square kilometre protected wildlife reserve in Mizoram. We selected five study strata: oil palm plantation, teak plantation, shifting cultivation (jhum), mature forest edge, and mature forest interior.

In March and April 2014, we surveyed for birds along twenty transects (100 metres long and 30 metres wide on either side) within each of the five strata distributed across multiple sites, constrained by logistics. We surveyed oil palm areas planted since 2007 and 15–25-year-old teak plantations established by the Forest Department near villages along the Tiger Reserve boundary. For shifting cultivation, we covered a range of sites representing the typical jhum landscape: recently burnt fields, fallows, and secondary forests with bamboo that had regenerated for 7–8 years, the typical rotation period. We looked for birds in moderately disturbed, forest-edge transects abutting the reserve and in relatively undisturbed tall, closed-canopy rainforests in the Reserve's core.

We recorded 107 bird species: 94 were birds of mature and secondary forests, 13 were open-country species of open and sparsely wooded habitats. In oil palm plantations, we found mainly 5 common and widespread open-country species such as olive-backed pipits and spotted doves on the ground and red-vented bulbul and common tailorbird in the few shrubs and trees. A mix of open-country and forest bird species occurred in teak and jhum sites, but the latter brimmed with birdlife from understorey babblers and flycatchers to canopy minivets and woodpeckers. Some forest bird species, such as yellow-bellied warbler, brown-cheeked fulvetta, and white-browed piculet, occurred more frequently in the jhum landscape and bamboo forests.

Mature forest sites had few open-country birds and a large diversity of forest birds, including 25 species not seen in other habitat strata. These included birds such as grey peacock-pheasant, red-headed trogon, blue pitta, mountain imperial pigeon, long-tailed broadbill, and wreathed hornbill.

Quantitative analysis reinforced our impressions. We recorded 10 forest bird species in oil palm plantations, a seventh of what we found in the rainforests. We found 38 species in teak plantations and 50 in jhum, compared to 70 in the rainforest interior. We also found that the jhum landscape not only supported more forest bird species, it had a higher bird abundance, and a more similar mix of species relative to mature forests than teak or oil palm plantations. Our vegetation data suggested that this could be due to the rapid recovery of secondary successional forests, including dense *Melocanna* bamboos, in jhum sites. Oil palm and teak plantations had few native trees and little bamboo, and oil palm also had sparser canopy. Our findings were broadly consistent with earlier research from Southeast Asian tropical forests.

Earlier research that had reported that mature rainforests supported more birds than jhum had compared them with individual fallow sites that had regenerated for a fixed number of years (1 year, 5 years, etc.). Our research presents a better picture by comparing a range of sites representing the jhum landscape with a forested landscape. Although mature forest interior was still better for rainforest birds, an impressive array of forest birds persists in the jhum landscape.

For forest and bird conservation, our results underscore the need to protect mature forests, but also attest that jhum cultivation is a better form of land use than oil palm or teak monocultures. Better land-use planning and practices such as retention of forest remnants and areas of high conservation value amid plantations are urgently needed to minimize the effects of ongoing conversion of forests and traditional shifting agriculture lands to oil palm monocultures.

If we go blind to these needs under the meteor shower of neoliberalism and market forces, the tropics may well succumb to the march of the triffids.

# How Green is Your Tea?

You could almost imagine this to be a forest trail. Fifty metres away, in the swampy valley, stands the dark hulk of a solitary bull gaur, over five feet tall at the shoulder, taut muscles and thick, curving horns, white-stockinged legs. Even as your eyes lock with his, your ears pick up the harsh bark of a great hornbill breaking through the cool mountain air from a patch of tall trees on the hill slope beyond. As you skirt the gaur and walk quietly down the trail, stepping past fresh scat of a sloth bear and dropped quills of a porcupine, a stripe-necked mongoose darts across, a flash of crimson bright against the background green. The green is not the multi-hued mosaic of a real forest, but the more uniform smear of a monoculture.

* This essay is written with Divya Mudappa.

Row upon row of neatly pruned metre-high bushes spread out in tight lines, punctuated by well-spaced and heavily lopped silver oak trees. In the mountains of the Western Ghats, at the edge of the Anamalai Tiger Reserve, you are walking through a large tea plantation.

Tea plantations, however, often get bad press from environmentalists as 'green deserts.' They tend to support far fewer wild species than comparable areas under native forests as large estates often grow only one crop. Still, when one takes a broader perspective, tea estates, too, can play a role in wildlife conservation. Recent field studies show that by modifying conventional land-use practices and protecting some neglected parts of the tea estate landscape, these plantations can help conserve wild species. Even a cup of tea reveals our enduring connections to nature, shows how conservation can be effective outside the boundaries of wildlife sanctuaries, and requires efforts that involve all.

Historically, tea plantations have been located in regions of great biological diversity. Over 90% of India's tea comes from West Bengal (particularly Darjeeling) and Assam in northeast India, and the southern states of Kerala and Tamil Nadu. While Darjeeling and the southern states lie within global biodiversity hotspots in the Himalaya and Western Ghats, respectively, most of Assam's tea comes from equally valuable conservation landscapes in the Brahmaputra Valley. These regions also hold the most significant remaining populations of endangered wildlife—such as Asian elephants, tigers, Indian rhinoceros, lion-tailed macaques, great hornbills, and many other species, including smaller and lesser-known reptiles and amphibians found nowhere else in the world.

Today, tea plantations, like those near the Anamalai Tiger Reserve, abut some of the most important wildlife sanctuaries and national parks in places like Kaziranga, Eravikulam, and Buxa

Tiger Reserve. Many of the protected areas are isolated from other forest tracts, and tea plantations are often embedded in intervening areas that are now large plantation districts. If threatened wildlife species can be enabled to survive in tea districts, it increases their conservation prospects as a larger population can persist in the wider landscape.

To appreciate the potential for tea estates to support wildlife, we need to envision the tea landscape as a space that includes tea fields and human settlements along with forest patches, grasslands, rock outcrops, rivers, and streams. The tea landscapes in the Western Ghats may have more than 250 animal species. This includes porcupine and gaur, which use tea fields; threatened species such as hornbills and lion-tailed macaques sheltering in forest remnants; and endemic animals such as frogs and small-clawed otters living in hill streams.

The critical first step for a 'truly green' tea landscape is to protect all natural ecosystems that remain, both within and around plantations. Tea estate maps detail tea fields, roads, factories, and housing, but often leave gaps marked as 'blanks' or estate jungles. These spaces, along with the streams and rivers (sometimes merely considered 'drains'), are the most interesting and valuable parts of the estate landscape. It is in these spaces that one finds evergreen forests, regenerating forest and scrub, grasslands, bamboo patches, swamps and water bodies that support diverse wildlife species. To place these forested areas firmly on estate maps and protect them—and not clear and convert them into a monoculture—is the first need of conservation. In addition, planters can retain or regenerate strips of native vegetation as buffers along rivers or as windbreaks on slopes; these would help protect water and crop while creating habitat corridors for wildlife. Planters can also connect with local conservation groups to carry out wildlife inventories and ecological restoration of degraded areas.

Protecting habitat and restraining land owners from further land conversion are, however, not enough. This needs to be accompanied by another form of restraint, one that's a critical second step: preventing hunting. In some plantation districts, although habitats remain, wildlife is scarce because of snaring and shooting. Besides poachers, some estate managers too, in a throwback to the colonial period, sport guns and shoot unsuspecting wildlife, especially where the Forest Department's patrolling and protection are lax. Still, support for wildlife conservation is increasing with the rise of a new breed of planters who have moved on to enjoy the more rewarding pastimes of trekking, wildlife watching, and nature photography.

A third measure concerns the tea fields themselves: the production area. Instead of merely seeing tea plantations as a 'green desert,' ecologists who have studied land-use practices find that small changes in conventional tea cultivation can make a big difference to conservation. Tea estates can avoid using the most toxic agrochemicals, and ensure that the chemicals are applied judiciously, by trained people wearing protective gear, and only on crop affected by pests such that the chemical drift does not enter adjoining natural ecosystems and water bodies. Shifting to organic cultivation is also an option.

Tea plantations can also do much better on the kind of shade trees used for tea bushes. For instance, in over 100,000 hectares of tea plantations in south India, planters mostly use a single Australian tree species: silver oak (*Grevillea robusta*). If shade tree species native to adjoining areas in the Western Ghats were used, more wildlife would thrive in the landscape, as such trees would provide food and foraging and nesting spaces. Bringing greater diversity into the landscape may also help improve the soil and control pests, thereby benefiting tea cultivation itself. Tea research institutes and planters have hundreds of native species to choose from, and they only need the will and foresight to begin trials.

Having wildlife in the landscape brings with it a fourth consideration, that of enabling their safe passage without harassment. The presence of wildlife in tea estates, by itself, does not signify a problem. But occasionally, it does lead to conflicts such as damage to property or unexpected encounters between wild animals and people. Small but significant measures can help avoid these: In the Anamalai Hills, some estates have left spaces for animals such as wild elephants and gaur to move through. Some have realigned fences to protect specific newly planted fields or property such as buildings and housing, while keeping substantial parts of the landscape open for animal movement. Conservationists along with plantation workers and managers and Forest Department staff operate early warning systems using alert beacons and mobile phone text messages to alert people to elephant movements and thereby avoid conflicts. Trained rapid response teams of the forest staff are also on standby to assist local people. These measures have helped reduce conflict while helping conserve these wide-ranging animals.

Finally, enhancing awareness enables coexistence between wildlife and people—from managers, workers, local residents to visitors. India's 0.56 million hectares of tea fields employ over 2 million workers and produce more than 1.2 billion kilos of tea annually, which millions of consumers relish. Yet, even the distant tea drinker, by learning about the landscapes tea comes from, and the ecosystems and people who sustain it, can help conserve wildlife by choosing to consume only tea that is truly green.

# Rhythms of Renewal

*2 January 2011*. Splashes of red dot the evergreen canopy, like blood on green canvas. The *Canarium*, stately white and tall, holds a red flush of new leaves above verdant, multi-hued forest. Skimming spectacularly over the trees, a great hornbill brushes grandeur onto the canvas. In the company of hornbills, a new year dawns on an unsuspecting rainforest.

The red flush is the flag of an ancient rhythm: a rhythm of renewal, carrying the cadence of nature's annual cycles. In the rainforest, the *Canarium* has endured months of sharp dry weather followed by lashing rains. It has stoically retained its space amid a thousand species, its leaves buffeted by many winds, aloft in sun and in rain, for another year of its decades' long existence. It has provided fruits for the hornbill, leaving seeds for hungry rodents or to germinate in a secure nook, and oozed

* This essay is written with Divya Mudappa.

resinous dammar from a cut. Drawing in the air with the breath of humanity, richer now in carbon dioxide, the tree has returned oxygen and thousands of litres of water to enrich the air and seed the clouds. As the second monsoon withdraws, leaving clear skies, spent clouds, and a winter chill, nature's seamless cycle enters another human year. There is now a renewed challenge of life in the environment, with other life forms of the forest, and with people in the wider landscape.

From the perspective afforded by the forests where the *Canarium* tree stands, here in the Anamalai Hills, one can take a sidelong look at events of the recent past and prospects for the year ahead. Local, national, and global change all have their imprint in this microcosm within a planet impacted by human action like never before.

Bolstered by a legal framework centred on conserving tigers, the state governments of Kerala and Tamil Nadu firmed up existing wildlife sanctuaries, declaring the establishment of the Parambikulam and Anamalai Tiger Reserves. Stretches of remarkable forest with threatened and endemic wildlife gain renewed recognition and, hopefully, better protection and improved management. In addition, valuable Reserved Forests, languishing in deliberate or benign neglect, are in the forefront as thousands of hectares are included within buffer zones. At the larger landscape level, these areas greatly add to the conservation potential of existing Reserves and help reduce the threat of forest fragmentation. Stung by past failures that aimed to exclude local people from conservation, efforts are being made to involve communities in the plantations and agricultural lands in the buffer zone. Overcoming suspicion and doubts to constructively engage these communities is essential to gain support for conservation and address pressing issues such as human–wildlife conflicts. This is no easy task, but efforts are under way, here, as elsewhere.

The people who share these forests of the *Canarium*, the tribal communities of the Anamalais, are also at a crucial juncture. Respected for their forest skills, the Kadar, in particular, have been partners in conservation of species such as hornbills, providing crucial support for wildlife research and forest protection. The Forest Rights Act (FRA) and the tiger conservation plan, both yet to be implemented, bring them promise and peril. The FRA was enacted in 2006 to recognize and restitute the traditional rights of forest dwellers, including individual rights to cultivated land and community rights to forest resources. It held promise to reverse historical injustice, but also provoked a section of conservationists to voice concerns that forests that were hitherto protected or excluded from human use will be lost, converted, or degraded. Over the year, even detractors of the FRA have learnt how invaluable the Act has been in fighting conservation battles against mining and forest diversion, where other environmental laws have failed. Can government, civil society, and tribal communities work together to deliver on the promise, while averting the perils of relocation, forest conversion and degradation?

The hills here are named for the Asian elephant, a species that better represents present conservation challenges. Elephant conservation implies thinking about swathes of land larger than our fragmented reserves, of corridors and agriculture, of people and property. The year 2010 saw a laudable initiative, the Elephant Task Force, of the Ministry of Environment and Forests (MoEF), culminating in a thoughtful report that promises to gently but firmly transform our view of the elephant and ultimately its conservation. The elephant has become, deservedly, our National Heritage Animal. A wider cross-section of society, good scientific understanding, and more transparent management shall be involved in elephant conservation. Movement routes and habitat fragments, including on private lands, shall gain additional attention, bringing benefits to myriad other species in the landscape including threatened hornbills and macaques, endemic amphibians, reptiles, and native plants. We shall no more be owners of captive elephants, only responsible guardians. Awareness of the need to phase out the demeaning existence and abuse of elephants in captivity is dawning. Now the elephant obtains a renewed place in our culture and consciousness. A position that recognizes and respects elephants as social, sentient, intelligent, and sensitive individuals and families, with whom we are privileged to share spaces.

Growing environmental consciousness is also driving changes in tea and coffee plantations in the landscape. Informed consumers are creating market demand for produce from farms that adopt responsible social and land-use practices. Consequently, certification programmes, such as Rainforest Alliance, require farms to protect natural ecosystems, revive native shade tree species, avoid toxic agrochemicals, and safeguard waterways. These promise to bring benefits both to the industry and environment.

Further downstream from where the *Canarium* stands, the ill-advised Athirapilly hydro-project, opposed for years on many solid environmental and social grounds, finally comes close to being scrapped. Partly, this stems from a welcome turn of events, with the Indian government finally appointing, in Jairam Ramesh, an environment minister who seems keen to uphold the environmental laws of the land. In a short span, he has transformed the rubber-stamp position of his Ministry to one that his detractors, even in more powerful ministries, are forced to take notice of. From aspects such as making the MoEF website one of the best government repositories of information, to taking clear executive decisions on dams, roads, airports, ports, forest diversion and exploitative industries, Ramesh's efforts have revitalized India's conservation movement and the dignity of his Ministry. His approach, mostly well-informed by ecology, is balanced by political pragmatism. The stance and strictures on preventing the proliferation of dams on the Ganga, on Bt Brinjal, Vedanta, POSCO, and coal mining, are battles that, if not won outright, are at least well fought. Like the stoic *Canarium* tree, he has many troubles to weather yet, to hold his present position.

Forces even further afield also impinge on the *Canarium*. Climate change is a decisive factor already affecting species, landscapes, and people's lives. The year 2010, poised to be the hottest year on record till date, was also marked by more heat than light in the aftermath of international climate conferences at Copenhagen and Cancún. Responses such as REDD (Reducing Emissions from Deforestation and Forest Degradation in Developing Countries), and voluntary, national, and international carbon markets are developing. Efforts are being made to recognize economic and other values of our natural capital and ecosystem services to move from an exploitative development trajectory riding on flawed and unidimensional measures such as GDP to sustainable development that values social and environmental goals. One can argue that these are too little, too late, or that forests are better REDD than dead, but time will tell if these are adequate responses to humanity's greatest global challenge.

Out in the Anamalai Hills, as the flag of the *Canarium* flutters red over the hill slopes, there is a sense of timelessness to the upheavals of life. And there are both storms and sunshine ahead.

*January 2019*. These eight years with more storms than sunshine, after Jairam Ramesh was moved out from the environment ministry in 2011, also saw a progression of five ministers, each vying to outdo the previous one in diluting environmental regulations. Many dilutions related to linear infrastructure projects such as roads and power lines through forests (see 'Long road to growth'). Even as the FRA remains to be implemented in the Anamalais, key provisions of the Act were diluted nationally by the central government, compromising the rights of forest dwellers in favour of industry, businesses, and private interests. In 2013, the requirement that projects involving forest diversion must secure the consent of local *gram sabhas* (a council of all adult members of the village) was scrapped for linear infrastructure projects. In December 2018, forest diversion projects were exempted from FRA compliance and gram sabha consent before first stage or 'in-principle' approval, creating fait accompli situations as projects approved in-principle mostly stood to be cleared. Rules notified in 2018 under the Compensatory Afforestation Fund Act 2016 also ignored gram sabha consent, so native forests with diverse plant species on community and common lands could now be replaced by artificial timber plantations, typically monocultures of species with little social or ecological value, to serve private or state interests. As for the recommendations of the Elephant Task Force, they remain largely unimplemented and ignored under the Modi government. In 2017, a nationwide elephant census—whose accuracy is anybody's guess as the methods used are obsolete and the data kept obscurely hidden—concluded that the population had dipped by about 10%, to around 27,000 elephants from the previous such census in 2012. The southwest monsoon of 2018 brought unprecedented rains and landslides scarring even forested hill slopes unmarked by such events, causing severe flooding in Kerala and Karnataka. Given climate change and the predicted rise in extreme weather events, this was perhaps only to be expected. As humanity's dependence on burning fossil fuels continues unabated, global atmospheric CO2 concentrations, which had crossed 400 parts per million in 2013 for the first time since weather data recording began, continue climbing to levels that the planet has not witnessed for millions of years. In the Anamalais, against a green backdrop on the mountain slopes, the new-year flush of *Canarium* trees now flares through atmospheric haze like red flags of a discordant nature.

# Conserving a Connected World

We live in a beautiful, strangely connected world. A world where a great argus pheasant dancing in the understorey of a Malaysian rainforest is linked by logging and timber trade to furniture in households in Coimbatore or Delhi. A world where an orangutan sleeping in its canopy nest in the rainforests of Indonesian Kalimantan is linked precariously to bars of chocolate and soap in Europe and cheap palm oil in Indian markets. It's a world where a person buying a packet of Indian coffee or tea anywhere is inextricably linked to hornbills and rivers, to threatened macaques and elephants of the forests and grasslands of the Western Ghats, Assam, or Darjeeling. In today's world, although perhaps unintended by the ultimate consumer, a purchase of a cell phone or laptop carrying coltan ore could signify slamming the door on equatorial forests and endangered gorillas of the Congo.

Nearly four decades ago, Indira Gandhi, in a famous speech at the first conference of the United Nations on international environmental issues at Stockholm in June 1972 declared that the 'inherent conflict is not between conservation and development, but between environment and reckless exploitation of man and earth in the name of efficiency'. Rejecting views that environmental problems were due to either population or affluence alone, she asked, 'Will the growing awareness of "one earth" and "one environment" guide us to the concept of "one humanity"?' The Stockholm conference was a global watershed, a harbinger of far-reaching changes in approaches to environmental issues, and in governance, international and national.

Twenty years later, at another international milestone in June 1992, the UN Earth Summit at Rio de Janeiro, many issues raised in Indira Gandhi's speech and in the developing concept of sustainable development began to crystallize. The Rio Declaration and Agenda 21 emerging from the Summit articulated the ideals of sustainable development, a form of development that gave environmental protection an integral part and placed human beings at the centre. Growing awareness began to influence various entities—from nation states and private companies to civil society groups and individual consumers—to incorporate concerns of sustainability in principle and practice. Sustainable development, or sustainability, became shorthand for a triple bottom-line approach that aimed to meld people, planet, and profit.

A more connected, globalized world also bloomed in the 1990s, with accompanying revolutions in access to information and communication through mobile phones and the internet. Bringing their own slew of environmental concerns over climate change and ecological footprints, they also brought unprecedented opportunity for sustainable development. When people's desire to know more about a product is linked to awareness on environmental effects and sources of production, consumer choices may change toward better alternatives.

The call for sustainability builds on accumulating scientific evidence of human impacts on the environment, and the communication and information revolution that allows consumers to play a conscious and proactive role in environmental conservation. This approach is gaining currency in sectors such as forestry, agriculture, tourism, and marine fisheries. It focuses on transforming land use and business practices and influencing consumer choices, through better standards and alternatives devised by civil society and environmental organizations in partnership with governments and businesses. Based on this, certification schemes such as Forest Stewardship Council in forestry, Rainforest Alliance, bird-friendly and organic schemes in agriculture, and the Marine Stewardship Council and Global Sustainable Tourism Councils now link businesses to environmentally conscious consumers through the shelves of markets worldwide, with the ultimate aim of fostering change on the ground.

For instance, in forestry, it remains critical to protect all primary forests and remnant natural ecosystems in timber concessions and in the surrounding landscape, and to curtail the severe soil erosion and negative impact resulting from roads and movement of heavy vehicles. India, along with China, now the largest importer of logs and palm oil from Malaysia and other parts of Southeast Asia, are driving extensive deforestation and destruction of the highly biologically diverse rainforests of this region. Oil palm, cultivated as enormous monoculture plantations after cutting and burning rainforest, is one of the most environmentally destructive land-use practices of today. Better practices, alternatives, and informed consumer choices are urgently required to save these remarkable tropical rainforests and the myriad threatened species, such as orangutans and argus pheasants, they shelter.

In agriculture, take the case of tea and coffee from India. Carrying the tag of industry because of the scale, tea and coffee cultivation involve agricultural practices with a large footprint in environmentally significant landscapes. Occupying around 120,000 hectares in south India and 460,000 hectares in the Northeast, in the biodiversity-rich hills of the Western Ghats, the Assam Valley and Darjeeling hills, the picturesque tea plantations are grown as intensive monocultures with vast swathes of tea under limited or no tree cover. Coffee, particularly arabica coffee, traditionally grown under the shade of many native forest tree species, in over 340,000 hectares in south India, has the potential to be a form of land use more gentle in its environmental effects in comparison to other crops such as oil palm, rubber, or tea. Still, inadequate treatment of waste water from coffee pulping and contamination of fresh water sources poses a continuing challenge.

Tea and coffee estates can do much for the environment by protecting natural ecosystems and wildlife habitats such as rainforest fragments and grasslands or by setting aside areas to revive such ecosystems or habitats that have been already lost, particularly along streams and rivers. By treating wastewater adequately, using shade tree species native to each region, and avoiding toxic and banned agrochemicals such as paraquat (gramaxone), endosulfan, and carbofuran, they can reduce their environmental impact. Produce from some certified Indian estates is currently making its way into export markets, with domestic consumers yet to connect with or benefit from ecologically friendlier tea and coffee. The billions of cups drunk every day may then leave a gentler imprint on the earth.

Yet, consumer choice via certification and corporate initiatives in social responsibility are not entirely fail-safe.

There remains an indispensable need for watchdog groups, for constant improvements in transparency and standards, and most important, for producers and consumers alike to be sensitive and adopt a spirit of caring for nature and the special needs of ecology and wildlife conservation, even if it entails reduction in production and consumption.

Now, it is worth examining what individuals can do in today's linked, consumerized society. Going back to the 'one humanity' of Indira Gandhi: 'It will not be easy for large societies to change their style of living. They cannot be coerced to do so, nor can governmental action suffice. People can be motivated and urged to participate in better alternatives.'

# Integrating Ecology and Economy

'One of the hardest things in politics,' former US President Barack Obama had said, 'is getting a democracy to deal with something now where the payoff is long term or the price of inaction is decades away.' Though he was speaking of the United States, Obama's words are equally relevant to the other great democracy—and its government—on the far side of the planet: India.

If applied to science, Obama's words most resonate with ecology. Its central concern is the long term and leaving a healthy environment for future generations. And within ecology, on a planetary scale, it is the science of climate change. So when the Modi government renamed India's Ministry of Environment and Forests (MoEF), appending 'Climate Change' at the end, it was a timely move. It signalled that even as the government pursues its stated policy of industrial and infrastructural expansion for economic growth, it places tackling climate change firmly on its agenda, along with the protection of environment, forests, and wildlife.

But a slew of media reports and actions of the ministry belied this interpretation. According to these reports, the MoEF—MoEFCC in its new avatar—so as to open more forests for mining, planned to redefine the criteria for inviolate forests. It also set about diluting environmental norms and procedures to bypass existing legal requirements for large infrastructure and defence projects. Through orders and guidelines, the ministry set about undercutting crucial provisions of the Forest Rights Act and loosening the environment, forest, and wildlife clearance process for new infrastructure or industrial projects. The intent was to make clearances for such projects speedy rather than sensible, after appraisals that were cursory and piecemeal rather than critical and comprehensive. Meanwhile, the MoEF has been slack on other pressing needs: acting to save critically endangered species such as the great Indian bustard (now down to less than 200 individual birds in the wild), or implementing proactive measures to combat climate change. Former environment minister Prakash Javadekar had stated that environmental protection and developmental activities need to go together. With pressures of poorly regulated development mounting, this is as good a time as any to recount five lessons from ecology on why environmental protection should concern India's government and 1.2 billion plus people.

Obama's words point to lesson one: ecology takes the long view. Development projects promoted for short-term gains may have unaccounted long-term costs for people and nation. The previous United Progressive Alliance government in its 10-year tenure allowed the conversion or loss of over 700,000 hectares of forest—an area the size of Sikkim—for development projects and non-forest uses. Natural forests of diverse native tree species function as watersheds, wildlife habitats, and sources of livelihood for tribal, farming, and fishing communities, contributing to long-term human wellbeing in ways not captured by indices such as annual GDP growth.

The science of restoration ecology attests that such diverse natural forests and the living soils they spring from, once destroyed, are difficult and costly, or infeasible, to restore, and appreciable recovery may still take decades to centuries. This is not adequately factored into the estimation of the net present value (NPV) of forests that tries to approximate economic losses over a 20-year period, by which time the losses are 'recovered' in compensatory afforestation sites. A project developer pays out the NPV—at 2014 rates, a maximum of Rs 10.43 lakhs per hectare for very dense forests in the most biologically rich regions such as the Western Ghats—and flattens football fields of forests for the price of a mid-range SUV. Furthermore, compensatory afforestation, if carried out at all, frequently involves raising plantations of one or few alien tree species such as eucalypts and wattles. Such artificial forests are no substitute for the more diverse natural forests of mixed native species, including centuries-old trees.

Lesson two is that ecology is a science of connections, of food chains from plankton and fish to sharks and humans, of energy flowing from sun through grass to deer and tiger. Pluck the hornbills out of their forest home, and forest trees whose seeds the birds disperse begin to decline. Destroy forest remnants amid coffee plantations, and farmers suffer as coffee yields dip from the loss of pollinating bees. Strip the oceans of sharks and predatory fish with industrial fishing, entire ecosystems and livelihoods of artisanal fishers unravel in what ecologists call a trophic cascade. So, the wholesale construction of 300 large dams in the Himalaya as proposed by the government would not just generate power, but have other negative consequences radiating down the chains and webs of life, including to people downstream. When these are taken into account, implementing fewer and smaller projects or alternatives appears more attractive.

The third lesson, the mandala of ecology, is that ecology closes the loop. Nature recycles, without externalities, wasting little. If the government applied this to everything from recycling municipal waste to curtailing pollution by industries, it could generate jobs and induce growth without leaving behind irredeemable wastes. Ecology is replete with such cycles. One sees it in the organic farmer practising rotational shifting agriculture on the hill slopes of northeast India, in the cycle of water from earth to cloud to rain and river, and in the dung beetles and fungi and vultures that help return dung and vegetation and carrion to the elements.

Fourth, ecological processes transcend political boundaries. We pump carbon dioxide and other greenhouse gases into the common pool of our atmosphere anywhere and affect people and the earth's fabric of life everywhere. The migratory warbler that picks the insects off the plants in our gardens may depend for its survival on protection of its breeding grounds in China or Central Asia. To conserve tigers and elephants in protected

reserves, we need to retain connecting corridors and forests, some spanning state or international boundaries. Development and infrastructure projects can be designed and implemented such that they do not further disrupt fragmented landscapes, but instead help retain remnant forests or reconnect vital linkages.

Finally, ecology teaches us that humans are not external to nature. Land and nature are not commodities to buy or sell recklessly or reduce to a packaged spectacle for tourists to gawk at. They form the community we belong to: we are part of nature, it is home. In the debate over ecology versus economy, we must remind ourselves that both words originate from the Greek word *oikos*, meaning 'home'. Ecology, the science of our home environment must inform economy, the management of our home resources.

What is often forgotten, in the debate falsely caricatured as environment versus development, is that for almost every destructive project, there are often alternatives and means of implementation that cause less harm to the environment and local communities and can provide overall long-term benefits. For instance, roads can be routed to avoid wildlife sanctuaries and provide better connection to peripheral villages, thus helping both people and wildlife. Decentralized village power generation systems that use biomass, solar power, and other renewable sources can help reduce reliance on mega power projects plagued by corruption and requiring long power lines that suffer transmission losses and cause forest fragmentation. Mining can be carried out avoiding areas valuable for conservation or to local people, after due environment and forest clearances, and keeping aside topsoil to ecologically restore even these areas later.

There are already many promising examples of ecologically sensitive development. If ecologists, engineers, and economists synergize their efforts, and governments choose to exercise their electoral mandates to take the long view, there can be many more. The integration of ecological considerations into economic development is vital if, in the quest for profit, we are to ensure the long-term wellbeing of our people and our planet.

# The Health of Nations

## *The Other Invisible Hand*

One of the perils of ignoring the environment is the consequent failure to notice that the environment never ignores you. Healthy environments support human health and flourishing even as conservation secures natural resources and livelihoods. On the flip side, environmental degradation rebounds as economic losses, while pollution strikes at the heart of public health. Can one afford to ignore the environment when it affects both economy and health?

This global question now confronts India, as a developing nation surging ahead towards its predicted destiny as the world's third largest economy by 2030. In its pursuit of a neoliberal growth model, focused on indices such as GDP, the country has accorded lower priority to public health and environment. The growth model presumes that social benefits will accrue via the 'invisible hand' of market forces, possibly mediated by increased public revenues and spending following economic growth. Meanwhile, environmental conservation remains predicated on creation of regulations and reserves, while public health is contingent on access to clinics and care. Governance systems consider economy, ecology, and health as different domains, ignoring their inescapable connections.

India cannot afford to let this situation continue longer. The country confronts unprecedented air and water pollution and environmental contamination and degradation. Connections among health, economy, and environment revealed by recent research need to urgently inform policy and praxis.

Take the air pollution crisis in the national capital. Implicated in serious lung, respiratory, and other diseases affecting its citizens including over 2 million schoolchildren, the crisis exemplifies a countrywide malaise. Over 660 million people, half of India's population, live in areas where fine particulate matter ($PM_{2.5}$) pollution exceeds the National Ambient Air Quality Standard, reducing life expectancy by an estimated 3.2 years on average. In 2018, the World Health Organisation's air quality database listed 13 Indian cities, including New Delhi, among the 20 most polluted cities in the world.

In 2011, $PM_{2.5}$ and other atmospheric emissions from 111 coal-fired power plants across India resulted in 80,000 to 115,000 premature deaths and over 20 million asthma cases. The economic cost to the public and the government was estimated at USD 3.2–4.6 billion. Agriculture, too, is seriously affected. One study estimated that rice yields between 1985–98 declined by 10.5% because of air pollution emissions from fossil fuel burning and the formation of atmospheric brown clouds with black carbon and other aerosols. Climate change and air pollution together (due to ozone and black carbon) were estimated to have reduced average wheat crop yields across India in 2010 by up to 36%, with yield loss up to 50% in some densely populated states and with most of the decline attributed to air pollution.

Indoor air pollution due to use of inefficient biomass-based stoves is another serious health issue. In India, it affects the health of over 400 million people, particularly women and children. A study published in 2019 in *The Lancet Planetary Health* estimated that air pollution led to about 1.24 million deaths in India in 2017—that is, one in every eight deaths in the country. This included about 0.67 million deaths due to ambient particulate matter pollution and 0.48 million deaths from household air pollution. Although less than one-fifth of the world's population is in India, over one-fourth of deaths globally attributable to air pollution occur here.

Similar concerns, connecting health, ecology, and economy, arise in water pollution and overdependence on chemicals in industrial agriculture. The Central Ground Water Board reported in 2015 that over half of India's districts suffered groundwater contamination, including with heavy metals above permissible levels in 113 districts across 15 states. A 2018 report by the Central Pollution Control Board found 351 polluted stretches along 323 rivers, with 38 rivers in 27 states in India having critically polluted stretches—besides the hundreds of polluted tanks and lakes and other water bodies. Water pollution does not only cause water-borne diseases and other direct health impacts, it can negatively affect diet and livelihoods due to the loss of fish and aquatic resources, contamination of soils, and loss of agricultural productivity downstream of industrial and mining sites.

A 2013 World Bank study estimated that the financial and social costs of environmental degradation in India amounted to about US$ 80 billion or 5.7% of the country's GDP. Of this, outdoor air pollution accounted for 29%, followed by indoor air pollution (23%), cropland degradation (19%), water supply and sanitation (14%), and pasture and forest degradation (15%).

How environment affects public health is often difficult to trace, but connections are evident and significant. A 2006 World Health Organisation (WHO) study attributed 24% of the disease burden (healthy life years lost) and 23% of all deaths (premature mortality) worldwide to environmental factors. The burden of environment-mediated disease and mortality is also higher in developing countries. Further, a large part of this is due to non-communicable diseases (NCDs).

In 2014, the WHO country profile for India noted that 60% of the 9.8 million human deaths were due to NCDs. The four big killers—cardiovascular diseases, chronic respiratory diseases, cancer, and diabetes—accounted for 48% of all deaths and 80% of deaths due to NCDs. NCDs are a global problem—they caused 68% of 56 million global deaths in 2012—that disproportionately affect low- and middle-income countries. Cardiovascular and respiratory health issues are caused because of air pollution, while cancers and hormonal disruption are known to occur from the many pollutants in the environment.

Scientists predict that the health situation will worsen under ongoing climate change. Increasing incidence of NCDs, besides other effects such as rise in vector-borne diseases and injuries due to climate extremes, is likely. In 2014, the Intergovernmental Panel on Climate Change concluded with very high confidence that climate change will exacerbate existing health problems over the next few decades.

The economic fallout will also be high. One study estimated that India's GDP in 2004 would have been 4–10% higher if NCDs were completely eliminated, while a 2014 report estimated that India stood to lose over USD 4.58 trillion between 2012 and 2030 because of NCDs. Economic policies that alter occupation, mobility, and diet can exacerbate these problems, such as when lifestyles become more sedentary or livelihoods change from rural or forest-based occupations to working in polluted industrial areas as wage labour. The health–environment–economy connect has become a vital concern in recent debates such as over mining in forests where forest-dependent communities live, land acquisition of farms for industry and infrastructure, and reducing pollution from coal and shifting to renewables.

As research findings accumulate, the connections between environment, health, and economy grow stronger. This has many implications for policy. For instance, health impact assessments must become a mandatory part of environmental and social impact assessments in industrial and development projects. The GDP-centric measurement of progress should make way for more holistic indices that include progress in health and environmental protection. Instead of viewing environment as a 'hindrance' or public health as a 'burden,' economic policy must consider these integral to human development and provide higher financial outlays. Finally, India's National Health Policy 2017 recognizes the need to integrate environmental and social determinants of health across all sectors, in keeping with the 'Health in All' approach, but concrete actions required in individual sectors are yet to be identified.

Ultimately, human lives and livelihoods, health and resources derive from the natural environment: humans are a part of nature. The environment is, in that sense, the other invisible hand that leads to a cleaner, safer, more alive, inspiring world where people can live and flourish. Environmental health subsumes and is connected to human health, just as the health of your body subsumes and is connected to the health of your heart.

# REFLECTIONS

## our place in nature

The same stream of life that runs through my veins night and day runs through the world and dances in rhythmic measures.

It is the same life that shoots in joy through the dust of the earth in numberless blades of grass and breaks into tumultuous waves of leaves and flowers.

It is the same life that is rocked in the ocean-cradle of birth and of death, in ebb and in flow.

I feel my limbs are made glorious by the touch of this world of life. And my pride is from the life-throb of ages dancing in my blood this moment.

—Rabindranath Tagore, *Gitanjali*

The heart is the seat of a faculty, sympathy, that allows us to share at times the being of another.

—J.M. Coetzee, *The Lives of Animals*

# The Wild Heart of India

Out of the parched forest flow the cool waters of the Charan Ganga. It is no insignificant stream this, weaving its course through the famed forest of Bandhavgarh, carving its signature across the land, quenching thirst of deer and tiger and langur, bringing life to the dry earth. Here in central India, in the middle of May, the sal forests and open grasslands bake in the tropical sun. The seasonal drought has turned many trees nearly leafless, their branches stark against singeing skies, and the browned grasslands crackle in the breeze. The heat of summer is hard to escape, here, in the heart of India.

Finding water, then, is key. The deer make their daily beeline to the waterholes through the grasslands. They appear, unhurried, unmindful even of lesser predators, such as the jungle cat who sits motionless, merging with swaying grass. Still, nearing the waterhole, the deer approach with cautious steps, eyes alert, ears cocked, noses twitching. They sense a tiger lying in wait. Sharing sentiment and presentiment with the deer, we make our own way in an open jeep, a group of friends heading to the waterhole, tourists in tigerland. The tiger we find, however, ringed by other vehicles with their gaggles of gaping tourists, is snoozing under the trees and the bamboo, behind a little rise and almost beyond our prying eyes. Sprawled on his back like a housecat, the tiger appears to be waving a disdainful paw.

The heart of India is tiger country. People come here to see tigers and be awed by their presence. The visitors quickly learn, from guides and tourists in other vehicles, that where there is water is a good place to wait to see a tiger. Some learn to mark the tiger's progress through the forest by the alarm calls of the deer, or by tracking paw prints on the dusty roads. Some begin to sense that the preservation of the tiger casts a protective shadow over a pyramid of prey and plants and other life of the forest. Others note that it is the tiger that needs such a forest to exist. But is this the primary message? Don't we need such a forest, too?

The soil is parched under stunning heat. And yet, the trees, as if knowing something we do not, or from habits evolved over the ages, are putting out fresh green leaves. There has been no rain—only an anticipation of it. Fresh leaves erupt on mahua trees, spent after an effusion of fragrant flowers now fermented into country wine, and on *sal* trees whose branches are laden with winged fruit. Perhaps there is an anticipation of wind, too. Even as fields lie fallow in the human countryside outside Bandhavgarh, the trees have found their moisture and are investing in growth, and in their future. And from the forest, the waters of the Charan Ganga continue to flow.

Deep in the forest lies a great idol of Vishnu, the Shesh Shaiya, a supreme deity signifying, pertinently, existence and preservation. Where the deity lies recumbent, the waters of the Charan Ganga appear to emerge from his feet. It is not hard to imagine, in a summer like this, that this source with such clear water, keeping the trees green here and for miles downstream, has a divine origin.

But if we could emerge high above the forest and soar on the wings of the Indian vultures that grace the cliffs and skies of Bandhavgarh, we may gain a different perspective. Then we'd see the vista of forest in the landscape around the spring where rests Shesh Shaiya.

From up there, it is the forest that appears to tap and soak and channel the water through aquifers to emerge as a spring. The forest is divine, in an aesthetic sense, but needs no divinity to perform this basic hydrologic function. Now, Vishnu, as a being signified by the idol, seems but a wise person who, like the tiger, found a good place close to water, to rest under the shade of the trees and the bamboo.

His presence, as a preserver, like the tiger's, is but a marker of what needs to be preserved.

The anticipation of the trees is not belied. The wind and the rain are coming. When we turn back towards our lodge as the day comes to a dusky end, a jackal lopes past into the growing darkness of the evening. And so the clouds gather, with gusts of wind, thunder, and lightning.

Under crackling, electric skies, the shimmering heat yields to cool draughts that precede the rain. Around the courtyard of the lodge, the waiting fruits of the sal trees take wing. Whirring like a fan, they disperse with the wind, carpeting the earth with winged seeds. The branches toss in tumult, the seeds skitter before the wind. From the verandah we watch the tempest unfold; our hearts skipping along with every gust.

Dark clouds grumble and crash, spilling grey curtains of rain. The rain sweeps across the sky like diaphanous drapes drawn by wind. The pre-monsoon thunderstorm has brought the water to nourish the soil where the sal may grow. And yet, the water is an unwanted burden on the fruit itself as it makes its own short but crucial spinning journey away from mother tree to moist earth.

Such is the economy of nature that, even as the parched earth soaks the water, the sal spins it off its seed.

From Bandhavgarh, in the middle of May, we travel on into another special landscape. A landscape of stately *sal* forests spreading to the horizon, amid sprawling meadows and plateaued hills. Here, the stage is set for a grand play of life and death. This is the land of the deer and the tiger, the quintessential prey and predator, a land that holds an essence of wild India: Kanha.

Kanha lies within a vast amphitheatre marked by the sweep of the Satpura mountains to the west and the Maikal range to the east. The soils and rocks seem old as the Earth herself—a piece of primeval Gondwana, the great land that sailed the early ocean. This land gathers the waters for the Narmada, flowing to the west, and for the Mahanadi, to the east. Here have lived the old peoples—the Gond, after whom the great land was named, and the Baiga, who live off the extensive forests and the deep soils.

It is special for us, too, being the landscape where the pioneering field biologist George Schaller roamed in the mid-1960s and studied deer and tiger. His landmark book, *The Deer and the Tiger*, introduced a whole generation of wildlife scientists in India to field studies of animals in their natural habitats, their behaviour and ecology through the cycle of the seasons, and the perpetual dance of prey and predator.

Kanha simmers in the summer heat; the monsoon is still some weeks away. Like green arms, the forests seem to hug the browned meadows that await the rain to spur another renewal of life. Along with Kanha's panorama of forests on view, the grand assemblages of ungulates on its meadows must rank among the great wildlife spectacles of Asia.

Dark groups of gaur, heading for water and forage, add grandeur to the landscape. Herds of swamp deer, the barasingha, weave through the grasslands. The stags, still sporting the many-tined antlers that give them their name, lounge among groups of prim and perfect does in the relative calm that comes after the rut. These are the so-called hard-ground barasingha, whose cousins of wetter turf live in the terai grasslands of north and northeast India. Here, in Kanha, these stately deer bring elegance to the meadows.

There are other deer, too, in Kanha: the diminutive and shy chevrotain; the cautious and excitable muntjac; the lithe and graceful chital; and that great deer of the forest, the sambar. The forests and grasslands resound with the bellows of chital stags, for this is the peak season of their rut. We watch, as Schaller must have watched, males displaying and sparring, pawing and preaching, fighting and mating. The sambar and muntjac afford us glimpses in the forest, but we fail to find any chevrotain, a more nocturnal species. Late one evening, we drive up to the Bamni Dadar Plateau in search of the four-horned antelope, the chousingha. Although the animal eludes us, we find a panoramic view of Kanha and stop to admire the landscape where, from open meadow to dense forest, each species of deer finds its space, and each its place in the cycle of life.

With the prey come the predators. There are tigers, of course, and in their shadow, other hunters: leopards, dhole or Asiatic wild dog, sloth bear, jackal, jungle cats, and other smaller carnivores such as civets and mongooses. With the help of the langur and a little luck, we are lucky to see some of them.

On our drive through the forest, a troop of langurs erupts with raucous alarm, bringing us to a halt. The monkeys closely watch and noisily track some movement down in the forest understorey. We wait in silence, all eyes, and ears; the langurs' alarm is almost contagious. Our patience is soon rewarded. At the edge of the road, a shape materializes out of the shrubs: a leopard. Wide-eyed, we watch as he quickly crosses over.

A bit later, a sloth bear with a grown cub scurries into the safety of the forest. And then, we spot a jungle cat resting in the shade of a little rock overhang to escape the heat of the afternoon. In the distance, a peafowl cries.

The sal forests swathe the landscape, and the *Bauhinia* climbers, bedecked with flowers, garland the sal. Yet, the really large, tall trees are few: a familiar story of past logging slowly transforming into a future progression of hopeful regrowth. The tree trunks are studded with the gems of orchid blooms and shoulder the burdens of strangler figs. Perched erect on the boughs, racket-tailed drongos sound their metallic clanging calls. Their glistening black plumage and tail extends into thin streamers tipped by black spatulae—the drongos attest the trees like exclamation marks.

The drama of the deer and the tiger will play out on the evolutionary stage, and shall forever mark this Central Indian landscape. Yet, it is sobering to recall that the present assemblage of wildlife is a truncated one, for the blackbuck, water buffalo, Asian elephant, and cheetah, which roamed in this region not long ago, are seen no more. They've vanished, yes, but are often forgotten, too, as human memory slides on a slope of shifting baselines: it is deer and tiger that one notes today. The vehicle that speeds past us on the unsealed forest road—hastening, perhaps, to the next tiger sighting—also misses the drongo and the orchid on the sal, leaves us coated in a light layer of dust.

It is not difficult to feel a part of this wild landscape, or feel a sense of belonging to a place where people have lived amid forests and wildlife for millennia on into the present day in the heart of one of the world's most populous countries. In a land of forgotten species and fading sensitivity, we can despair at what we have lost, yet exult at what we can experience, and strive for what may lie ahead. As we depart from Kanha, we cannot shake off an awareness that, deer or tiger, *sal* or stream, the land has a value inestimably greater than the sum of its parts. Parting from the landscape, we feel a tug at our hearts. But that only seems natural, here, in the great landscape of forests and meadows in the heart of India.

# Close Encounters of the Third Kind

For me, close encounters with animals in the field have been of three kinds. The first, and most frequent, was when the creatures slipped past me unseen. One morning, in a watchtower camp in Mizoram, I stepped down to discover, neatly imprinted on the soft earth, the fresh pugmarks of a tiger. Another time, in Kerala, I opened the backdoor of our family home one afternoon to find stretched beside the porch the metre-long, scaly, sloughed-off skin of a rat snake. Such encounters were almost daily occurrences. The animals left only traces of their passing.

Then came the close physical confrontations: direct encounters that thrill or terrify—the sort often recounted in wildlife magazines and promoted in adventure tourism. Turning a bend on the forest road in a vehicle, I would thrill at the sight of a crossing sloth bear or elephant, knowing that for a person on foot or for an animal on the path of a speeding vehicle a similar encounter could be life-threatening. Like other wildlife biologists, I, too, have sat for hours on bamboo platforms up on trees or in camouflaged ground hides to spy on wild animals up close. When, in Dadara, Assam, a greater adjutant stork landed on a nearby branch to gaze at me with his pale, sky-blue eyes; when, in Guindy, a chital doe with twitching nostrils passed the shrub I crouched in at touching distance; and when, in Ladakh, from icy stream to rocky slope a silver snow leopard darted just metres away, the encounters brought an irrepressible rush of excitement. In conversations with other people—journalists, family, colleagues, and casual acquaintances—these were the encounters they were often most curious about.

And yet, the most memorable encounters were not those that left subtle traces or provoked thrills and terrors. It was those in which, as witness or participant, I felt transformed, filled with wonder and new perception into the lives of animals.

## Elephant Crossing

On a clear February morning, 10 feet from the highway's edge, behind a whispering wall of rainforest plants, a great grey shape waited. She stood among tall trees, her forehead bulging above an attentive, honey-coloured eye with dark lashes. She fanned her ears, sounding a leathery *slap slap* against her neck. A shiver of shrubbery behind her revealed a smaller shape in the understory; there they waited, mother and calf, to cross the road.

Ahead, the highway sliced a 4-hectare fragment of tropical rainforest and coursed through vast, waist-high tea plantations towards a nearby town in the Anamalai, or 'elephant mountains,' in southern India. I had been driving at speed when I saw the mother by the road—elephants always give me pause—so I backed up to park and stepped out to watch her. She was already swinging her ponderous bulk back into the trees. It must have been my driving.

Traffic sounds punctured the morning quiet: the chatter of passengers and schoolchildren rose above the growl of a town bus; an open lorry rattled past carrying a standing cluster of workers on their way to pick tea leaves in the plantations. None of them could hear the soft slap of ears and the deep, low rumble from the trees.

I flagged down a tea-estate tractor that was chugging past. 'There are elephants ahead,' I told the driver. 'Watch out.'

'I know. They passed by our housing lines around midnight. The rest of the herd is still down there,' he said, pointing to a nearby valley.

Then, with a nonchalance that estate workers here often have, he added, 'If we leave them alone, they will soon go their way.'

When people and elephants share a landscape, not all interactions are this trouble free. Wild Asian elephants sometimes break doors or push down walls of storerooms to access grains, salt, or lentils. They quest for food in a landscape—once a vast tract of rainforest and perennial rivers—now swathed in a monoculture of tea scattered with fragments of forest and a lattice of grassy swamps. Elephants, ranging over hundreds of square miles, inevitably encounter people, cultivation, or human settlements. In a close and unexpected encounter—the kind that makes headlines—a person on foot or on a bike can be injured or crushed to death.

At a lull in traffic, the elephants hurried across the road on padded feet and melted into the trees. Ahead, a man in a bright white shirt appeared riding a motorcycle; I could see that he would pass just a few feet away from the elephants. To alert him, I waved and pointed. He didn't see me or failed to understand, and kept on coming. From where I stood, I could see the mother turn. She stepped forward, placing herself between her calf and the road, and curled her corrugated trunk up to test the air. The rider came closer, and closer. Less than a dozen elephant paces from the animals he had not yet seen, but whose eyes and ears were tracking him, he passed without pause.

Farther up the same highway, a conservation signboard marked a forest corridor: ELEPHANTS HAVE RIGHT OF WAY. There is something heart-warming in the idea that one can make space for elephants on the roads, as the tea-estate workers have in their lives. And yet I can't shake the feeling that the more remarkable phenomenon is the rarely noticed quietude of the animals themselves.

## Kicking the Habit

One day, around noon in Chennai city's Guindy National Park, I was walking along the edge of the road keeping close to the shrubbery, scanning the thorn forest ahead for deer and antelope, when a rustle of leaves at my feet brought me to a sudden halt. There, by the side of the road slithered a sinuous whipcord, gleaming black, banded with white rings, long as my arm.

I reacted instantly. I jumped back a step. And then, as if by force of habit, I swung my leg to kick sand and stones at the recumbent snake.

I do not know if I wounded the snake, although some of the stones must have struck him. As I watched, the snake glided slowly away under the shrubs, further into the leaf litter, and was gone.

It was a common krait, one of India's 'big four' venomous species, implicated in a large number of deadly snakebites across the country. Common kraits are nocturnal, hunting small prey like rodents after dark. By day they are rarely seen and can be quite sluggish and I had never seen one before in the Park. This snake had just been lying there, perhaps resting, perhaps waiting for an opportunity to climb onto the road nearby to bask in the sun for a while. The snake had neither attacked me nor meant to.

Walking in a National Park, engrossed in my self-important task of field research, I had failed to recall that this was one of the few spaces in the city meant to be a safe haven for species such as the krait. I had not paused to note that the snake could not have bitten me through the shoes I wore. I had not hesitated even a moment before kicking stones at the snake. I felt foolish and ashamed.

Whenever this incident comes to my mind, I recall the lines from D. H. Lawrence's famous poem 'Snake'. In the poem, as a snake comes to a water trough to drink, he evokes in the poet a sense of honour and horror. Honour because the snake sought the poet's hospitality and came to his water trough to drink, but horror, because of an inculcated prejudice, which urges him to attack the snake. In the end, the poet hurls a log, chases the snake back into a hole. And yet, he is left disconsolate:

> And immediately I regretted it.
> I thought how paltry, how vulgar, what a mean act!
> I despised myself and the voices of my accursed human education.
> And I thought of the albatross
> And I wished he would come back, my snake.

For he seemed to me again like a king,
Like a king in exile, uncrowned in the underworld,
Now due to be crowned again.
And so, I missed my chance with one of the lords
Of life.
And I have something to expiate:
A pettiness.

My snake had meant me no harm. And I had missed my chance, too, with a beautiful animal in the wild and have a pettiness of my own to expiate. If anything, it was my habit of reacting from prejudice that needed to be kicked and not the snake.

## Two in the Bush

On a dark December night at our field station in the mountains of Kalakad, a young herpetologist friend of mine returned from a night survey for reptiles in the surrounding rainforests with an announcement.

'I found a strange bird sleeping in the forest. Do you want to see it?'

I was studying birds as part of my doctoral research in the same forests and he knew I'd be interested.

Minutes later, joined by another researcher, we were off. Down the narrow trail we went, our headlamps and torches bouncing with our steps, swinging and casting diffuse beams through the mist. We soon reached the spot. A mass of shrubby evergreens stood under tall rainforest trees and lianas. And just a few steps away, on a thin, horizontal branch barely higher than my head, there she was.

Perched on one leg, the other folded in, she looked like a small, cosy fluffed-up ball of feathers, like a dumpy sparrow. A chestnut patch marked her throat and breast above a white belly and reddish tail. Her greyish head was turned and tucked in by her side against rufous wings. She slept a deep sleep, apparently undisturbed by the lights and whispering voices.

She was a female white-bellied blue flycatcher, a species endemic to the rainforests of the Western Ghats. I had rarely encountered one so close or watched one motionless for so long. In the forest, by day, I'd mostly hear a nervous ticking if I spooked the birds or the soft notes of their ventriloquial songs that made it difficult to locate them in the undergrowth. The males, plumaged in brilliant blue on their upperparts, heads, and breasts, also have white bellies like the females.

My field studies were revealing that this flycatcher species preferred the undergrowth of mature rainforests, and declined or disappeared where forests had been disturbed, logged, or converted to single-species plantations. But I wanted to know more. I had applied for a permit to use mist nets to catch birds like the flycatcher, slip rings on their legs, to help identify individuals and map their territories. The State Forest Department had, however, denied me the permit.

And yet, that night, there the flycatcher sat, within arm's reach.

I slunk up to the bird, reached up from behind, and grabbed her in my fist.

A fraction of a second before my hand closed around her, with a terrified chirp the little bird shrieked awake. I held her gently, but as the fluffed-out bird fluttered awake, she shed dozens and dozens of her feathers. The soft feathers settled on my sweatshirt or drifted off into the darkness. I took the palpitating bird to the field station for measurements, then quickly returned and released her in the same place. She flew to perch on a high branch and that is where I left her, in the cold, that night.

The incident has haunted me since. What was it that made me grab the bird? A quest justified in the name of science—an arrogant belief that conflated the taking of that one useless measurement with taking the measure of a wild bird? Was I just venting on the bird my frustration at being refused a bird-catching permit? It was just a joke to show off how I could outsmart a blameless little bird.

Another day, walking a nearby trail, I came upon a pair of white-bellied blue flycatchers. The female flew out from the base of a large tree, where I noticed a snug little nest tucked into a crevice among the tree's buttresses. Woven with thin dry rootlets and fibres, the small cup nest was lined inside and tasselled below with finely-veined leaf skeletons. The nest had four pale, bluish-white eggs, capped in brown around their broader ends. From a distance, I watched the pair and listened to them singing in soft trills and quavering high-pitched whistles. They seemed at home in the deep rainforest. Whoever claimed a bird in the hand was better?

# Who Gives a Fig?

The ground shook as the lorry thundered past. I felt the blast of diesel fumes and smelled the hot rubber from the tyres. The lorry's colourful carriage was shrouded by a drab tarpaulin thrown over an apparently heavy load. The beat of music from the driver's cabin was drowned by the sound of the vehicle's shrill air-horn. A swirl of dust arose in its wake and wafted here and there; some of it settled on me.

The road was not a large one, but was considered an important connection through the hills linking two cities. The cities considered themselves important enough to have this road, too. Here and there, the blue asphalt was broken by potholes—not big ones—and the edges of the road were ragged. Yet, it was a good road and wide enough to let two buses pass by each other at top speed and with space enough between them to avoid terrifying the passengers by the windows. There were little dips and rises in the road as it took the contours of the countryside. It wound its way through field and fallow, farm and forest—as the wheels churned the miles in the incessant rush to the city.

All around was a landscape of gently undulating plateau. By the side of the road grazed cattle and goats, herded by village boys and girls with sticks. In the tree across the road—a stately banyan—were a troop of monkeys, a hornbill, and a couple of active, ardent squirrels. The monkeys and the hornbill were looking for some choice fruits to eat, for the fig fruits were ripe, red, and luscious. The squirrels, for the moment, were only interested in each other.

The tree was large and thrown into folds—it was ten times as thick as a grown man and perhaps ten times as old, too. Its canopy formed an archway across the road. A multitude of roots came down from the branches. Those by the side of the road almost reached the ground and were thick enough for village children to swing on and play. Those over the road, with the constant pruning from vehicular collisions, never made it that far, but never stopped trying either—the roots hanging high over the road were still growing. Framed by this arch-like ceiling of root and branch, the road could be seen in a unique perspective.

The canopy cast a deep, cool shade over the otherwise blistering hot tarmac. Under the tree, the breeze seemed to matter, and to refresh. Out on the road, in the sun, it had seemed only a desiccating gust of hot air. A bit further on, another fig tree had enticed a man to escape the hot afternoon by catching a few winks in its cool shade. I looked around and saw some mango and jamun and tamarind trees too, planted by some blessed soul decades and centuries past. They did not have any fruits now; if they did, there would be people looking to collect any fruits that had fallen to the ground. A farmer had scaled another fig tree to lop the branches as fodder for his livestock; his child stacked them onto a bicycle below. On the road, unmindful of all this, the vehicles roared on back and forth.

I savoured the scene. It would not last long. A few kilometres away, the road was being developed. It had been given an even more important status. The road was being widened, they said. What no one said was that they were felling all the centuries-old ficus trees, even those that would not lie in the path of the wider road. Hundreds had been cut already—their huge dismembered corpses cut into pieces and lying by the hot, dusty roadside. A few sorry Australian acacias and eucalypts were to become their replacements here and there. The road was being made better, they said. What no one said was that they were divesting it of all character and all connection with the surrounding landscape. The rises were flattened, the dips filled, the curves straightened. More vehicles could ply, they said, and they could go faster and spend even less time in this countryside in that incessant march to the city. And the trees? Oh, the trees were a traffic hazard—what if a speeding, swerving vehicle didn't miss the trees? That's a bad situation to be in. Well, soon, there will be none of these trees by the road. Will anyone now, I wonder, still miss them? Is that a bad situation to be in, too?

My thoughts went to another road, another fig tree. That tree's branches had reached across the road, too. On the opposite side, a tree that had earlier locked branches overhead with it had already been cut. Only a row of low shrubs and weeds now ran parallel to the road. Yet, I'd watched a squirrel cross the road through that canopy. He moved through the branches over the road, until he was above the row of shrubs, but there was still a gap of a couple of feet. Tentatively, he moved a bit more, and then, his own weight bent the twig downwards, letting the squirrel virtually step onto the shrubs. At that moment, I had something like a vision, as if the fig tree had assumed a larger persona—a munificent one that, with its long arm, had held and gently placed a little creature safely on the other side. I blinked, but the vision did not go away.

On this banyan, the lead squirrel was not yet ready to give into her partner's attention. Down the trunk of the old banyan she came, hesitated a moment, and then darted across the road. Intent on his mate, the other squirrel followed. An instant later the follower seemed to realize the import of his move, for he saw a car coming at good speed. In mid-run, a sudden reconsideration brought the squirrel to a halt, a quick turn, and a bolt back. The vehicle, alas, had no such doubts on its direction and speed. The squirrel had no chance. The sight of that crushed body, with the eyeballs bursting out of the skull, bore almost no resemblance to the delightful bundle of life of just a few moments ago.

The first squirrel called shrilly. Perhaps she was calling in alarm at the plight of the other, or in relief at her own safe crossing. Maybe she was just announcing her new location to the surroundings that were now one squirrel poorer.

From the banyan, the hornbill took wing. It now seemed a long haul to the next tree and in dipping flight she went her way. After a little while, it was the turn of the troop of monkeys to cross the road and they started descending from the branches. Suddenly, it seemed it was time for me to leave as well—there was some important work to do elsewhere. And besides, who would have believed me if I'd said that I could not bear to see some monkeys cross the road?

# Welcome Back, Warblers

Every year, as the southwest monsoon fades across our land, a sense of restlessness and upheaval brews in the high Himalaya. The grey skies of August transform into the clear blues of September and a developing chill marks the air. The landscape and trees are gathering the colours of autumn; winter is not far. Then, in the high mountains, in ravines with willow and rhododendron, in lichen-encrusted forests of fir and birch, millions of little birds prepare themselves for a great journey.

The birds are so small that they can nestle snugly in the palm of your hand, or even fit into a loosely closed fist. They are most unassuming and drab, dressed in pale greens and humble olives, or in dull browns with scarcely a dash of yellow or orange, sometimes dabbed as pale wing-bars and stripes. They merge so well with the leaves that were they not so active and restless—flitting their wings and calling regularly to announce their presence—it would be hard to even spot them. And yet these wispy little birds, weighing around ten grams, can stake claim to great achievement. Every year, millions of them migrate hundreds to thousands of kilometres—in a matter of days even—flying south from the high Himalaya, the Caucasus, and mountains of Central Asia to winter in the foothills, plains, plateaus, and hill ranges across India. And here, after a lull of many months, when the trees and shrubs are aflutter with lively chirps and twittering song, we know that the leaf warblers are back.

The leaf warblers, as one may guess, have a close association with the foliage of shrubs and trees where they restlessly search for their insect prey. Their restlessness is heightened in the days that precede migration. The birds feed in the foliage as if in a frenzy, to load up for their journey a crucial stock of fuel: a few grams of fat. That hundreds to thousands of kilometres can be efficiently travelled on a few grams of fat is one of the primal wonders of bird migration. Burning fat is more efficient than burning sugars or proteins, and it produces as a byproduct water, another key need for those long hours on the wing.

Although many birds, including small ones like warblers, fly non-stop between their breeding and wintering grounds, the leaf warblers may make brief stopovers en route. Thus warblers heading to the southern tip of India may be recorded on passage at sites in northern India or the Deccan in August–September and then again in April–May on the return journey.

In southern India, the most ubiquitous of the leaf warblers is the greenish warbler. This species has three forms that differ slightly in plumage and call, which ornithologists have in recent years separated into three species. They are found in a range of habitats from urban gardens and plantations to tropical forests, preferring the canopy of trees.

When these warblers leave the subtropical and temperate forests of the Himalaya or the mountains beyond for the tropical deciduous and evergreen forests of the south, it is not just the tree species and habitat they use that changes. They make a fundamental change in their lifestyle. Up north, these warblers live and breed in pairs during the summer, each pair defending its territory from other pairs for its valuable trove of insect food. Yet, when they come south for the winter, the males and females separate—each individual maintains his or her own territory. A female sings and defends a territory from other members of her species, just as a male does. Following the monsoon rains, insect prey are rich enough in the foliage to attract the warblers, but scarce enough to warrant staking out a territory to defend it from the warbler multitude.

The territories the birds defend are not large. A single hectare of tropical forest may pack two to four birds holding territories. By marking individuals with numbered and coloured rings on their feet, ornithologists have shown how the same individuals return to the very same quarter hectare of forest after their long journey every year—a great feat of fidelity and orientation in so small a bird.

When the warblers arrive in our forests and gardens in September and October, they come and settle with song. These are territorial songs staking claim to their all-important grove of trees or little segment of forest. During the first few weeks, the trees are busy with songs and territorial skirmishes as some warblers settle down in their winter turf and others are chased out of it. The songs then give way to simpler, short call notes serving to merely announce their presence (a boon to birdwatchers to detect and identify each species). Then, over a relatively quiet period, the warblers moult into a new set of feathers, as if to greet the new year.

As April arrives carrying the promise of Himalayan spring, the relentless forces of nature and instinct turn the birds northward. Once more, the birds feed briskly to load up on fat. There is a flurry of song, as if in preparation for the territorial battles to be waged shortly on their breeding grounds. And then, one day, the tree where you have watched the warbler for several months is silent, and the bird is gone again.

And yet, when the warbler departs, the bird leaves behind a new awareness. An awareness, stirring deep wonder, and strangely uplifting, that a tree in your garden may be linked to a specific, even if unknown, corner of the Himalaya by one individual bird. A renewed sensibility that your garden and that Himalayan corner, and a range of stopover sites along the warbler's route are all needed to keep alive this tiny linker of worlds. The warbler's journey then seems a brave voyage of survival and connectedness, surmounting artificial boundaries and national differences in a way that transcends our best-intentioned bilateral efforts at co-operation. Softly and unobtrusively, as the bird has done for millennia, the little warbler continues to tie us to other lands and peoples and nations far away.

# Musician of the Monsoon

Mystical whistles float out of the depths of the dark rainforest, through mercurial mist and morning air. The notes are continuous, almost breathless, yet gracefully slow and pleasing. This is no repetitive tune—it is an entrancing melody of ever-varying notes that rise and fall in an unpredictable series of undulating tones. In this forest, the notes announce, lives a flautist of the most unbridled creativity.

Nearby, a rippling stream of cool, clean water—fed by monsoon rains and deep aquifers—courses down a ravine of rounded pebbles and stubborn boulders. The sparkle is now stolen from the water by the towering forest canopy, but the water and its murmuring voice flow unhindered. The fluty whistles continue, too, as the secret songster fills the crystal morning with music.

As the sun climbs and the mist dissipates, the songster comes to light. Perched on a boulder by the stream, he first appears a simple, unremarkable black; only a closer look reveals his sleek, glossy plumage. The bird, unconcerned by this simplicity, sports a proud, erect bearing, chin up and chest out. He bobs smartly to attention, fanning his tail open and closed, up and down. He surveys the stream with a stern eye, like a general scanning his territory. If the bird had heels to click, he would. His military sprightliness is strangely accentuated when sunlight catches his black uniform. Suddenly, the drab black acquires an iridescent purple-blue sheen, and on the forehead and shoulders glisten—like epaulettes of high rank—spangled badges of ultraviolet finery.

Birdwatchers know this bird as the Malabar whistling-thrush, a species found in the Western Ghats and associated hill ranges of India. The name, adequate though it is, does little to describe the landscape of hill forests and streams that is the bird's home. Here be a bird—the name seems to say—thrush-like in appearance, if not in ancestry, for the species is more closely related to Old World flycatchers in the Family Muscicapidae. The birds whistle, for sure, as one can attest, and by christening it Malabar a certain perfunctory attention is paid to their provenance. A more casual tag is sometimes affixed, stemming from an interpretation of their carefree and rambling whistles—the 'whistling schoolboys.' And yet, when you wake up on monsoon mornings to whistling-thrush symphonies, the name seems inadequate, and you wish there was greater tribute to pay. In the great traditions of Hindustani classical music, the Raga Malhar is associated with the rains; among our birds, then, this is surely the Malhar whistling-thrush.

The whistling-thrush has a fondness for flowing waters on hill slopes. There he hunts aquatic snails, frogs, and crabs, staying open to what opportunity may offer, including worms and bird nestlings. Holding the prey firmly in his bill, the thrush batters it lifeless on a rock before consuming it, concluding his predatory bout with a piercing whistle, perhaps, or a dipping flight down the stream in search of more. With the approach of the monsoon, as the streams are recharged with waters, his song acquires a new zest, and the bird begins to breed, even as other bird species in the rainforest are done with their nesting and are out with their young. Pairs build nests in little nooks and crevices along streams, among rocks and cut banks. When forests give way to plantations and rocks to buildings and bridges, the whistling-thrushes, fortunately, are forgiving and may adopt a space under the eaves or a hole in a wall to nest. Yet, the streams and rivers are never far.

As long as the streams are alive, even with a vestige of flowing water, the whistling-thrushes may survive in the ever-changing hillscapes. One may see them in coffee, cardamom, and tea plantations, swamps, and rocky, wet slopes, and hill towns. Along hill roads, they are often seen on culverts, a habit shared with their Himalayan cousin, the blue whistling-thrush. Seeing culvert after culvert with its resident bird, a friend in government service coined an official designation—'culvert-in-charge'.

When streams wane and disappear with human-caused transformation of land and water, and the last vestiges of forest are swept away for human ends, the whistling-thrush numbers decline. Clearly, there is a measure to the bird's forgiveness, too. And where the whistling-thrush has disappeared, the mornings are now bereft of beautiful song.

Today, the whistling-thrush persists in these hills. Amid a litany of ugliness—pesticides and poisons, check-dams and catapults, pebble pillage and boulder burglary, sand mining and dynamite fishing, drainage and diversion—the bird survives and fills the air with songs of rare beauty. As the rivers and streams transform to dreary drains and smothered swamps, as rank grass and sedge and weed take over spaces where verdant moss grew and wild balsams bloomed, the whistling-thrush hangs on to the land like a question hanging in the wind. Will we cherish the whistling-thrush as a marker of our hills and streams? Or will we let it slide, like the dead waters of fouled rivers deflected into our modern reservoirs of stagnation? If we do value these birds and treat their homes and haunts with greater respect, then, perhaps, there will be a different future. One where we can still dip our hands into the cool, clear waters of many a hill stream and raise it to our lips to slake our seemingly insatiable thirst, even as the song of the whistling-thrush quenches an intangible inner human spirit.

I cannot imagine these hills without the whistling-thrush. The day the bird's song fades from these hills would be a day of unbearable pain, as of unbridgeable loss. And when the monsoon arrives to revive the parched hills, with its thunder and clouds and lashing winds and pelting rain, there will be a strange emptiness. For in the hills of the Western Ghats, it is not just the rain, but the voice of the whistling-thrush that sings the music of the monsoon.

# The Caricature Monkey

The road points like an arrow towards the hills. Amid fallow fields and coconut farms, flanked by rows of large tamarind trees, it takes a curve at a little rise. From here, the wide panorama of hills ahead is an inviting blue grey. From the city, the vehicles stream in with loud jumbles of people—people talking and laughing above the din of DVD players or stereo music. They bring food and personal effects, electronic gadgets and favourite toys, in the metal cocoons that carry them and bits of their city. They share an anticipation, almost unconscious, a desire for something different, above the mundane, that might yet shake the city off them. As they draw near, they watch the hills through tinted glasses.

First, there is the mandatory stop at the inevitable dam at the foothills. Here, they will pick up soft drinks and colourful balloons, or eat their fill of cotton candy. Manicured lawns and clichéd blossoms ornament the garden by the dam. For a while, the open expanse of water of the river-that-was piled up against stone and concrete seems somehow refreshing. But soon, it is time to move on.

The road climbs through the forests of the hills. A little distance from the road a troop of bonnet macaques forages in the branches of a *Sterculia* tree, the monkeys picking fruit from crimson pods and stuffing them into their cheek pouches. The distended pouches store their takeaways for a leisurely meal. Adequate it may be, yet the macaques are alert for more food. The macaques in the troop—there is a full twenty of them, young and old—walk and leap across branches with little effort, along their three-dimensional path towards the road—the road that holds easy pickings.

Along the way, two females pick at the young leaves of a tree, pluck remnant ripe figs from a large *Ficus* tree, and stop to groom each other. The females—one with an infant clinging at her breast, nipple to mouth, eyes luminous and glancing sideways—sit in a close huddle. They are friends, perhaps, or relatives sharing a close bond. Close by, a male follows a young female, red-faced and in oestrous, with apparent disinterest—there's a larger scar-faced alpha male sitting on a rock nearby. The troop appears watchful but relaxed. The forest around holds myriad tree species, shade upon shade of green, flushes of colour of leaf and blossom. Now, through the canopy and over the ground, the troop arrives at the browned roadside, and the steel-grey asphalt of the highway.

In the approaching tourist van, the driver is thumbing the incessant horn and slowing down. The vehicle stops by a bridge and a little cascade of water from the forested slopes. A board proclaims this place as Monkey Falls, linking the monkey to the water, and obliterating, by its sheer superficiality, a deeper appreciation of all that it entails.

The macaques, meanwhile, on the sidewalls and the road, watch the people emerging from the van. The passengers are weary: the men stretch, and rub their faces; the women straighten tousled hair and crumpled sarees; the child is hungry, clamouring for food. Reaching into her handbag, one of the women gives the child a packet of biscuits. Yelling over the noise of the stereo, she asks him why he needs to shout for his snack. The men are arguing, too.

A few macaques converge upon the vehicle. The mood of the people changes instantly. Wide-eyed, they point: Look, monkeys here! One of the macaques stands on his hind legs, arms outstretched, peering into the vehicle, attention caught by a handbag slung on a woman's shoulder. The people watch two macaques that are still on a tree—boisterous youngsters, pulling at each other's tails, playing tag on the branches, like the monkeys of caricature. The child is absorbed in his biscuit, locked eye-to-eye with a macaque on the road. The latter approaches, nervously, looking around at the other people, the other monkeys, and the biscuit packet.

And then, in a game-changing instant, the little boy flicks a biscuit to the monkey.

The macaque rushes to the biscuit, but even as he darts forward, even as he picks it up with one hand, two other macaques rush at him. There is a tussle, snarls and grimaces, as one drops and another picks and runs aside with the biscuit, stuffing it into his mouth to evade another challenge. The boy is delighted; he is clapping his hands and laughing. Smiles dawn on the other human faces.

With another flick, the boy tosses the remaining biscuits. There is chaos on the road. Now, there are more monkeys rushing for the snack, posturing and flashing their signs of dominance or submission, intelligible only to themselves. They dart this way and that, narrowly avoiding speeding traffic. They're merely monkeying around as far as the people are concerned—*if*, indeed, they are concerned. The people are all laughing now, their weariness gone, their tension evaporated. The macaques are tense, having to face the momentary attrition of well-worn relationships.

With just that flick of wrist, the boy has transferred the social tension from his family to the family of the macaques.

A while later, refreshed from the encounter, cleansed of city grime in the cascade, the family boards the van. Spuming black exhaust from the tailpipe, they leave behind the squabbling and scattered troop, along with their soap and foil and plastic.

One imagines them later, in the city, being asked by a friend how their trip was. Their answer will be heartfelt, honest, 'Wonderful! It's like we left all our troubles back in the hills.'

## Postscript

If you like monkeys, don't ever feed them. If you see others doing it, stop them politely. Feeding monkeys affects their health, stress levels, and social relationships. Monkeys habituated to human foods frequent roadsides, where they are often killed by speeding vehicles. Or they enter houses and shops where they may be hurt badly if people retaliate. The result: monkeys with broken bones, festering wounds, lost limbs. Feeding monkeys—and littering—creates a cascade of problems, including the so-called monkey menace. It's really the menace of uninformed tourists and littering that needs to be controlled. Instead of literally killing monkeys with kindness, learn to enjoy them in their various habitats—urban parks and gardens, countryside banyans and village groves, wildlife reserves and deep forest—eating natural foods from the trees.

# Turning the Turtle

At the edge of the foaming sea, behind the spent waves on the beach, shapes materialize in the night. These ancient shapes have appeared countless times over millions of years. They slowly pulse towards the shore, their domed shells barely breaking the surface. Under a waning gibbous moon, scaly flippers strike the sand. Unblinking eyes, on beaked heads stretching from wrinkled necks, survey the beach. Like time travellers from some primeval epoch, a great wave of olive ridley sea turtles lands.

The turtles are all pregnant. Impelled by the timeless purpose of reproduction, they seek their ancient tie with land, where they will leave their eggs. Beached like soldiers off a flotilla, domed and armoured, the turtles advance upon the shore. Though they seem to move as a group, each is driven by her individual purpose: crawl up the beach past the high tide line, find the right spot for her nest.

One of the turtles is ahead, leading, though not leader of, the group. That she is not the first tonight is marked clearly on the sand. A row of scythe-like scuff marks leads from and loops back to the sea: a sea turtle had emerged and returned. Perhaps she found no suitable spot. Perhaps she decided to return later. The time is right, now, and the lead sea turtle trudges ahead. Walking with difficulty on sand, she finds a spot and begins digging with her flippers. She gives up after a while, moves ahead, and digs again. This time the place seems right, and she locks into old instinctive habit. She digs the jug-shaped nest with her hind flippers, smoothens the sides, and lays over a hundred eggs, one by one, over an hour. Job done, she pats the nest down using her flippers and her hard plastron undershell, weighted by her body. She swishes the sand around with her flippers to hide the nest. Her eyes brim with salty tears. Her reptilian visage is inscrutable. It is time to head back.

Normally, she would turn instinctively to the brighter horizon over the ocean, but now there are lights inland, and the land horizon is beguilingly bright, too. She pauses in inner turmoil. Finally she turns, disturbed and befuddled, heading towards town and village, dogs and roads—towards certain death.

Fortunately for her, help is at hand tonight. The kind arms of volunteers coax her, guide her with friendly lights, even heave her 40-kilogram bulk, turning the turtle around. She is brought to the edge of the surf, where she effortlessly steps into the waters, heading east, into the ocean, where dawn will soon break the darkness of night.

Seven weeks later, from the nests that have escaped poachers and sand mining and beach erosion, little sea turtles emerge by night into their new world. Along with other early terrors of survival—dogs and nets and litter and kites and poachers—the baby turtles, too, face the serious quandary of light.

Millions of years of evolution have honed their perception of the kind and quality of light over the sea that will lead them back to it. But that guiding light is now masked by the shimmer of lights from land. Farther up the shore, the lights of the village-become-town and the town-become-city beam. The horizon has a dull artificial glow. The clouds over the land of the people gleam in reflected light.

The moon is behind clouds. The hatchlings on the beach are inch-high specks on the dunes under the great sweep of sky, now blurred by unfamiliar lights. They are all at sea on land. On their very first steps, the hatchlings face a higher risk than their mothers of turning to the wrong horizon. A turning that is usually fatal.

Lights that show the way for one species may be the death of another. Now, the sea turtle—that great and ancient navigator of the oceans and its currents—is disoriented by the new current that runs at the flip of a switch, lighting up the horizon.

How difficult is it to give sea turtle mothers and babies a better chance at life on the ancient shores? Shores that we have only recently obtained the privilege to share with them. How difficult is it to dim or obscure the seaward lights for a few weeks during nesting season? Will individuals living along the coastline oblige and turn down their lights? Or does this need combined effort as a human society? Together, we could designate lights-out periods, night closures for vehicles on beach roads, and create shades or allow appropriate natural vegetation along the seaward side of towns and cities to obscure our lights. Perhaps such collective purpose can only emerge from each person's awareness of the problem and motivation towards its solution. The annual tide of turtles on the beach offers a parable for us, for their great *en masse* nesting phenomenon is still driven by deep and motivated individual purpose.

Sometime later, farther down the beach, new shapes emerge from the sea, boats returning with the day's catch. The fishers draw their catamarans and skiffs to higher ground and moor them with stout ropes tied to stakes driven into the beach. The nets are hauled and tossed, then the fishers walk to their huts. Where their feet strike the moist sand along the edge of the beach, thousands of tiny lights spring up briefly and fade away. These are the lights of sea sparkles—tiny one-celled creatures, dinoflagellates—that bring biological luminescence to the seas. They are named *Noctiluca scintillans*, the night lights that scintillate. In the night, then, the subtle lights in the footsteps of the fishers flash a different message. On the shores of our planet, the human footprint can be beautiful, too.

# The Deaths of Osama

*May 2011*. Osama bin Laden was killed in Abbottabad in Pakistan, say the news reports. How could it be? He's been killed twice in India already! Once in 2006 and again in 2008. Osama's first death was in December 2006 in a tea estate in Assam in northeast India at the hands of a hunter, a hired gun. As if that wasn't enough, he was killed again in May 2008, in Jharkhand, at the hands of an empowered mob sanctioned by the government—the Forest Department and the Police. The second death wasn't easy. It took twenty bullets to silence him this time, but going by the news, even that didn't work, after all.

The painful truth—literally—is that the first two deaths of Osama referred not to the terrorist mastermind and leader of al-Qaeda but to two separate individual Asian elephants, a species with the contrasting reputation of being the gentle giants of our forests. These individuals were named for a feared human on the most-wanted list of a distant superpower. They were labelled serial killers and raging bulls, as rogues and as terrorizers. And yet, when people came to see their prostrate corpses, they placed flowers on the bodies even as many asked whether the right animal was killed or if it was yet another innocent elephant victim.

Now, as before, it is open season on the Asian elephant. The character of the elephant is on public display in the media, interpreted to us by all manner of people. There are journalists and filmmakers, naturalists and scientists, politicians and hunters, mahouts and zookeepers, temple priests and elephant 'owners.' Everyone knows, or seems to know, the elephant.

From the forests come stories of great tuskers and makhnas and their roving lives of ranging and musth and disproportionate peril. There are tales of tenderness among mothers and calves, and of itinerant family herds led by rugged matriarchs over familiar routes across vast and varied landscapes. The stories speak of communication by unheard sounds, unfelt vibrations, and undetected pheromones, and of elephant memory and cognition. They speak of individuals that are self-aware and social, that can be doting or depressed, loving and forgiving. We learn of stable yet sensitive societies, and begin to know sentient and intelligent individuals. These stories proclaim an understanding of elephants and elephant cultures that is barely beginning to grow.

From crop fields and human habitation come tales of rogues and raiders, marauders and mayhem. There is an image of a lone tusker, wilful and vicious, or of a huge herd on a rampage of raiding and pillage. The elephant tramples, the tusker gores, snuffing out the lives of the people whose path has converged tragically with the elephant's own path. The elephants are not on old routes anymore; they are said to be straying herds, individuals on trespass. The words say it all. Each elephant and its action, known or unknown, is judged and placed within the ambit of a common belief. Pinioned by belief and judgement, claims and media reports, the elephants, unaware, must await retaliation—retaliation and pain at the hands of the self-aware, social, sentient, and intelligent human.

What does it take to cause pain to an elephant? That great beast, the ponderous pachyderm standing tall on sensitive feet, but still dwarfed by the immensity our worldly landscape and its perpetual perils. Will it take a land mine, planted in a contested forest by warring people, that tears away a leg? Or the final body blow delivered by a speeding train that brooks no obstruction to its path? Will it be a flaming torch flung at the elephant by an irate farmer whose ire has overwhelmed his tolerance? Or perhaps the sting of an electric wire strung deliberately across land that someone now claims as theirs, and only theirs? It could come, too, from a bullet as it bursts its way into the elephant's heart or brain—from the poacher who wants only the animal's teeth. Any of this may bring pain, and yet, the deepest pain to an elephant may come from the loss of one of their own. A pain we barely sense, far less understand, as we watch the elephant visit and caress the bones of the dead.

We have arrived at a grim moment, where we are asked to rethink our tolerance and veneration towards the elephant, a relation spanning millennia. We are asked to find ways to deal with the elephant, as we would with a troublesome pest, spawned by an interaction between people and landscapes gone awry.

And so, the ecologists, wildlife scientists, forest managers, judges, and administrators are coming up with their answers. Protect the reserves and the movement corridors, they say, and the elephants will find their way through 'our' land. Erect this kind of barrier, not that, and here, not there, this way, not that, they say, and the elephants can be kept at bay. Compensate the people for their loss justly and quickly, for everything today has a price, and perhaps people's love can be bought, too. Understand the elephant, they say, strapping a collar on her neck or probing her DNA and her habits, for this will inform us, and information is power. Capture and relocate the elephant, or kill him where he lives, say the pragmatists, for we can then evade the elephant as easily as we evade seeing the brutality in ourselves. We can even mark our broken tolerance by filling elephant camps with broken elephants. By and large, these methods and solutions have one character. They treat the elephant as an object, a commodity even, to be valued or traded, upon whom, in the words of a leading student of elephant behaviour, G.A. Bradshaw, 'things and people act to produce a programmed response'. As J.M. Coetzee writes, in *The Lives of Animals*:

> The heart is the seat of a faculty, *sympathy*, that allows us to share at times the being of another. Sympathy has everything to do with the subject and little to do with the object .... There are people who have the capacity to imagine themselves as someone else, there are people who have no such capacity (when the lack is extreme, we call them psychopaths), and there are people who have the capacity but choose not to exercise it.

Aren't we missing something? Will it not help to bring in an element of empathy to elephant individuals, societies, and cultures? Should we not aspire to a higher understanding of the psychology of elephants, whose selfhood, rights, and emotions should matter to us but are relegated to the dustbin of false anthropomorphism or misguided pragmatism? Have pragmatism and experience really provided solutions, or has our direction suffered from a shallowness of understanding? Manuals and action plans have been written on how to understand and stave off conflict with the elephant-object in India. Why is so little said about the elephant-being with whom we share so much of our true nature? As Bradshaw notes in her fascinating *Elephants on the Edge: What Animals Teach Us about Humanity*:

> Elephants are merely mirroring the circumstances in which they have come to live.... Under such conditions, human-elephant conflict (HEC) takes on a very different meaning ... issues surrounding elephants are 'not about the animals'. Rather, they are about humans: human-elephant conflict revolves around questions of social justice and human introspection. Much like other cultures that have refused to be absorbed by colonialism, elephants are struggling to survive as an intact society, to retain their elephant-ness, and to resist becoming what modern humanity has tried to make of them—passive objects in zoos, circuses, and safari rides, romantic decorations dotting the landscape for eager eyes peering from Land Rovers, or data to tantalize our minds and stock the bank of knowledge. Elephants are, as Archbishop Desmond Tutu wrote about black South Africans living under apartheid, simply asking to live in the land of their birth, where their dignity is acknowledged and respected.

One wonders what the future holds for the human-elephant relationship, a relationship between two intelligent, sentient species. Will it remain a perception of elephants as objects of conflict seen through the coin of economics and the lens of science, when it could lead to coexistence if passed through the prism of humanity?

# An Apology to the Iyerpadi Gentleman

He was standing behind the building when we first saw him. Dignified and stately, yet aware and watchful, for he had some business of his own. We had come to see him unannounced, but he held no wish to meet us.

We waited on the road, watching the traffic go by. Behind the building, we saw him move. He was a young tusker, his left tusk slightly curved forward while the right pointed, asymmetrically, straight down. With dignity and grace, and with all senses alert, he walked down towards the road.

The road was not very busy by the standards of a large city, but for the little hill town of Valparai, here in the Anamalai Hills in southern India, it was arterial. The tusker was in a little plantation of eucalyptus above the road. Below the road were the Iyerpadi tea estates, the Iyerpadi factory, houses of estate managers and workers, a swamp-stream, and beyond that a patch of rainforest close to the Anamalai Tiger Reserve. In the landscape of plantations and rainforest patches on the Valparai plateau, elephants often come for water and forage, or to pass from one part of their range to another, as they have done for long.

Besides us, some people from the estate were watching the elephant. On the road, vehicles plied back and forth and a few people went walking past, hardly 50 metres from the elephant. The elephant could see, or sense, all of us; with his trunk up, he monitored our scent and presence. Nervous, he let out a short blast of a trumpet. Yet, it did not seem that he trumpeted from anger; it seemed a brief warning to get us off his path, and let him pass.

While others watched, I moved closer for a better look and a photograph. A little skid-trail came down the slope to the road and below the road a path led through the tea estate. The elephant seemed to be moving that way. I stood right there, at that intersection, and sure enough he emerged, less than 30 metres away.

He looked grand in the evening light. I was awed and clicked away, trying to get a photograph that would record his grandeur. And yet—I stood right in his path. He stopped, alert, and looked straight at me.

Divya and the others with me who were watching from some distance away urged me to move. But I stood as if transfixed. Perhaps it was that wholly unnecessary photograph that kept me. Or a falsely superior rationalization: This is not the way he should go. There are houses and people down there. Maybe if I block this path, he'll go around taking what I think is a better route.

He gave me a few moments to reconsider my stupid, irrational decision. As I did not budge, he did. He turned away gently, to swing down, taking a more inconvenient, steeper, rocky slope. A group of women coming to collect firewood was casually walking towards him, thinking he was one of the domesticated elephants being used in tree-felling operations. We convinced them that he was wild and urged them not to go in that direction. On the steeper slope, the tusker turned back and swung down towards the same route he would have more easily taken had I not stood in his way. He kept moving, now forced to cross the highway a little further ahead of where he had intended.

We tried to halt the traffic on both sides for a few minutes to let him cross. It was scarcely necessary, for he knew how to deal with traffic and crossed the road without a hitch, without disturbing anyone in vehicles or on foot.

He was heading in the general direction of the houses and the factory and anyone watching him, who did not understand the elephant in him, could have thought this spelt trouble. The tusker wanted no trouble, however; he just wanted to be on his way. And it was wonderful to watch how gracefully he moved, carefully avoiding the proximity of the houses. He needed to go that way, because beyond the houses and factory in Iyerpadi was a rainforest fragment and the Anamalai Tiger Reserve, and, perhaps, respite from others like us.

He quickened his step. He walked down. He swung away from the houses. He avoided a car coming up on an estate road, though he was close enough that the passengers may have got a scare. He turned down the valley, past the temple, into the swamp, and reached a path that would take him, without crossing any other road or colony, towards the forest patch.

An excited bunch of kids appeared from the vicinity of the houses and tried to follow the elephant. We dissuaded the children; with a little persuasion, they stood to watch him from a safe distance. We'd come to the colony to inform the people to watch out for this tusker on the move, but, again, it was not necessary. The people had seen him and, moreover, the tusker had no interest in the houses. He really knew where he was going.

I will remember him as a gentleman of Iyerpadi and I will remember my foolishness of that evening. This is my apology to this gracious and peaceful elephant. I am sorry I stood in your path. I am sorry for thinking I knew better than you.

# An Enduring Relevance

*16 June 2014.* It is tough for a single publication or its author to have an impact across nations, cultures, genres, and disciplines. It is tougher still for such a publication to spark a social movement—one on the global stage, at that—to rekindle human values and awareness, and create new mandates for action. And toughest of all is when the author is a woman, a scientist, who must overcome the prejudices of her time—of gender, of notions of progress, of the omnipotence of untrammelled industry—to articulate a clear-eyed, renewed vision of a better world, a cleaner environment, where people do not merely live, but flourish.

If I had to pick one such exemplary work from the environmental canon, it would be the one that burst on the scene on this day, 16 June, over 50 years ago, in the United States of America, and then swiftly encompassed, in its scope and sweep, the rest of the world. The book, *Silent Spring*, and its author, marine biologist Rachel Carson, are widely credited to be the sparks that lit the fire of the global environmental movement. Carson, whose 107th birth anniversary came and passed quietly on 27 May 2014, with little fanfare other than a commemorative Google Doodle, died 50 years ago after a battle with breast cancer. Why should we bother to remember Rachel Carson and *Silent Spring*? What could a woman, a book, from over five decades ago have to do with the enormously changed world we live in today? Yet, over the last few weeks, during fieldwork and travels in India's northeast and the Western Ghats mountains, I thought frequently of Rachel Carson and her prescient words in *Silent Spring*.

*27 February 2014, Chawrpialtlang peak, Dampa Tiger Reserve, Mizoram.* Slicing through the air over crackling-dry grass, a black-tipped arrow streaks past, plunges down the sheer cliffs, swerves around the mountain, and is gone. For one rushing moment, the ripped air appears to shimmer, as if in sudden clarity, then closes in the fleeting wake of the bird. A peregrine falcon. Windswept and breathless, I stand on the peak and think of Rachel Carson. For it was in *Silent Spring* that she described—and I learnt—how the chemical pesticide, DDT, sprayed or dusted into the environment, entered water and soil and animal tissue as a persistent organic pollutant, and travelled up the food chain, accumulating from pest to predator to top predator, into birds like peregrine falcons and bald eagles, thinning their egg shells, making the brood crumble instead of hatch in the nest, bringing down populations, endangering the species itself. Only when awareness of this issue soared after the publication of

*Silent Spring* and concerted efforts, including a DDT ban, were made did raptor populations recover, so that the birds could wing and scythe through the air again.

*20 March 2014, Mamit District, Mizoram.* On the outer wall of bamboo hut after hut, in village after village, in one of the most remote and malaria-prone corners of India, I see inscribed in chalk: DDT 15/03/14. The date varied a little from village to village, but it took me only a moment to realize that this was just a marker that each of those huts, the homes of Mizo and Riang tribal peoples of the state, had just been sprayed with DDT. DDT, the one chemical for which Rachel Carson's work is most known for and most frequently and unjustly vilified. Carson, using a growing body of research, highlighted the environmental and human health consequences of excessive DDT use in *Silent Spring*. The book, along with the growing tide of awareness led to a ban on DDT and consequently, or so the accusation goes, it became unavailable for use in malaria control and led to the death of millions.

In reality, DDT was banned for use only in agriculture and unrestricted aerial spraying. It is readily available, and continues to be used, for malaria control across the world. And here was evidence, decades later in Mizoram, that this accusation is untrue. In Mizoram, as in other states in India, the government has an Indoor Residual Spray programme of DDT, usually twice a year, coupled with distribution of deltamethrin insecticide-impregnated mosquito nets. DDT was banned for agricultural use in India in 1989, but even this was not a complete ban, as it carried a rider allowing the use of DDT under 'very special circumstances' for plant protection from pests, under the supervision of the state or central government. In 2006, an Indian government order permitted the use of up to 10,000 tons of DDT annually for public health and vector control measures. In Mizoram, where I knew malaria was still frequent (it knocked me down for two weeks during my fieldwork here in 1994), there were more pertinent issues than these false debates and the vilification of Rachel Carson. Loss of effectiveness of DDT from overuse and overdependence, the need for better public healthcare facilities, and the fact that more than 90% of the local people prefer insecticide-impregnated mosquito nets over indoor DDT sprays, all seem more important issues to be discussed.

*23 May 2014, coffee and tea estates near Sakleshpur, Western Ghats.* As our car speeds past the gate of the coffee estate, I think back to my previous visit to that plantation, in 2011, when I was doing a diagnostic audit for a company that was planning to apply for Rainforest Alliance certification of their coffee production. There, beside a small pond, a group of workers had been preparing a pesticide concoction for spraying on the coffee bushes. In the group, helping mix and spray the chemical on coffee bushes, without any protective equipment to cover her face or exposed hands, was a 12-year-old. Even as she was exposed to the chemical, the pesticide tub overflowed and spilled into the pond. Decades after *Silent Spring,* and despite knowing the effects of pesticide pollution on the natural environment and learning more and more about how pesticide exposure affects human health, it is a pity that in many of our plantations and agricultural fields, so little is done to reduce or prevent pollution, to minimize or avoid exposure to agrochemicals.

Later, in a tea estate, I listen to a manager describe how their chemical sprays had failed to control a pest, the red spider mite, because, he said, the sprays killed the natural predators of the mite such as ladybird beetles. Again, I recall how in *Silent Spring* Rachel Carson had explained how insecticides had the counterproductive effect of increasing spider mite infestation: by not affecting them directly, by killing instead mite predators like 'ladybugs,' and by scattering mite colonies that now focused on increasing their reproductive output as they had no need to invest in defence against predators because the people with the chemicals were inadvertently doing this job on their behalf. I suggest to the manager, like Carson did to her readers, that perhaps the best way ahead is to change cultivation practices, foster more biological diversity in the farm landscape, and reduce their reliance on agrochemicals. He nods, but I am not sure he is ready, as yet, to agree.

*26 May 2014, Highway to Valparai, Anamalai Hills.* All along the highway, the vegetation on the sides of the road lies slashed. Beautiful ferns, orchids, wild balsams, and a number of wildflowers that lent grace and beauty to the road, now lay withering on the tarmac, crushed under the spinning wheels of speeding vehicles. The Highways Department had been 'cleaning' the roadside again and scraping the soil, leaving brown strips beside grey tarmac and concrete. Soon, the exposed earth would be taken up by invasive alien weeds, changing the roadside aesthetic from the lush green of small native plants and wildflowers to dour greys and browns and weeds. Again, Rachel Carson's words in *Silent Spring* came to mind, for she wrote also about the beauty of wildflowers along the roads, criticizing 'the disfigurement of once beautiful roadsides by chemical sprays' and 'the senseless destruction that is going on in the name of roadside brush control throughout the nation.'

Whether it was wildfowl or wildflowers, Rachel Carson's insistence that scientific understanding of the environment should integrate ethical and aesthetic values struck a chord with readers. *Silent Spring* did not merely inform them, it *affected* them, and spurred them to act, thus catalysing the birth of a movement.

The environmental movement, as philosopher Arne Naess once remarked, was one of the three great movements that marked the twentieth century; the others being the ones for world peace and social justice. Of the three, the ecological or environmental movement is relatively nascent. One can trace the roots of environmentalism, at least in its modern form, to early concerns over nature conservation and vanishing species, but it was really in the latter half of the last century that the movement really took off.

In the aftermath of World War II, with the development and testing of atomic weapons, concern over the perils of nuclear war and radioactive fallout was widespread. Still, there remained unmitigated optimism over the promise of new and powerful technologies. At the same time, rising pollution of air and water following industrialization and consequent effects on human health spurred early efforts to curb pollution beginning in the 1950s, culminating in laws enacted over the ensuing years and decades in various countries. In the 1960s, the great phase of dam-building was also in full steam. As the environmental historian J.R. McNeill recounts, on average one dam was built per day around the world during that decade. The construction of dams and the displacement of thousands of people by reservoirs also brought growing awareness of the alteration of entire landscapes by human action, and of its harmful impacts on the environment and livelihoods of people living in the catchment area and downstream.

Still, this was a period when the industrial juggernaut rolled on, backed by a specific vision of development based on technology and large, so-called infrastructure projects. It was a period, in India and elsewhere, when impacts on environment or the lives, lands, and livelihoods of local peoples could be brushed aside on the basis of a grandiose, little-questioned development trajectory. Besides, India and the other countries stood at the cusp of a major transformation of agriculture into intensive cultivation dependent on a slew of chemical fertilizers and pesticides: the Green Revolution.

It is in this context that we must view the publication of *Silent Spring*, first serialized in *The New Yorker* magazine from 16 June 1962 onwards, and later published as a book by Houghton Mifflin on 27 September of that year. The book burst on the scene with a telling and convincing account, based on scientific evidence, of the perils that chemicals used as pesticides and fertilizers brought to human health and the environment. Carson, a skilled writer, explained in clear but compelling detail the various kinds of chemical poisons used in agriculture and pest control, such as DDT, chlordane, and lindane, organophosphates, and carbamates. With care and clarity, she collated research findings published in scientific papers, recorded personal experiences of people around the US, and described the effects of the chemicals on human health, their persistence in the environment, and build-up (bio-accumulation) over time in the bodies of people and wildlife. She explained how pests developed resistance to the chemicals, how that ultimately led to resurgence of pests, and to a vicious cycle of more potent poisons being created, among other things.

> The current vogue for poisons has failed utterly to take into account these most fundamental considerations. As crude a weapon as the cave man's club, the chemical barrage has been hurled against the fabric of life—a fabric on the one hand delicate and destructible, on the other miraculously tough and resilient, and capable of striking back in unexpected ways.

By ignoring ecology, the agrochemical industry appeared poised to fail in finding long-term solutions. Rachel Carson did not stop with careful explanation and evocative descriptions of the problem of increasing dependence on chemicals. She went further and described a way forward to sustain productive agriculture without recourse to the 'chemical barrage'.

> A truly extraordinary variety of alternatives to the chemical control of insects is available.... All have this in common: they are biological solutions, based on an understanding of the living organisms they seek to control, and of the whole fabric of life to which these organisms belong. Specialists representing various areas of the vast field of biology are already contributing—entomologists, pathologists, geneticists, physiologists, biochemists, ecologists—all pouring their knowledge and their creative aspirations into the formation of a new science of biotic controls.

There were several reasons why *Silent Spring* was so effective. Rachel Carson drew upon her earlier experience as a biologist with the US Fish and Wildlife Service, where she served as an editor in the Division of Information, reading scientific publications and transforming them into readable and informative articles for citizens. Today, she would be called a leading science communicator in biology and the environmental sciences. What was remarkable about her writing was that even as she explained science to the citizen, she did not flinch from simultaneously interlacing into her writing moral values and the ethical consequences of environmental harm, which she was convinced was equally significant to her readers.

> Why should we tolerate a diet of weak poisons, a home in insipid surroundings, a circle of acquaintances who are not quite our enemies, the noise of motors with just enough relief to prevent insanity? Who would want to live in a world which is just not quite fatal?

Rachel Carson was a dedicated writer. She had always wanted to be a writer since her early childhood. When she joined the Pennsylvania College for Women (later Chatham College) in 1925 as an 18-year old, she enrolled for an English major, until a biology course in her junior year reawakened her 'sense of wonder' for nature, another fascination since childhood. Later, she obtained her Masters degree in zoology from Johns Hopkins University, Baltimore, following which she taught zoology in Maryland and worked at the famous Woods Hole Marine Biological Laboratory, Massachusetts. Her biological knowledge as a trained scientist, her field experience as a naturalist and keen observer of nature, and her literary talent came together as a potent combination in her books.

Although Rachel Carson is perhaps most known for *Silent Spring*, she wrote other books including a trilogy on the sea and marine life, *Under the Sea-Wind*, *The Sea around Us*, and *The Edge of the Sea*; a book for children titled *The Sense of Wonder*; and a number of magazine articles. She won the National Book Award in 1952 for *The Sea around Us*; it remained on the *New York Times* Best Sellers list for 86 weeks.

The success of Rachel Carson's books such as *The Sea around Us* and *Silent Spring* was at least partly from the way she managed to meld scholarship and literary talent. As she said in her acceptance speech for the National Book Award:

> The aim of science is to discover and illuminate truth. And that, I take it, is the aim of literature, whether biography or history or fiction. It seems to me, then, that there can be no separate literature of science.
>
> ... If there is poetry in my book about the sea, it is not because I deliberately put it there, but because no one could write truthfully about the sea and leave out the poetry.

Still, there was more to *Silent Spring* than just scientific rectitude or literary flair. Carson recorded and used many case studies and personal experiences of people who had witnessed the effects of aerial spraying and pesticide overuse. The *Silent Spring* metaphor itself, referring to a spring that goes silent as songbirds decline and disappear because of pesticide use, was inspired by a letter from a friend who noted dead birds lying around her house after aerial spraying of pesticides in her area, and who now wanted it to stop. Taken as a synecdoche, it suggested that people were sensitive to environmental destruction and it had reached a point where they had had enough.

*Silent Spring* and its author were (as one would expect even today) attacked by government agricultural scientists and companies with high stakes in the agrochemical industry such as Velsicol, a major manufacturer of DDT, and Monsanto. Velsicol threatened to sue Houghton Mifflin and *The New Yorker*. Detractors and vested interests made personal attacks on Rachel Carson, asking why 'a spinster was so worried by genetics,' and disparaged her as hysterical, emotional, unfair, one-sided, and as given to inaccurate outbursts. But, ultimately, the science behind *Silent Spring* withstood public scrutiny, including a congressional hearing where the author stood calm and dignified with her research, credentials, and explications, and the book, instead of being pulped as her opponents may have wished, went on to sell millions of copies and make history.

The reactions and desire for change that the book triggered influenced environmental legislation and policies worldwide. The years that followed the book's publication saw the first Earth Day celebration and the formation of US Environment Protection Agency in 1970, the gathering of representatives from 113 countries at the United Nations Conference on the Human Environment at Stockholm in 1972, the enforcement of a ban on DDT in 1972, and other efforts around the world overtly inspired or tangentially influenced by *Silent Spring*.

In that period, India, too, made several legislative and policy efforts, as the country enacted the Insecticide Act in 1973, laws to prevent water and air pollution and protect forests and wildlife in the 1970s and 1980s, and the Environment Protection Act in 1980 that also created the State and Central Pollution Control Boards and other authorities with environmental mandates.

The appearance of *Silent Spring* was one of the defining moments in the history of environmentalism, one that would irrevocably shake the complacency and complicity of state and industry in environmental harm. Today, one may quibble over the details of *Silent Spring*, over what the author chose to write about—such as her focus on biological control, a measure not without its own problems—or over how she wrote about it. But what one must acknowledge is that much of what Rachel Carson wrote about, and the scientific, literary, and moral clarity she brought to it, remains relevant to this day. From Maryland to Mizoram, then as now, the problems she described and the solutions she offered remain valid, apposite, and vital. In that respect, Rachel Carson and *Silent Spring* continue to be of enduring relevance.

# River Reverie

The river flows on, but sluggishly. Its surface is calm and smooth. It turns a bend at a clump of bamboo, gently passes a grove of coconut, and now drifts along with scarcely a murmur. It is wide, too. The engineers had asked for unspeakable amounts of concrete and money to build a bridge across, and when that went into disrepair with age and neglect, they had demanded even more to build another.

The only signs that the river is flowing are the little waves lapping at the pillars of the bridge and at the rocks on the riverbed, and the slow movement of a few floating leaves and twigs. On a little island in the river grows a clump of screw pine. A cormorant is perching on it, wings held open to dry in the sun. Near the banks, slim and gangly water skaters briskly glide back and forth on the surface, even as energetic dragonflies zip around overhead. They somehow accentuate the river's stillness. And yet, an enormous volume of water flows past here every minute.

A white flock of river terns appears. The terns energetically flap their pointed wings but mill around in an effort to go slow with the flow. They swoop and pick off the surface of the river small silvery fish floating strangely immobile on their sides. It is easy work, for the fish are already dead. Dozens of dead fish follow, sprinkled and sparkling on the river, killed by poison or by the shock of a dynamite blast upriver. Some feed the terns, others drift here and there and below the culverts and into the nearby fields.

The waters have travelled far to get here. Blown by winds from across the ocean, meeting the great escarpment of the Western Ghats, rising as vapours and clouds, and bringing wafting mists and torrential rains, they drenched the slopes of the mountains a hundred miles away. Not all the rain travelled the ocean, though, much had arisen from the forest itself, ascending through millions of roots and stems and transpiring through billions of leaves and leaflets. The forests pump hundreds of thousands of litres of water into the air, and the air returns some of it, falling as rain, condensing as dew.

In the dense forest, the rain drips from the tree tops, down onto more leaves—leaves with pointed tips that gently drip the water down towards the soil. The water also courses down the tree trunks in shining sheets, reviving verdant moss and epiphytes, gathering bits of bark and leafy debris on its way to the soil. The soil itself—that quiet player in the catena of water supply and transport—is almost unseen. A thick blanket of leaf litter and fallen twigs and branches covers it, holding it in place, along with an incredible meshwork of fine roots. Even as the fungi, bacteria, beetles, and termites eat into the litter blanket, the plants top it up with new, spent material. The litter blanket is crucial for the soil.

Some water flows overland, much sinks in, sponged by the leaf litter and soil. Below the surface, the water travels through pipes and aquifers far and wide, recharging ground waters, emerging as springs, and draining into streams feeding the wide river.

The clear waters from the forest join other waters. Waters that gather the dust and carry the soil from the road-scars and the mine-wounds on the hill slopes. Waters deadened by passage through dams and reservoirs, through stagnant pools and ponds with hyacinth and algae. Waters carrying earth from furrowed and exposed soils under alien plantations of acacia and eucalyptus and from forests whose litter blankets are harvested to enrich nearby fields with nutriment. Waters course in with the wastes of villages, towns, and cities, the effluents of factories, and the oil and fuel spilled from lorries washed on the banks. The lorries shine after the wash and a glistening film of oil glides away on the river surface. Outside the paper factory close by, more lorries line up with their loads of bamboo taken from the forests.

The river passes a rice mill. The mill faces away from the river, with a neat garden in front and a mound of waste dumped at the back, on the banks. Farther on, there are houses, too, with their backs to the river. Sewage and farm run-off, effluent and debris, pesticide and poison, the river accepts all impartially, dilutes it, and covers the ugliness under a shimmering surface of quiet beauty. The flow of the river goes one way, our vision of the river remains perpendicular.

The story of the river seems so familiar. I can almost hear my school teacher, her voice emerging shrill and powerful from her corded throat: The river gives us water for irrigation, drinking, washing, bathing, navigation, and power. It provides us fish and fertile plains, reeds and recreation.

But, does the river really give to us all this or do we just take it? And what do we give back, if anything?

We take the water but turn away from the forest-covered watershed that could bring more. We exploit the fish and the fertility but turn in our sewage and our wastes. We derive inspiration from its serenity, its space, and its glorious flow, but scarcely miss a chance to kill that very flow with a dam or barrage, a realignment or link, or a straightening of a curve with dynamite, concrete, and metal. We then proudly gaze upon the river thus engineered and canalized—only, the river is there no more.

Over the water, the river terns wing ahead. In handsome grey and white, with shiny black caps, red legs, and yellow beaks, they contrast sharply with the turbid green-brown waters. In the evening light, to the gentle drone of insects, they fly purposefully forward, their heads pointed down scanning the river. The birds make effortless progress even when their vision is perpendicular to the river. If our vision remains perpendicular to the direction we must take, will we make progress, too?

# Behind the Onstreaming

Upward, behind the onstreaming it mooned.
—'Tlön, Uqbar, Orbis Tertius,' *Fictions* by Jorge Luis Borges

'You all know what a river is,' the biologist says, standing on the banks of the Cauvery. Behind him, the river mumbles and roils over low rocks, gleams slick silvery flashes in noontime sun. The man, who has spent a good part of his working life studying rivers and the animals like otters who live in them, is talking as a resource person to a group of scientists and conservationists on a field trip after two days of a conference on river otters in Bengaluru city. 'The river's upper course begins in the mountains, the water comes down the slopes, becomes perennial in the middle course,' the speaker gestures to the river behind him, 'and then flows through plains in the lower course before finally entering the sea.' In the audience, listening, he thinks the biologist seems self-assured and competent: the man is an expert on rivers, he must know, he must be right. And yet, it seems too pat, too succinct, too simple, that a great river like the Cauvery weaving its way through southern India is described thus—as neatly organized and fulfilling as a three-course meal. It seems to suggest that a river is but a line—though a watery, purposeful one—drawn from mountain to sea. A line.

Standing in the shade of a sprawling banyan tree beside a small riverside shrine near Mavinahalli village—located in Karnataka state about a third of way down the river's 800-kilometre length from nascent spring in the Western Ghats mountains to yawning mouth in the Bay of Bengal—the group looks across the river. They look past waters gushing through the innards of a centuries-old check dam, spewing white from between stones dislodged by the last monsoon flood, past calmer waters slipping smooth around boulders in the river bed, through little tuft-like islets with sedges and grasses and leafy *Polygonum* tangles waving their heads with wind and water. They look across water looping around elliptical islands and ephemeral sandbars, past a dry stretch lying in the lee of a small hydroelectric dam conterminous with the check-dam, towards the opposite bank over 300 metres away. From the reservoir brimming behind the dams, a canal draws water through the fields of paddy and sugarcane, onion and tomatoes, pastures and stands of coconut trees. Downstream the river slides, roils, slips, glides, waves, shimmers, effervesces, and chatters—down a long, gently-curving stretch, with little channels curling and purling between islets—before it swings out of view. An hour ago, a lone smooth-coated otter, swimming, head held above the water, had suddenly turned, dived, and vanished under the marmoreal surface. There is so much the river reveals, so much it hides.

He tries hard, after listening to the expert, to imagine the river as a line, but fails. In the sparkling sweep of water, fish wink in little splashes at the surface, radiating ripples subsumed by the silent current. Then a cluster of ebony rocks, marked with white acid splashes of bird droppings, rises a couple of feet above the water—enough, it seems, for the darter standing motionless and coloured like the stone itself, wings held open to dry in sun and wind. Water again, a tar-black cormorant skimming ahead of a leafy twig bobbing and turning slowly as it drifts—to what unknown destination? And land, again, an islet edged by a strip of fine, dark sand carrying the calligraphy of wader feet, and crested with sedges and shrubs into which a little tricoloured munia flies carrying a long blade of dry grass to tailor his nest among reeds. Then, water again, with Asian openbill storks walking—on stilt legs, like old people, heads bent in contemplation—wading in inches-deep water in the company of stately grey herons frozen in ambush, fierce eyes behind dagger beaks. Land again, a sandbar, tossed like a throw cushion on the satin sheet of the river, attended by a line of snow-white egrets, onto which a common sandpiper sweeps in on stiff, twitching wings, even as a cloud of three hundred pratincoles flares into the skies, wheeling, quartering, rising, twisting, dipping, flashing now white now brown, in a stupendous, heart-stopping, aerial symphony of movement, like one giant bird exploded into hundreds, yet alive. Then water, then land, water, marsh, and land again.

The river is a braid of land and water, he realizes. A braid, like the ones that the village women washing clothes by the riverside have made, hair parted neatly or drawn back from their foreheads, pulled into long bunches, woven over and across each other, a tassel at the end; the bunches braided tight together, gleaming neatly down their backs, just as land and water and marsh and rock come together in the river shimmering its way ultimately south. Now, the egrets and otters and storks and plovers are but *mallige* and *kanakambara* flowers decorating the braid.

Still, he is not satisfied, although the idea of a river as a braid seems better than calling it a line, or describing it with that ambiguous, ambivalent, amphibious word: wetland. In the river, the waders, the storks, the otters—do they not belong to water and land? And do we not have to see but once the great wheeling flock of pratincoles to know that they belong to land and water and air? And if they are not part of the life and energy of the river—then what is the river?

No, what he sees before him now constrains his vision: an all-too-human constraint. He needs to look further. He thinks of what that river expert said of the river's origin in the mountains: the tiny spring of sweet water that somehow became this great, wide river. The origin, in a sacred pool, springing from the belly of the mountains in Kodagu.

He has been to the sacred place, the hillside temple at Talacauvery in Kodagu. In the austere heights of the Western Ghats, at an elevation of nearly 1,300 metres in Kodagu district in Karnataka, a tank and shrine of the eponymous river goddess mark the origin. He has been there with her, a girl of Kodagu, her middle name Cauvery, too. Soon after their wedding, the two of them had travelled with her parents on the traditional visit, like so many others from Kodagu, to worship and be blessed by the river. Modest offerings at the shrine, mediated by a priest, a few sips of sacred waters, a purifying walk through the pool a few yards below. Now, as then, it seems that wherever crystal springs of sweet water emerge, as if from the earth's navel, no one, religious believer or scientist or atheist, can be faulted for heaping such places with spiritual value.

From source to sea, the Cauvery basin occupies more than 81,000 square kilometres and sustains nearly 33 million people: an area about the size of Austria, but with four times its population. The basin—about 245 kilometres at its widest and 560 kilometres at its longest—spans the states of Karnataka and Tamil Nadu, with small portions in Kerala and Puducherry. At 400 people to the square kilometre, it is among the most densely populated river basins in India. Around 70% of the population is rural, two-thirds of the basin is agricultural land. It is a land with a rich history going back thousands of years, a crucible of the cultures that make south India what it is today. The river is the circulatory network of water-filled arteries and veins of the land. As the river's watery tendrils weave through the land, its vitality weaves the tapestry of landscapes and peoples.

Yet, the river is no closed circulatory network, it is more open and dynamic. Downslope from Talacauvery, the water joins a first-order stream emerging from the dense forests and swaying grasslands on the mountains, becomes a clear stream chittering over smooth pebbles and rocks, along banks shaded by evergreen trees. There, they had bathed in the waters, in public with other visitors and pilgrims, the women entering water almost fully clothed—a dip, some splashing, water cupped lovingly in hands and poured over heads, a ritual cleansing of body and soul. He had felt then how strongly the community identified itself with the river, and the river's significance for their land: as birthplace, as mother, as source of pride, as progenitor of water and life. Like the two of them had done, all Kodagu couples would visit these waters, he thinks, if it was within their means. They would bring their children, their children's children. At the end of their lives, many would, like her father, have their ashes consigned nowhere but in the river.

Perhaps, like a bloodline, the river is a community of waters. A coming together of freshets, a commingling of streams traversing earthly life and landscape, surmounting boulders, cascading down cataracts, sweeping through calm interludes, receiving, sharing, giving, branching out into little distributaries that join the vast oceans of life. Through Karnataka, then Tamil Nadu, the Cauvery gathers the waters of the Hemavathi, Kabini, Shimsha, Arkavathi,

Bhavani, Moyar, Noyyal, Amaravathi, and more, before splaying out into the great delta of over 14,000 square kilometres across Thanjavur and its neighbouring districts. At every confluence of rivers, one is likely to find shrines, or ghats, or great trees, silent markers of religion, culture, and ecology.

He ponders, now, over the two streams joining near the origin, the water lapping against as-yet unsullied banks. Something that a hydrologist said at the conference comes to him: the daily flux of water with the breathing of the trees. He imagines now the water level dropping subtly by day as the trees in the watershed draw water up through their roots, breathe them out through their leaves, the level slowly rising again as the trees close their stomata in the quietude of night. He imagines the tree as a river itself, waters drawn from the earth, coursing through the trunk, branching out, breathed out into the atmospheric ocean, the air then burdened with moisture, condensing as mist and cloud, rushing back into the mountains, then falling as rain, the water drenching the trees and the earth, shimmering down the tree trunks, sponged by the leaves on the soil, percolating through soil pores and root tubes, then drawn out again, into the tree, into the river. Now, the river is a dendritic network of water melding with the trees.

A yell snaps him out of his reverie back to Mavinahalli. Standing on a bamboo coracle, wooden paddle in hand, a fisher shouts across to another in a coracle downstream asking, perhaps, about the day's catch. On the banks, the conference group splits and climbs into half a dozen coracles and two inflatable rafts to experience the river, discuss field survey methods—their task for the day—and look for otters. The coracles and rafts are full and he waits on the banks with her for another coracle from the village. A boy takes off at a run to the village for the purpose and returns soon with the coracle. He carries it inverted over his head, only his legs are visible, feet slapping bare earth: he looks like a walking mushroom. Behind him, a fisher follows, holding a paddle.

The sun blazes furiously over their heads, and there is no breeze. The shimmering surface of the river reflects the light, the water is barely cool to the touch. Sweating, squinting against the light, squatting at the three points of a triangle in the coracle for balance, the two of them and the fisher make their way slowly downriver, the coracle swinging now this way, now that, to the rhythm of the rowing.

Out in the river, the fishing is under way. In the water, a pair of smooth-coated otters gambolling in the current, skimming the surface with heads over the water, plunging suddenly as if diving for a fish, reappearing somewhere else. A fisher on a coracle swinging his net, heaving it out, the net fanning out in a circle, dropping, then hauled up and checked for unsuspecting fish caught. Another setting out a net line from the edge of an islet out into the river, weaving his coracle around the unseen net in the water.

Trained as an ecologist, he sees the river as an ecosystem, one that sustains life and livelihoods; he wonders how the fishers perceive the river.

In *A Sand County Almanac*, Aldo Leopold had noted:

> There are two spiritual dangers in not owning a farm. One is the danger of supposing that breakfast comes from the grocery, and the other that heat comes from the furnace.

To this, keeping rivers in mind, one could add as corollary:

> There are two spiritual dangers in not knowing a river. One is the danger of supposing that fish come from the market, and the other that water comes from the tap.

The history of the use and abuse of rivers in India is a sorry one. Few rivers have been left undammed, their dynamic flow, their open throbbing engines of life, left untrammelled. After 1947, independent India embarked on a phase of dam building that reached a crescendo in the 1960s onwards—a time when around the world there was one dam being built every day on average. The archaic and obsolete ideas of development based on large-scale impoundment of water for agriculture and electricity generation has spared few of India's rivers. The Cauvery and its tributaries already has a series of major dams—at Harangi, Hemavathi, Krishnaraja Sagar, Kabini, Mettur, Lower and Upper Bhavani, Avalanche, Emerald, Kundha Palam, Pegumbahallah Forebay, Pillur, Porithimond, Parson's Valley, Nirallapallam, and Amaravathi, and smaller impoundments—65 dams already, plus a series of irrigation canals drawing water out—Devaraj Urs, Mettur, Kodivery, the Grand Anicut of the Cauvery Delta, Lower Coleroon, and more—all of which feed the prosperity and eternal discontent of human needs in the basin. Over a hundred dams, beguilingly named mini-hydels, built or being built, generating power and profits for privateers, are set to scupper what is left of the river. The 2012 *River Basin Atlas of India* proclaims that about 90% of the Cauvery basin's average water resource potential of 21,358 million cubic metres is 'Utilizable Surface Water Resource'. The live storage capacity of completed reservoirs has already usurped half that pot of water—a disputed and contentious pot, the sharing of which Karnataka and Tamil Nadu have fought over for more than 150 years. Another large hydroelectric dam, recently proposed at Mekedatu, already has the political powers-that-be jousting, indicating the bitter fight is not over.

And then there is potable water, too, he remembers. He travels often between the cities of Coimbatore, Bengaluru, and Mysore, where nearly 11 million people depend substantially on the Cauvery or waters from its basin for their drinking water supply. How much of all that water has he imbibed over the years? If over half his body is water by composition, is the Cauvery now a part of him, too? Is he then not part of the river of life that is the Cauvery? And if the river declines, will it only presage his own?

Meanwhile, as the river silently gives, it silently receives: the wastes of cities, the washing and leavings of humanity, agricultural chemicals and poisons, fertilizer runoffs and residues. Invasive alien plants like water hyacinth and giant salvinia choke its flow, as native riparian vegetation along the banks is stripped and replaced by unsparing farms and alien *Eucalyptus*. Alien fish fed into the waters proliferate, as native fish beat their retreat under the hooks and looks of Fisheries Department officers and anglers. The fishers still eke out their living, on a fluctuating catch from a changing river, taking recourse to poison and dynamite to fill their nets with stunned and shattered fish. All along the river, lorries queue up in the hundreds, gouging sand from the bed, destroying the fabric of the river from the bottom up, even as the sand helps erect the concrete framework of houses and buildings in the growing cities and towns. And farther down, at the end of the river—if any river can have an end—down in the Cauvery Delta, every drop of water that enters the sea is considered wasted. Not as waters that forever build and sustain fertile plains, not as waters that sustain flourishing ecosystems well past the Coromandel coastline—no, any water entering the sea, so the blinkered view goes, is water wasted. And so it goes.

To utilitarian human eyes, the river is no line, or braid, or network, or tree of life. The river is only a pipe from tap to flush-pot, with feeder pipes to drain or dump at will.

The next day, a hundred kilometres downstream in the Cauvery Wildlife Sanctuary, he is again by the river, sitting quietly on the banks. A welter of broken forest-clad hills, having somehow resisted farms and dams, have left a stunning stretch of river winding past stately *holematthi* and mango and *jamun* trees towering along the banks. The others have left on their coracles, but he prefers to sit silently and watch the river. Past the disembodied darter, swimming with only snake-neck and spear-beak showing above the surface, a mugger crocodile basks on rocks across the river. On a high branch in a tall tree, a lesser fish-eagle waits, eagle eyes on the river, suddenly plunging into the water, rising with the merest splash, back to his perch, a fish in his talons.

In the quietude, he wishes now for a good book about the river. Perhaps the translation of *Eternal Kaveri* by T. Janakiraman and Chitti, to learn, as the subtitle suggests, 'the story of a river'. Or another book by the same name edited by George Michell, documenting the historical sites along the river. Even,

*It Happened Along the Kaveri*, touted as a delightful travel book replete with the requisite anecdotes and trivia. Would he have understood the river a little better then? Would he have known the river in its essence, as the otter and fisher on the coracle, perhaps do, every day? Maybe. But the book he carries with him, the book that he is absorbed in, now, by the river, is *Fictions* by Jorge Luis Borges. Fictions—what can one learn about a river from mere fictions?

He remains dissatisfied with all his metaphors, distressed even at his attempt to describe the river thus. Line, braid, network, tree, tap! Is that all he can come up with? He is disgusted with his ideas, his scrawls in his notebook.

The river is a great engine of life, energized by the sun's fire, the wind's breath, the cloud's sweat, the pull of earth, and the touch of mountains. An elemental, spinning circle of water, air, earth, and fire, that distils the alchemy of ethereal life. His eyes on the ever-moving current—water in him, in the air, in the earth—all moving, melding, dissolving, he wonders how one can pin down in words something so fluid, so full of movement. It's like trying to hold a fistful of water.

Finally, it is in Borges that he finds his answer. In 'Tlön, Uqbar, Orbis Tertius,' where Borges writes about an imaginary realm of idealistic nations, Tlön, a world whose language and everything derived from language is suffused with idealism. For Tlön's people, Borges writes, 'the world is not an amalgam of objects in space; it is a heterogeneous series of independent acts—the world is successive, temporal, but not spatial.'

Consequently, in their language, the 'conjectural Ursprache' of Tlön, there are no nouns. There are 'only impersonal verbs, modified by monosyllabic suffixes (or prefixes) functioning as adverbs.' There is no noun like moon, Borges writes, but there is a verb that in English would be 'to moonate' or 'enmoon.' In Tlön, one does not say 'The moon rose above the river.' One would avoid nouns altogether, and say: 'Upward, behind the onstreaming it mooned.'

Imagine that, he thinks, a river not as a noun, or an object of human consumption, but as a verb, a ceaseless flowing part of a community of ever-flowing life!

It is so the onstreaming engines on, sunblazing, windbreathing, condensing, earthpulled, and hillblocked. It is so in circlespinning, watered and aired and fired and earthed. It is so the onstreaming ethereally alchemizes into lifesparking.

As it sundowned, watching: downward, behind the onstreaming, it flamed.

# Earth-Scar Evening

The road winds through a disfigured landscape of tea plantations. It skims the contours over the open reservoir with its sloping banks of naked red earth. It passes the check post with the inevitable tea stall, and only then does it plunge down. Down towards the rainforest, our destination for the evening. The Nilgiri langurs—black shapes with long dangling tails—sitting on the tree near the tea stall, watch us go. There is a hint of rain in the air. Clouds hang dark over the landscape.

We come upon the fallen trees a short while later. Twenty-two of them, many towering giants felled as if by an invisible blow, scattered along less than two kilometres of road through the forest. In their fall, they have snapped some of the neighbouring trees leaving crownless, leafless boles standing like wooden pointers at the sky. These trees were not felled by axe or chainsaw; their fall appears natural.

The trees must have all come down at roughly the same time, and not so long ago either: the leaves are still attached to the twigs and just turning brown. A thunderstorm with downpours and gusts of wind and even hail, typical of this pre-monsoon season, would be the most obvious, immediate cause. There wasn't one, but two recent storms, during the third week of April. The ground is littered with leaves, twigs, and branches, much of the latter has clearly broken off in the wind and rain. The tree falls seem only natural.

All but one of the trees that have fallen are large, over a metre in girth, some more than twice that. Several are white dammar trees, true giants of the dipterocarp family. Upright, their crowns would have dwarfed the rest of the forest canopy, drinking in the bright sun, but exposed to every buffeting wind. Their disproportionate misfortune—if one may so label the almost instantaneous end to their centuries-long existence—seems natural, too.

Almost a third of the trees had fallen on a short stretch of road, less than half a kilometre long, which climbed a little rise—a small, exposed hill crest—before it dipped into a stretch of bamboo and drier forest. The forest here had clearly received the battering of the wind and rain, in sharp contrast to a more sheltered valley a little distance away. It was the storm that must have done it.

And yet, a nagging thought tugs at us: Why are all these trees along the road? Is it because we could not see far into the interior of the forest, where doubtless other trees have also fallen? Or, is it the road itself, this earth-scar cleaving its way through the forest that in some insidious, silent way brought down these rainforest giants?

The road takes a sharp bend, and we see the opposite slope above the earth-scar. A tree has fallen on that slope, amid hundreds, and it is just over the earth-scar. The fallen trunk has been axed and sawed and moved out of the way of vehicles. The earth-scar brooks no obstruction.

All along the road, the earth has been scraped or gouged off the sides, to fill potholes. Even as these quick-fixes have exposed the roots of tree after tree, they cling to the sides, trying to hold back what is left of the earth. The earth-scar feeds on itself.

Punctuated along its length are deepening furrows where, with the open sky and the slope, the pelting rain can now directly strike the earth and carry the soil away. The gullies cut the sides and more roots show. The road goes one way, the soil another. The earth-scar spawns scars.

The forest churns with dynamic life, more complicated than anything human-built, and it can clothe and heal itself. As it tries to, through a succession of forest ferns, shrubs, and trees, its innards are ripped again by the repeated, thoughtless slashing of vegetation along the road. The canopy, once fully covered overhead, is now rent; the streaming light feeds the weeds. Now the weeds have to be controlled by slashing, again. The canopy that kept the weeds at bay and clothed the earth with beautiful ferns and orchids for no extra charge is ignored by people who, for wages paid by the government, slash away under the arc-sky over the road. The earth-scar craves the sun.

The weeds that now stifle the rainforest seedlings have travelled along the road with the vehicles and the dust and the people and their plastic and debris. The *Mikania* is here, and the *Lantana*, as is the chromolaena. With the fall of the giants, light can now stream into the forest, and the weeds, too. The road has also brought a plantation nearby; the seeds of the robusta coffee grown there have now spread into the rainforest. The understorey is a beguiling green—every fourth or fifth plant growing among the future forest is a robusta. The earth-scar brings visitors.

Like the vehicles, the wind, too, can speed along the earth-scar. It can sway the branches and gently pitch and shiver the leaves. It can lighten the humidity and desiccate the earth. It can bring moisture to the forest even as it lifts it from the leaves. It *can*—and *does*—blow the trees over. The earth-scar funnels the wind.

Is it Nature that felled these trees? Perhaps. Is it the road? Or is it I, who, getting into my car, ride the earth-scar back home? As we reach the check post, the langur are still watching. There is a hint of rain in the air. And clouds hang dark over the landscape.

# The Butchery of the Banyans

The roads from Mysore, leading west into Kodagu, and south towards the Biligiri Rangan Hills, are old roads. We know they are old, not from the road itself, or the people, and certainly not from the speeding vehicles. We know it from the magnificent trees growing by the roadside for mile upon mile. These are grand *Ficus* trees, the fig trees we know as banyans, metres in girth and sprawling in canopy, planted and nurtured to life by some blessed soul centuries past. Today, they bestow an uplifting aesthetic and rejuvenating shade to the otherwise bare and dour tar road. And yet, all along the roads, these huge, ancient, centuries-old banyan trees are now being hacked.

Winding through a picturesque countryside, taking little dips and turns hugging the contours of the Deccan plateau, towards the Western Ghats and other hill ranges, these roads, until recently,

* Written with Divya Mudappa.

seemed to sit gently on the landscape. There has always been ample space for vehicles, even large ones, between the trees on either side. And even as the vehicles plied back and forth, the trees were full of life. Indian grey hornbills and barbets and mynas come to feast on the luscious red fruits of the banyans, as do monkeys and squirrels. Myriad creatures feed, roost, mate, sing, rest, hunt, play, and sleep in the trees.

It is not just the animals that benefit. These are trees planted by people, primarily for people. From the scorching sun of the Indian summer, these trees offer dense, cool shade, the only respite from the heat in the open landscape. Many travellers—there are many who even now travel on foot, bicycle, cart, and without air-conditioning—rest in the shade and move on refreshed. And who cannot appreciate, in the heat of noon, the joy of a nap under the shade of a tree? Village children often swing from the roots of the banyan or scale the branches to lop some fodder for the livestock on their farm. When the trees are many, the lopping seems a minor matter, and the trees have perhaps borne the

children and provided for livestock for centuries. But now, the trees are few, and even as you read, becoming fewer. A massacre of the awe-inspiring trees has been underway along these roads for some time, and continues even now.

This is a requiem for the grand banyan that we saw dismembered along the Mysore-Madikeri road. This great tree is now gone, along with many others. Now the authorities plant obnoxious Australian *Acacia* and *Eucalyptus* trees—alien species that can never muster even a fraction of the ecological importance or aesthetic grandeur of the banyan. These alien species may detrimentally affect soil and ground water, besides contributing little to the local people, wildlife, or environment.

This is a requiem for the great banyans now being destroyed on the Chamarajnagar-Asanur and Chamarajnagar-Gundalpet roads near Mysore. Dwarfed by the massive stumps of the destroyed giants, the vehicles and people pass—apparently untouched and unrepentant. And all along the roads the logs pile up but will not stay here for long—even when dead, the trees are too valuable and the lorries are busy collecting the logs—the spoils of slaughter.

We stop to talk to the people cutting one great tree. They tell us that the order is passed by the Highways and Forest Departments to cut the trees. *The order is passed*—what a passive statement of active slaughter! They say the road will be made wider—another order has been passed, perhaps. They *also* think the trees are over five hundred years old. They continue their work—swing their axes and pull at their saws, taking turns to rest, and to hack. Two men hold a rope tied to the top of the tree and pull taut, away from the sawyers at the base of the tree; it should not fall on them, or harm them, even in its fall. They saw away with zest.

It is just a day's wage labour to obliterate the growth of centuries.

The extraordinary value of fig trees is something the entire world of ecologists, particularly those from tropical countries, has come to appreciate. Fig fruits are a favourite food of many animals. Research has so far identified over 1,200 species of animals that eat fruits of different *Ficus* species around the world.

Studies have also highlighted how, by fruiting copiously, producing tens of thousands of fruit on a single tree, often during seasons when other foods are scarce, figs are a critically important resource, labelled 'keystone resource' or 'keystone species' by ecologists. The remarkable relationship between the tiny fig wasps and the magnificent fig tree is the stuff of ecological legend and fascinating natural history. Anyone who has spent an hour under a fruiting banyan can attest to the life that such a tree brings to a landscape.

Why, then, do we need to cut these trees? Yes, we need roads, good roads; that is something most of us would not dispute. But what does *a good road* really mean? Something that is wider, more open, more homogeneous, and more barren in appearance, and, coincidentally of course, also requiring bigger contracts? Or something that is well surfaced, well marked with road signs, safe, and well integrated into the landscapes that it passes through? Studies have shown that roads and avenues with beautiful and pleasing vegetation, with stately trees on either side, even have positive, restorative effects on driver behaviour, reducing frustration on the road and, perhaps, making it a more enjoyable journey.

We need to keep the banyans we have, and where they have gone, bring them back as the flagships of our countryside roads. Is it too much to ask that trees such as these, which are markers of our country's great natural and cultural history and heritage, be saved rather than sawed?

# Of Tamarinds and Tolerance

For centuries, long rows of majestic tamarind trees have marked our roadsides, particularly in southern India. Around Coimbatore city, in particular, innumerable tamarind trees flanked the roads radiating from the city. One could see them on the road to Mettupalayam and the hazy blue mountains beyond, on the road to the sacred hill of Marudhamalai, towards the Sathyamangalam hills and Mysore to the north, through the expansive plateau and plains to Salem, and southwards past Pollachi to the ancient hills of the elephants, the Anamalai.

The trees have stood like old sentinels, serene and solid through the rush of years. Their sturdy trunks and strong branches have towered over and across the roads, quite unmindful of buffeting rain and searing sun. Their twigs, festooned with dark green leaves, each with its paired row of little leaflets, have provided impartial and unstinting shade and shelter for all. In return, the trees only needed a little space by the side of road, to set their roots in, to stretch their arms.

They stood like this until the men came with axes and saws for their slaughter. The men brought bulldozers and earth movers—construction equipment powered for destruction—to gouge the ancient roots of the tamarind trees out of the earth. Trees that had stood for centuries were brusquely despatched in a matter of hours.

The tamarind tree is an old and dignified citizen of our city avenues and gardens, our countryside and farms. Its name, derived from the Arabic *tamar-ul-Hind* or the 'the date of India', finds mention in written historical accounts of India going back centuries. There is irony in this, for the tamarind is native to Africa and not a species that grows naturally in India's forests. Despite being alien to India, the tamarind has not run wild and become an invasive pest, becoming instead what biologists call a naturalized species. Embraced by a deep tolerance and cultural acceptance into Indian cuisine and culture, the tamarind is today a familiar and inseparable part of Indian life and landscape.

Before the men and the machines came, the tamarind trees seemed to have an abiding presence, like torchbearers marking a productive countryside, like the enduring blue mountains in the distance. Their wide trunks rose above stout roots that pushed into the soil, like muscled and flexed thighs gripping the earth. Their fissured bark was thick and brown, aged and toughened and weathered, like the wrinkled face of the old woman selling mangoes in the patch of shade below.

Under the dense canopy, thousands of pedestrians and riders of two-wheelers found quick shelter from rain. Or, in scorching summers, a refreshing coolness cast by the tiny leaflets—how many leaflets does a tamarind tree have, a million, ten million? Even the air-conditioners seemed to waft easier and cooler in the metal cocoons of parked cars that escaped roasting in the sun. The trees seemed to abide, they granted benefits, and their beneficence was taken for granted.

Every year, the twigs were weighed down with hundreds of lumpy brown pods, with skins like coarse felt covering pulp, tart and tasty, and disc-like, shining seeds. The fruits were there for the taking. The adept and nimble climbed the branches to knock down the fruit. Their friends darted around to grab the fallen pods, dodging traffic. On the roads, many tamarind trees had managed to rise above anonymity: each tree, even if not named, was numbered; each individual claimed by negotiation or auction by someone from the village or panchayat for its fruit. Collected, dried, and packed, the fruit of the tamarind trees would eventually find its way into a thousand dishes, enrich the palate of millions, and become inseparably incorporated in people's cuisine, in their lives, in their very bodies. And no one could stop the children, who needed only a handful of stones to claim their share. The trees brought utility, food, cash, plain fun.

And yet, there is more to the tamarind. Beyond its utility and benefits, there is something intangible, overlooked. It seems to emerge as a touch of beauty—an enlivening green in an increasingly dour landscape. A beauty fragile from the prospect of its loss a chain-saw away.

It seems to emanate from the trees, too, from the sounds where a few still remain. The soughing of wind through ten million leaflets, in mournful restlessness, carrying the delicate aroma of the tamarind's modest, finely marked flowers. The creak of branches and the click of twigs holding the tamarind's pendant fruit. Or, when the wind abates, a calming susurrus pierced only by the occasional screech of parakeets. And when dusk descends, the tamarind trees darken to the chuckle of mynas, the chatter of shy owlets, and the hoots of somnolent owls rising with the stars. The trees are silent but full of sounds, and one who hears them may find things worth listening to.

Then the old roads were labelled tracks, the tracks became streets, the streets became roads, and the roads became highways. And yet, we are not satisfied, we need super-highways. This idea brooks no questioning, no obstruction. The trees must make way for tarmac. The people who stood in the shade must make way for the cars that proliferate. The vitality of a living countryside must make way for the deathly artificiality of the city, spreading like a virus down the arteries.

The tamarind trees are now painted with broad waistbands in white and black, so that they are more visible to the highway motorist who can then avoid them. How effectively we mark something to be more visible and to be more ignored at once!

So the tamarind trees drift into wayside anonymity, from anonymity to disuse, disuse to neglect. The fruits fall and are crushed under the tyres of vehicles. The road surface is studded with hard, shining seeds driven into hot tar, eyes without eyelids gazing at sun and sky. Shade and greenery are replaced by heat and grime. The songs of birds and the sighing of wind in the branches are replaced by the cacophony of vehicles.

Now the tamarind trees are but old fixtures in the landscape, like old people, grandparents and elders, suddenly out of place in a redefined world, suddenly unwanted. And when the old trees fall, the countryside is bereft, like families broken.

It does not have to end this way. Engineers and ecologists, citizens from the city and the countryside can join hands to find better design and transportation solutions. Solutions that retain the old trees, such as tamarinds and banyans, as essential components of roadsides for their varied and indisputable uses, and as representing a more refined aesthetic sorely needed for our cities, roads, and countryside. What call do we have to deprive those who come after us of the public utility and beauty of these astonishing trees?

Even now, many stumps lie metres away from widened roads: one wonders why they had to be felled at all. Natural landscaping, planning service lanes around trees, traffic regulation and public transportation solutions need to be found before engineers and bureaucrats wield the axe, albeit from behind their desks, distanced and disconnected from land and landscape. Taken as a matter of wide public importance, decisions to retain or fell such trees should be based on democratic and public debate and consultation with and concurrence of citizens and citizen groups, and involvement of representative local administrative bodies and the judiciary.

Widening roads at any cost represents a one-dimensional view of progress, one that compromises other human values, capabilities, and needs that are all not really fungible. Our increasing disconnect with these values and capabilities only erodes the deep wells of tolerance and breeds alienation between people and nature, land and culture. There are better roads, so to speak, to take, and there is time yet to take them.

# Forest of Aliens

Like the proboscis of a malarial mosquito the Andaman Trunk Road pierces the Jarawa forest. The road carries a steady stream of vehicles, bunched into convoys with guards. By the road are heaps of stones and the claw marks of heavy machinery: the road will soon be wider. Just beyond, on either side, stretches the home of the Jarawa—lush rainforests with tall dipterocarps and *padauk*, myriad trees and lianas, palms, cane, and bamboo. If the forest bears the mark of the Jarawa, it is subtle and difficult to discern.

Up in the trees, a flock of birds is busy hunting prey. Dressed in smart black, the Andaman drongo forages in the canopy with long-tailed Andaman treepies. The forest thrums with the territorial drumming of the black woodpecker of the Andamans, even as the cries of a spectacular dark serpent eagle shrills from the skies. Towering above the other trees, an emergent *Tetrameles*, smooth-trunked and leafless, holds a dollarbird on a high exposed branch.

The endemic Andaman birds mark the uniqueness of the forest, but the dollarbird suggests an ancient commonality with lands across the ocean, for one can see it similarly perched atop tall trees in the rainforests of the Western Ghats, in northeast India, and in southeast Asia.

The road hurtles on, like an arrow of time, past the island of Baratang, into a more open forest. Huge logs lie by the roadside. 'Welcome to Middle Andamans' proclaims a Forest Department signboard. The signboard is only half green—the other half is red. This forest bears the mark of a different kind of human.

Here, the tall trees are few and scattered. Amid remnant evergreen trees are many that are deciduous. The undergrowth is dense with palms, shrubs, and saplings, in dense tangles with weeds and vines. Through the canopy shredded by logging sunlight streams to feed the light-hungry weeds in the undergrowth. Alien weeds thrive: the *Chromolaena* in dense clusters, the *Mikania* woven into green shrouds over saplings. The forest is criss-crossed with logging coupe roads. Some are overgrown, some erode away, but some remain, tenacious scars marking an old, unforgotten wound.

In the forest itself, the ground is thrown up into little mounds. The mounds are covered with a fine sort of soil that termites conjure from earth and wood. Little seedlings germinate on the mounds. There is *Ficus*, of course, but ferns and other plants, too. The mounds are rounded at sawing height off the ground. Theirs is a strangely haunting presence in the forest, like ghosts of trees past. On the forest floor all around are dotted seedlings and saplings of forest trees—pioneers, deciduous, and evergreen—a tenuous cohort presaging an uncertain forest of the future.

At either end of the road are altered landscapes of settlement, agriculture and forest remnants, seeming destinations—end points—not just in space, but in time as well. Here, alien mynas and native starlings share and contest space, in the continuing biological tussle of introduced and indigenous so unfortunately frequent on islands. Crows and bulbuls, chital and elephants, many animals have been brought and released here, subsequently thriving as feral populations. By the roadside in Port Blair and Wandoor are rain trees, another alien, festooned with bird's-nest ferns and orchids, growing luxuriantly in the humid tropical climate and soil. The land has welcomed and accommodated the people and life forms that have arrived, providing resources and succour. How those who arrived have accommodated to the land is another matter.

After a long spell of logging and a brief reprieve, the forests are on the cutting block again. The island forests rise behind a skirt of dense mangroves whose aerial roots claim purchase at the very edge of land, forming a shelterbelt from the surges of the sea. The mangroves now give way to desolate wastes and burgeoning resorts with the all-important sea view. The sand beaches that hold the nests of turtles and the roots of *Manilkara* trees are mined away for the homes of people and the foundations of buildings. The soils from slopes and crop fields erode into streams and into the sea to smother with silt the coral reefs—the ones that aren't already bleached and crumbling from ocean warming or extraction. A tsunami came and went, but the tsunami of this type of development continues—yet, it seems only a promise to squander in years what peoples such as the Jarawa have sustained over millennia.

Will the spread of alien plant and animal species into the sensitive landscape of the islands ever abate? Will the tussle over space and resource, over lifestyle and culture, among the indigenous and the settled peoples amicably resolve? And yet, isn't alien and native a matter of perspective? Seen with immigrant eyes from the streets of Port Blair, the introduced myna and house crow appear more familiar than the Andaman teal or treepie. To the native Jarawa still embedded in the island ecosystem, whose name for themselves, *eng* means 'people'—to them, we are the alien, people from another world barely known or understood. But to us, as people bereft of intimate connection with nature, it is the Jarawa—our name for them meaning 'the other,' 'the stranger'—who appears alien. And so it may remain. The Jarawa live a world apart. A world they can scarcely construct for us without somehow losing it in the process.

Unbidden, a strange feeling then appears on the journey down the road. A feeling, as if we are destined to always be second-comers, carrying an atavistic insecurity originating in early human migrations from the African savanna into new lands. As aliens forever, we cope with insecurity by revelling in alienness, seeking shelter in superiority, making it aspirational, a developmental goal. It is our proud red against the darkling green of the Jarawa, who are people like us but who arrived in ages past, taking a path towards a destination altogether different.

Our road could yet lead to a different sensitivity and perception. A sensitivity that allows us to make space for diversity—biological and cultural—on the land itself, in our hearts, our minds. A perception that we simultaneously inhabit different worlds and that a more powerful world should not trample a weaker one to the earth. By making space for survival and recovery of other peoples and other species in their natural homes, the forest of the future may be, not a forest of aliens, but a forest of the human and the humane.

# The Tall Tree

The tree stands tall, head and shoulders above the rest. Its long round bole reaches straight to the sky. Its branches hold out firmly, even as the leaves sway and whisper with the wind. With its first branch at over a hundred feet and the uppermost leaves nearly half as much higher, the tree is one to reckon with; even a monkey would need to work hard to climb it.

High above, the tall tree's branches hold clusters of red-brown, two-winged fruit. A gust shakes the high branches and a couple of winged fruits with their package of seed take to the air and go whirring in the wind. In evening light, they are like fiery butterflies pirouetting in an aerial ballet.

The tree is a landmark, for those who choose to see it as one. In the distance, the weaving tributary of the Brahmaputra courses through a winding dip in the land. The forest around is dwarfed by the tall tree. Across farms and scrubby undergrowth tangled with vines, only a smattering of trees meets the eye, and there are none so large. The tall tree is special. What does it stand for, even if it stands alone?

From fallen seed to stately landmark, the tree has been here long. One imagines the year it is born, when its first leaves emerge in a little sun fleck below a verdant forest. Years pass in dense shade, before a thunderstorm, or perhaps the lashing of a monsoon rain, brings down a nearby tree and gives the seedling a new lease of life. With sunlight streaming into the forest, the seedling surges upward, along with nearby clumps of bamboo and pioneers. Herds of elephants pass by, crunching on the bamboo; more softly pass the deer, nibbling on herbs and fruits.

As years turn to decades and then centuries, the seedling grows into a tree, bold and straight, up into the canopy. The trunk is magically cylindrical as the lower branches give way to the higher, leaving scarcely a trace. High above, within the dome-like canopy where a branch has broken off, the fungi are carving a hollow. A hollow that will perhaps be a calling-hole for the tokay gecko, or a home for a hornbill pair, a flying squirrel or civet, or an eagle-owl. The elephants still pass by the tree and the trunk is smoother and moss-free where they enjoy a good scratch.

Who is the first human to see this tree? Perhaps it is one of the subjects of the king who ruled the land when the elephants first established their scratching post. A barefoot forest-dweller, who also scratched the bark, but with a machete, and waited for the resin to collect. He would burn it as incense and use it in medicine.

Or, the first man to see the tree is a native of the colony, as the land came to be called later with the rulers from the west. He needs wood for his hut—fuel, poles, rafters—but there are many other trees around, easier to cut and haul. The man stops, briefly, scratches a little resin, and goes his way.

Or, maybe, it is a forester, the span of whose career is a small fraction of the tree's lifetime. He stops, admires the straight bole and frowns at the flanging buttresses, and decides that the cut can wait a bit longer. The path of the elephants is widened now and is called a road. Along the road, his bungalow is not far, and neither are the creeping farms and plantations, nor the saw mill.

Finally, the man is transformed from subject and native and colonist to citizen, and the landscape around slowly changes in the new nation. The air is different, too, as the new people vent their wastes into it. Molecule by molecule, growth-season by season, the tree faithfully assimilates the collective breath of humanity into the silent history in its wood. The tree and the land around is a history book, but there is no one to read it.

Now, the botanists and ecologists are here. With rapt attention, they try to place the tree, which they call a dipterocarp, within the bounded scope of their understanding. They try to unravel its mysteries, its overwhelming significance in the shrinking forests. Yet, they are just scratching the surface, like the resin tapper, only he has been doing it for centuries.

In a distant court, the strike of a gavel marks a defining moment. It decrees that the forests with the tall trees, what is left of them, shall be felled no more. A wave of protest rises over this seemingly unjust ruling. What will happen to the saw mills that cut the trees? To the elephants who haul the logs, who may now miss their loads, their chains, and their flogging, only to beg on the streets for an ounce of humanity? To the people, who are said to depend on forests for their livelihoods? To the state, which prefers to measure its progress with metal, mortar, and money?

The forest all around is gone, slashed for agriculture, clawed out for minerals, drilled for oil, smothered by plantations, submerged by dams, logged by industry, burnt by fire and human ire. No hornbills venture here, nor deer—they keep to the distant hills. The occasional befuddled herd of elephants tries to make use of their historical passage, on the path that has become a road and through the forest that has become plantation and farm. Their great journeys have now become mere trespass. They are chased this way and that, by one group of people or another.

Are these the final years of the tree? Or will it span another minor human lifetime? The wood from the other tall dipterocarp trees has gone around the world. People live in homes of dipterocarp wood, with dipterocarp floors and cots and chairs and tables. Families mark their human years, decades, and generations, with the wood of trees that had watched centuries pass. The wood is of lasting quality—even when dead, it will outlast the pitifully short human lives.

The tall tree, somehow, has escaped the axe. Gusts of wind still carry its seeds, like twirling beads of hope, over the dwarfed landscape. Where the tall tree still stands, it stands alone to mark the forest that once was.

# The Pigeon's Passengers

There is a modesty in their conquest of mountains. From the heights, they commandeer vistas of rugged mountains covered in forest or countryside dotted with large trees. From tall trees on high ridges, they scan the landscape, their heads turning on long, graceful necks. They have scaled peaks, even surpassed them. Yet, they speak in soft and hushed tones that resonate among stately trees. The imperial-pigeons are a dignified lot, keeping the company of great trees.

Down in the valley, the pigeon's voice throbs through dense rainforest: a deep *hu, hoo-uk, hoo-uk,* repeated after long pauses, like the hoots of an owl. In the dawn chorus of birdsong, it sounds like a sedate basso profundo trying to slow the tempo of barbets and calm the errant flutes and violins of babblers and thrushes. The calling pigeon, in a flock with others, is in a low *Symplocos* tree whose branches shine with dark green leaves and purple-blue fruit. They are all busy picking and swallowing the ripe fruits, each with fleshy pulp around a single stony seed.

These large birds, neatly plumaged in formal greys and pastel browns, are mountain imperial-pigeons—a species found in the rainforests of the Western Ghats and the Himalaya in India. In more open forests and on huge banyan and other fig trees along the roads through the countryside, one can see their cousins, the green imperial-pigeons shaded in more verdant sheen. As a group, the imperial-pigeons have a penchant for fruit that necessitates roaming wide areas in search of food. Weeks may pass in a patch of forest with no sign of pigeons, but when the wild fruits ripen, the nomadic flocks descend from distant sites and the forest resonates with their calls again.

Like other birds such as hornbills and barbets in these forests, imperial-pigeons eat fruits ranging from small berries to large drupes, including wild nutmegs and laurels and elaeocarps (rudraksh). Yet, the pigeon's bill is small and delicate in comparison with the hornbill's horny casque or the barbet's stout beak, which seem more suited to handling large fruits with big stony seeds. The imperial-pigeon's solution to this problem is a cleverly articulated lower beak and extensible gape and gullet that can stretch to swallow the entire fruit and seed.

Lured by the package of pulpy richness in fruit, the pigeon becomes a transporter of seed. Many seeds are dropped in the vicinity of the mother tree itself, scattered around with seeds from rotting fruit fallen on the earth below. The concentrated stockpile of seeds below elaeocarp and nutmeg trees are attacked by rodents and beetles, leaving little hope for survival and germination. But when the pigeon takes wing, some seeds become passengers on a vital journey, travelling metres to miles into the surrounding landscape. Voided eventually by the pigeon, the dispersed seeds have an altogether greater prospect of escape from gnawing rat and boring beetle and—when directly or fortuitously dropped onto a suitable spot—of germination. By carrying and literally dropping off their passengers where some establish as seedlings and grow into trees, the pigeons become both current consumers and future producers of fruit.

Still, it is the quiet achievement of the trees that seems more impressive. Rooted to a spot, the trees have enticed the pigeons to move their seeds for them. Deep in the forest, one discovers a seedling where no trees of its kind stand nearby, bringing a rare pleasure like an unexpected meeting with an old friend. The pigeons are plied with fruit and played by the trees. The birds' modest conquest of the mountains is trumped by the subtler conquest of the pigeons by the immobile trees.

In speaking of the pigeon's passengers, one recalls with misgiving the fate of passenger pigeons. The passenger pigeon was once found in astounding abundance across North America in flocks numbering tens of millions—flocks so huge that their migratory flights would darken the skies for days on end. Yet, even this species was exterminated by unmitigated slaughter by hunters and by the collection—during their enormous nesting congregations—of chicks (squabs) by the truck-load. Within a few decades, the great flocks and society of passenger pigeons were decimated in vast landscapes transformed by axe and plough, plunder and profiteering. By 1914, the species—that not so long back was perhaps one of the most abundant land bird species in the world—had been reduced to a single captive female. The last known passenger pigeon, Martha, died in Cincinnati Zoo in September 1914, closing the page on another wonderful species, in another sorry chapter of human history on Earth.

Our imperial-pigeons are more fortunate, but in many areas they, too, are dying a slow death. Some fall to the bullets of hunters who take strange pride in their dubious sport or skill. Some roam large areas of once-continuous rainforest that now have only scattered fragments. The mountain imperial-pigeons are still seen winging across in powerful flight from one remnant to another, over monoculture plantations and still reservoirs. Their forays are getting longer, and their journeys often end fruitless. Our countryside, too, is becoming bereft of their green cousins, as banyans and other fruit trees vanish along our widening roads, and diverse forests of native trees are replaced by wretched Australian wattles and eucalyptus, if they are replaced at all. As their homes are whittled away, the hornbills, barbets, and other pigeons vanish silently. With them vanish subtle splendours and prospects of regeneration. On the roads, the vehicles speed along on their wheels of progress, carrying passengers of a different kind, barely aware of the majesty and opportunity for renewal left behind.

From the valley, the imperial-pigeons take wing and—in a minute—fly high and swift over the mountain to distant rainforest. There, sometime in the future, new seedlings will perhaps still emerge in silent testimony that it is possible to fly high and strong forever, if you only consume what you also regenerate in perpetuity.

# The Mistletoe Bird

It is one of those little plants that you barely notice in the rainforest. It perches on tree branches, like a sea fan on a coral boulder, like a Christmas decoration. At a glance, it seems like another tassel of the tree's twigs and leaves. But look closer and you see that these leaves are smaller, paler green tinged with coppery yellow, unlike the tree's longer, parrot-green leaves. On the tree's brown branch, powdered with white lichen, the little plant rises out of a swollen bulb-like base, holding out dark brown twigs dusted with white spots, chocolate sprinkled with sugar. Clusters of pinkish-red berries and buds line the smaller plant's twigs, on the tree bereft of fruit or bud.

The clutch of leaves, berries, and flowers are on the tree, but are not of the tree itself. The little plant is an epiphyte: a plant that grows on other plants. It is a mistletoe.

In the company of mistletoes lives an unassuming little bird that you'd barely notice in the rainforest. A tiny bird, small enough to hide behind a leaf or hold in a closed fist, drab enough to escape the attention of everyone save an ardent birdwatcher. An undistinguished little bird, dull olive brown above, rather dingy white below, with sharp eyes, glinting dark and attentive, and a sharp beak, gently curved to a point to poke among the flowers. A metallic, fidgety *tick-tick-tick* announces her presence as she darts through the boughs. You have to be quick to spot her. In keeping with her modest appearance, birdwatchers call this species the plain flowerpecker.

I've travelled far from my home in the Western Ghats to see this flowerpecker. And not just any plain flowerpecker, but a particular one—a bird of that species flitting among the mistletoes on the same trees where I had seen it two decades earlier. I am seated on the steps of the Dampatlang watchtower in Dampa Tiger Reserve in Mizoram. Twenty years ago, I spent many quiet, contented hours watching birds from these same steps while camping here for fieldwork. To the south, steep cliffs plunge to Tuichar valley. An evergreen forest with many trees adorned with mistletoes surrounds me on three sides.

Alongside the watchtower grow two small orange trees and a straggling *Holmskioldia* holding bunches of scarlet cup-and-saucer blooms. Against the wild forest backdrop, the planted orange and cup-and-saucer plant marked what seemed a very human temperament to cultivate and ornament the lands we live in.

Seated two stories high, I am almost eye-to-eye with the flowerpecker. The bird flits from branch to branch, dives into each mistletoe cluster, peeking, probing, seeking with eye and beak. Flowerpeckers remain closely tied to the mistletoes on the trees within their territories, which usually span a few hectares at most. The birds consume mistletoe flower nectar and fruits, but this is a two-way relationship. The plant, too, gains when the birds pollinate its flowers and disperse its seeds.

Many mistletoes have tube-like flowers that, when probed by a flowerpecker beak, part like a curtain or pop open, furling the petals down and thrusting the stamens out to dust the bird's head and face with pollen. After the bird sips the sugary nectar with a special tube-like tongue—who needs a straw when your tongue itself is rolled into one?—and flies over to probe other flowers of the same species, some of the carried pollen may rub off on receptive female parts, triggering the plant's reproduction.

Despite this penchant for flowers and the bird's name itself, the flowerpecker remains, at heart, a fruit-lover. Mistletoes often have long and overlapping flowering and fruiting seasons so there is always food for a hungry flowerpecker. Ripe mistletoe fruit never fails to attract flowerpeckers.

Mistletoes represent a group of over 1,300 plant species worldwide belonging to five families, chiefly Loranthaceae and Viscaceae. As the latter name suggests, the fruits are viscid, the usually single seed surrounded by a sticky pulp, often enclosed in a rind-like peel.

The plain flowerpecker and its close cousin in southern India, the Nilgiri flowerpecker, manipulate mistletoe fruits in their beaks to gently squeeze the seed from the pulp. They swallow the sugary, nutritious pulp and wipe their bills on twigs to remove the sticky seed. If the flowerpecker swallows the fruit, the seed passes rapidly through the bird's gut to be excreted out. To remove the still sticky seed, the birds wipe their rears on twigs or tree branches. In either case, these actions have the same result, which biologists call 'directed dispersal': the mistletoe seed gets planted where it is likely to germinate.

Mistletoes are also partial parasites. They synthesize their own food through photosynthesis, but their special roots draw water and nutrients from the host tree. Extreme infestation of trees by mistletoes is rare in natural forests, occurring more often in degraded or managed forests and monoculture plantations. Still, foresters and others concerned with production of timber or fruits from trees sometimes call for mistletoes to be removed or eradicated.

Recent research suggests that this may be unwarranted. In forests, falling mistletoe leaves add vital nutrients to soil under the trees on which they grow. Experimental removal of mistletoes causes a cascade of harmful impacts, including declines in soil nutrients and populations of other species. Besides flowerpeckers, mistletoes sustain a large number of other species worldwide. The barbet-like tinkerbirds of Africa, the mistletoebird and honeyeaters of Australia, the sunbirds and white-eyes of Asia, mouse lemurs and sifakas of Madagascar, tyrant and silky flycatchers and colocolo opossums of the Americas, the eponymous mistle thrush of Europe, myriad insects and other creatures—all find food and spaces for hunting or nesting in mistletoes.

Back at the watchtower, I watch the feisty flowerpeckers defend their mistletoes, darting at intruders who enter their territories, zipping between branches with rapid ticking calls, giving chase. Fighting flowerpeckers have been known to fall to the ground while grappling fiercely with each other. One imagines their raging little hearts palpitating, as they flay and peck at each other to defend what they perceive as their own.

Together, the flowerpecker and mistletoe epitomize an irreplaceable vitality of the forest.

An hour later, as I leave the watchtower, I sense that there is more to it than just a symbiotic evolutionary link between bird and plant in a forest webbed with ecological connections. Perhaps, behind the gleam of that flowerpecker's eye, there resides, too, a temperament to cultivate and protect what she consumes, an aesthetic to adorn the trees in her forest with the prettiest little plants she can find.

# The Walk that Spun the World

It starts as a walk in a forest in Vermont that takes me, strangely enough, into the high Himalaya. On a balmy July afternoon, with hesitant clouds massing out west, I set out on foot down the road that passes through the village of Craftsbury Common. I leave behind the public library and the silent church whose spire towers over the open meadow of the commons and the white clapboard houses in the village. Ahead, the forest appears, flushed green and dense and dark from summer rains. Open fields, loon lakes, and lush farms adorn the landscape, but it is the tranquil forest that entices. Almost involuntarily, I am drawn into the woods, up the winding trail that disappears into darkness.

On the trail, the dark, wet earth carries the tread of tyres and the stamp of boots overlaid by tracks of deer and spoor of weasel. The imprints attest a land that the animals share with people. Shafts of evening sun brighten the canopy with amber light, but little reaches the shaded undergrowth. Much is still hidden from my eyes.

I pause where the deer tracks swerve into the forest. The print is now no more than a suggestion. I follow, stepping off the trail into the trees. The ground feels soft underfoot, matted by pine leaves and litter, sprigs of tiny grasses, a coterie of creepers, mounds of moss. Standing shoulder-to-shoulder with me in the undergrowth, saplings of red maple and sugar maple jostle basswood and hemlock in the eternal quest for light.

In a gap, a tree stump stands testimony to logging—not the rampant clear-cutting of a more reckless past, but the mindful selection of a woodsman. The forest is still used by people. If the trail and boot prints are not evidence enough, the pails and pipes strapped to trees reveal the touch of people tapping the syrupy sweetness of maples. And further down, a stack of logs waits beside the trail to be hauled and hewn into canoe or board or burnt for warmth in kitchen and hearth.

As I look up at trees that stand tall without being colossal, I sense the passage of decades rather than centuries. Yet, this land has witnessed a great flux of trees over the ages. As ice-age glaciers retreated around 11,000 years ago, forests appeared on the open lands. In the warmer world that followed, vast tracts of forest arose in which Native Americans lived and hunted for thousands of years. Colonial settler populations surged from the 1760s, and in the ensuing decades clear-felled millions of acres of forests for timber and to open land for farms in New England. By the 1850s, around three-fourths of Vermont was open land: farms and sheep pastures bounded by long walls of stacked stone slabs. Then, in the mid-nineteenth century, the tide in people and land turned, as civil war and migration whittled the settlers and rested the farms, bringing a great resurgence of trees to clothe the landscape. Today, forests once again cover over three-fourths of Vermont. In the forests, assisted by conservation efforts, resident wildlife species such as moose, grouse, beaver, and American marten are staging a comeback in a propitious rewilding of the New England landscape.

From where I stand, the river on which the timbers were once floated, like giant rafts of huge logs spanning bank to bank, is distant and hidden. Nearby, the moss-felted, lichen-blotched stone wall of an old farm crumbles into a linear tumulus weaving through trees.

The land still carries the marks of history. If I could read the rings in the recumbent trees or discern the trajectory of the resurgent landscape, I would understand better the revival of forest on exploited lands. Perhaps then, I would also perceive what it was that brought respite and respect for trees, in this momentous transformation in America's environmental history. In 1864, at the cusp of that great turnaround from clear-cuts and farms to forest, as the nation underwent the upheavals of civil war and the civil rights movement, Vermonter George Perkins Marsh's *Man and Nature; Or, Physical Geography as Modified by Human Action* appeared, a book that was to deeply influence the conservation movement. Now, a century and a half later, embracing the insights from environmental history, conservationists are better placed to resolve the 'great question' with which Marsh concluded his book: the question as to 'whether man is of nature or above her'.

If nothing else, the tracks on the trail suggest the human footprint can now be a soft one, too.

Looking at deer tracks, I ponder over my own journey. I have travelled from India to Vermont to join a workshop on writing about the natural world. Here, I know, I am a stranger still. Yet, that is not what I feel in the forest. Watchful, wistful, I meander through tall conifers, modest maples, beech, and ash. Taking a turn, my walk through the trees becomes a trek through stately forests of conifers, ash, and oak on great mountains. In a forest above a landscape of farm terraces and pastures, I am walking in the Garhwal Himalaya.

It is a landscape and people with a special place in India's environmental history. For over a century, the colonial British government, followed by Indian foresters, exploited these Himalayan forests, taking timber and tapping pines, for products ranging from bats and boards to resin and railway crossties. Even as the state felled forests for commercial and industrial uses, it restricted access of local people to trees and forests for livelihood needs. In 1970, severe erosion and floods raised concerns about the widespread impacts of deforestation. Three years later, near the village of Mandal, government authorities curtailed access of local people to ash trees but allotted felling permits for the same trees to Symonds, a sports goods company. This drew the simmering tensions to a point of taut confrontation and more than a century of resistance boiled over into a new movement of non-violent action: *Chipko*, meaning 'hugging.' By defending and hugging trees marked for felling, supported by sustained protests, villagers, particularly groups of women, saved forests and livelihoods from being destroyed by logging. As the Chipko movement spread and captured public imagination and attention, it triggered regulations that restricted deforestation, safeguarding forests on the mountain slopes.

The Chipko movement inspired a generation of environmentalists in India and elsewhere. The peaceful resistance of Chipko, named for the gentle, human act of hugging, signified a new environmentalism, an environmentalism of the poor affirming livelihood needs but rejecting overexploitation by industry, renewing people's connections to land, reaffirming the human place in nature.

I come to a halt before an ash tree. I gaze at the grey-toned bark, at the rising ridges and falling furrows on the rough surface, at the trunk soaring up into a canopy flaming in evening light. Now, the soughing shiver of broad-leaved trees, the sibilant whispers of conifers, and the incessant keening of mosquitoes become all too familiar. Now, too, the strange songs of vireo and chickadee and warbler are transmuted into known voices of yuhina and tit and flycatcher. Spruce becomes *raga,* cedar becomes divine *deodar*. Raven remains raven. Suddenly, I am struck by a sense of belonging, as if the ash and cedar are to me what the ash and *deodar* are to people in the Himalaya, the forest a space for solace, succour, and veneration.

Now, I am in India and it is the Vermont forest that has travelled around the world.

# Aesthetics in the Desert

It was a week of stark contrasts: a week spent in the cities, countryside, and wilds of Rajasthan. Not just the banal contrasts of old palaces towering over congested housing and modern squalor, or the civic contrasts pitting city against village or desert. These were contrasts among four divergent ideas of the aesthetics of parks.

From Jodhpur city in Rajasthan, where we visited Rao Jodha Desert Rock Park and Machia Biological Park, we travelled on through an arid park-like countryside to the sprawling dunes and spare expanses of the Desert National Park. This vast dry landscape extending into the Thar Desert carries the imprints of four contrasting, yet intersecting, aesthetics: the aesthetic of care, of control, of the countryside, and of the wilderness.

## Caring for the Land

It is a warm August afternoon; clouds of wailing kites swirl overhead as a small group of us enters the rocky valley lying in the shadow of Jodhpur's spectacular Mehrangarh fort. White-eared bulbuls gorge on white berries of *ghatbor* (*Fleuggia leucopyrus*) bushes nearby as francolins cackle in the distance. Ahead of us sprawls the Rao Jodha Desert Rock Park, a landscape of ancient rhyolite—volcanic rocks that formed around 700 million years ago—overlaid by pink sandstone. Here, amid formidable rocks and arid, thorny desert vegetation lies a remarkable story of ecological revival.

Rao Jodha, the ruler after whom the park and city are named, established the fort in the middle of the fifteenth century. The Desert Rock Park itself was established by the Mehrangarh Museum Trust with the support of conservationists only recently, in 2006. Today, the 72-hectare park embodies one hallmark of attentive human care: restoration.

Both architectural and ecological restoration are under way in Rao Jodha Desert Rock Park. The park entrance lies at the restored Singhoria Pol, an arched gateway through the old city wall. About two-thirds of the 10-kilometre-long city wall has been restored with stones held in place by lime plaster rather than cement. Sandstone slabs marked with ancient wave patterns, the imprint of geological time, form the walkway. Integrated into the Singhoria Pol, along the stone steps and narrow corridors are well-designed interpretive displays about the rocks and geology of the park, about desert ecosystems and their uniquely adapted native plants, and about the park's history and restoration.

The Trust worked with a team guided by environmentalist and tree aficionado Pradip Krishen to ecologically restore the park by painstakingly removing thousands of mesquite (*Prosopis juliflora*) shrubs, an invasive species introduced from Central

America that had proliferated among the rocks. The restorationists employed the local Khandwaliyas, expert at detecting hidden cracks and crevices in rocks, to chip away at the stone to remove the invasive mesquite. Over the next seven years, they worked to carefully bring back a multitude of native plant species.

An urge to see the transformation brought about by ecological restoration had impelled a small group of us to visit the park. Arriving at Singhoria Pol, we pick up informative, illustrated booklets on the park, its rocks, animals, and plants, and meet two of the park's trained naturalists, Sachin Sharma and Harshvardhan Rathore. The two men, dressed in sand-coloured uniforms, lead us on one of the trails into the park.

Near the entrance, in a small raised bed walled by stones, plants burst forth in a living display of arid-land microhabitats—sandy soil, rocks, calcified or saline soils—each miniature habitat holding a selection of native plant species attuned to that particular environment. We walk along a cool, dark gully, an old aqueduct cut from rock to carry rainwater from a catchment in the north to the Padamsar Lake at the base of Mehrangarh fort.

After the monsoon rains, it is hard to visualize this as a harsh, arid landscape. A low stream of clear water flows down the gully. Beyond, over open terrain, low carpets of coarse grass sway softly in the breeze. The grasses hold spiky flower heads or wispy inflorescences that, in one species, hangs like a diaphanous mist over the green blades. The dark pink *missi* or cowpea witchweed (*Striga gesnerioides*) spikes upward holding delicate pink blossoms. The plant lies ensconced among rocks at the base of a 5-foot tall candelabra-like plant, the succulent, thorny, leafless spurge or *thhor* (*Euphorbia caducifolia*).

Over the pale, sandy or gravelly soils, herbs like *Tephrosia, Indigofera,* and creeping *Launaea* sprinkle small, attractive purple, red, and yellow blossoms. From sandy areas and rocky outcrops grow wiry green shrubs like *kair* (*Capparis decidua*) dangling red flowers and milkweeds like *kheer kheemp* (*Sarcostemma acidum*) holding white flower clusters, and *kheemp* (*Leptadenia pyrotechnica*) with their subtle, velvety yellow flowers. Small trees thrust their branches out over the trail: the *peeloo* or toothbrush trees (*Salvadora persica*), the gum arabic or *kumatiyo* (*Acacia senegal*) with curved thorns, the desert date or *hingoto* (*Balanites roxburghii*) with spike-like thorns, and the *bordi* or jujube tree (*Ziziphus nummularia*) with twigs bearing rows of paired thorns—a long straight one and a small hook—little spears and scimitars that pierce and snag our hats and clothes.

Yet, the harshness and aridity of the landscape are evident. The plants are low, shrubs and trees are scattered and sparse. The stones, hot to the touch even on a late monsoon afternoon, will be blistering in the summer. Most plants have small leaves, some waxed or sandpapery like the *goondi* (*Cordia sinensis*), others protected by thorns. To conserve water and survive the heat, many plants have dispensed with regular leaves altogether, and photosynthesize through green stems: wiry as in the *kheemp* or thread-like in the gymnosperm *Ephedra*. Below, the plants put out deep, wide roots to access the little water found in the landscape that receives only around 600 millimetres of rainfall over thirty rainy days in a year.

The naturalists point to planted saplings, raised in the park's native plant nursery from seeds sourced from mother plants in the wider desert landscape around Jodhpur. Some saplings were planted in the same earthen pits or rock crevices from which the alien *Prosopis* was carefully uprooted with the help of the Khandwaliyas.

The ecological restoration in the Desert Rock Park has been thoughtfully executed, avoiding the ill-advised tree planting that is often carried out under the guise of 're-greening' the desert, dune stabilization, or afforestation. The restoration does not regard the arid, rocky terrain, grasslands and thorn scrub as 'wastelands,' as state agencies are wont to do. Instead, the restoration effort recognizes the desert and arid-land vegetation as natural ecosystems in their own right.

The transformation is stark: across the city wall lie stone quarries, disturbed soil, and a monotony of *Prosopis*. Inside the park, in the carefully restored, sparse yet vibrant vegetation, a multitude of native arid-land species flourish among ancient rocks. Here, the earth and the revivified landscape bear few scars, reveal only the healing touch of intelligent care and nature recuperating and resurgent.

By dusk, swarms of little swifts titter and careen in the skies above. Loud cackles of grey francolins subdue the cooing chuckles of laughing doves—birds of earthen tones merging with the rocks. An Indian crested porcupine emerges from behind a thorny bush, her cloak of quills rustling and clicking as she shuffles along on her crepuscular sortie. Out of the calm evening—in the company of rock-loving plants and birds and porcupine—arises a sense that the Rao Jodha Desert Rock Park is a place where, with a little care, people, too, can blend with the landscape.

## Land Under Control

The diversity of species and sensitivity to land that makes the Rao Jodha Desert Rock Park distinctive are completely missing in the presiding aesthetic of Jodhpur's Machia Biological Park. Managed by the State Forest Department, the 41-hectare park carved out of a 600-hectare protected forest promises, at first, to be a centre both informative and recreational. The former objective, which one expects in a place named a Biological Park, soon proves elusive.

We enter Machia Biological Park through an unremarkable entrance that has, appallingly for an arid landscape, plush, heavily watered lawns and a fountain. Past the inevitable parking lot near the entrance, you immediately see the same unremarkable ornamental plants and trees that are seen in thousands of other gardens across the country—yellow oleander, bougainvillea, *pongam*—none of which belong to the desert or the surrounding landscape. The same tired species and other similar ones are found in the visitors' area, along the trails, even in pots inside the office compound.

Where Rao Jodha Desert Rock Park had narrow foot trails beautifully integrated into the landscape, leaving as much space for native plants as possible, Machia has garish, wide roads, paved with quarried sandstone. You can drive two cars abreast here, or even a large lorry. The roads, an eyesore already, additionally have ditches, iron railings, and wire fences running alongside, and scattered shelters made of cement and concrete (each, mercifully for the heat, capped with thatch).

A huge building, built with taxpayer money no doubt, rises like a blister—this is the visitor interpretation centre. Kept locked, a man employed by the park opens it for us to view. Inside, in a room ostensibly meant to provide information about local plants and vegetation, we find walls adorned, bizarrely, with framed dry herbarium specimens: dead plants likely to evoke the enthusiasm of only a diehard botanist.

Another room holds a selection of spectacular photographs of a handful of desert wildlife: the endangered great Indian bustard and lesser florican, mammals like desert cat and chinkara, and a couple of reptiles, the spiny-tailed lizard and desert monitor. Below each photograph, a small metal plaque offers a tidbit of information and the name of the species in English and Hindi. Still, the well-appointed room with a rare species captured within each frame seems more suggestive of photographic conquest than a serious introduction to the region's peculiar desert wildlife. A small group of local visitors walks in as we're looking at the photos, does a swift walk around the room, and exits. Shortly after, as we depart, the man locks the door behind us, leaving us wondering about the privileges and purpose of the centre. One comes out hardly informed or wiser about biology, leave alone the unique geology and ecosystems of the region.

And then it gets worse. Machia includes a miserable little zoo. Behind fences topped by razor wire, wild animals languish in cages and enclosures, baking in the singeing heat. In a small, green, scummy concrete pond hangs a listless gharial crocodile. There are other animals, too, we are told: sloth bear, lion, leopard, fox, wolf, even a porcupine—but we give them all a miss. The fierce sun rains its energy on the place; the hot, wide sandstone road bounces the heat right back at us.

Yet, there will be shade here in the future. The Forest Department, in its undiscerning fixation with trees, has considerately planted saplings alongside. None of the trees are typical species of the surrounding arid landscapes; instead, there are neem, banyan, peepal, *pongam* and other naturalized familiars of urban gardens and parks and the countryside in other parts of India. While one cannot fault the planting of tree species such as neem and peepal, common in the Indian countryside, there is a notable lack of attention in the Biological Park to many valued and interesting species of the regional arid-land vegetation.

And still, the worst part comes a bit later. Seated imperiously in his office compound, under a neem tree in which a Eurasian collared-dove sat brooding at her nest, a forest officer regales us with the challenges of beating the Machia landscape into the shape of the Biological Park. It was all a 'wasteland' here before, just rocks and few plants here and there, he says. If he knew he was describing the landscape and vegetation of a naturally arid area, he shows no trace of it. Not only did we have to build all this here, he says, we had to bring trees into this rocky 'wasteland.' And the way we did it: with dynamite.

Boom! A dynamite blast to shatter 700-million-year-old rocks! And then, bury more dynamite a couple of feet deep for a second blast! With pride in his 'double-blasting' method, the officer describes how they set aside the shattered stones, and then filled the gaping cavity with soil and manure trucked in from outside: all to plant banyan and peepal saplings which were then supplied daily with piped water. 'Look how much they have grown in 1 or 2 years,' he says, pointing to saplings 8 feet tall.

I hesitate to say that they could have just as simply made concrete troughs (painted Forest Department green), placed them on the rock, filled it with soil, to pot their trees. It seemed excessive to have destroyed a rock face that could have told its own geological and earthen story to every visitor, to make artificial pits, filled with imported soil and manure, to grow trees that never occurred here before.

From framed photograph to caged animal, to potted tree to watered lawn, to gushing fountain to quarried roads and wire fences, everything in Machia manifests the same idea: an idea of dominion, that nature, to be tolerated, must be controlled. An idea that the rest of nature, physically and emotionally distanced and separated by fence or frame or human fiat, could then be gazed upon by an appreciative citizenry. And only nature subjugated, objectified, and controlled by the brute force of government machinery, to be turned into spectacle or entertainment or manicured civic park, could be publicly enjoyed.

A sense of relief washes over me as I leave that double-blasted park behind.

## Countryside Aesthetic

In the latter half of the fifteenth century, in the countryside not far from where Rao Jodha's Mehrangarh fort had just come up, a new land ethic arose. It was an ethic that would foster the protection of the trees and animals of the countryside as a matter of faith and as a facet of the lifestyle of the local Bishnoi people.

The Bishnoi sect arose as followers of Jamboji, a religious leader born in 1451 CE. The 29 (*bish-noi* in the local language) precepts he laid out also directed followers to protect nature, not kill animals or cut trees, in particular the *khejri* (*Prosopis cineraria*). The ethic, possibly built upon a reverence for nature and an existing connect between people and land, caught on and spread to occupy a swathe of villages in the arid tracts in the region. In 1731, when a local ruler, Maharaja Abhay Singh, who wanted wood for his palace, ordered *khejri* trees to be cut from the Bishnoi village of Khejarli, southeast of Jodhpur, the villagers rose in protest. Accounts of the incident detail how Amrita Devi, a Bishnoi woman, faced the tree cutters, hugging the tree to protect it, and for her action was beheaded. Her three daughters, who came to her side, were also killed. As more and more Bishnoi stepped forward to protect the trees, they, too, were brutally beheaded. The incident, called the Khejarli massacre, is commemorated today by a memorial in the village. A memorial to the 360 people who lost their lives defending trees and their ethic of reverence for nature.

The influence of the Bishnoi ethic manifests in the countryside aesthetic even today. Near Bishnoi villages, herds of chinkara and blackbuck graze in the open savanna and fallows among scattered *kair* and other shrubs and desert trees, including *khejri*. *Khejri* and other trees, protected or planted, stipple the landscape. The Bishnoi tolerate the occasional antelope herd that enters a crop field to forage. They fiercely oppose poachers—in one famous case, the Bollywood film star Salman Khan—who try to kill the animals.

In the wider landscape of Rajasthan's arid tracts, around Bishnoi and non-Bishnoi villages and towns, the landscape manifests the relationship between people and land. Driving from Jodhpur towards Jaisalmer, we find the highway throbbing with thousands of people—on foot, packed in lorries and other vehicles—on their annual pilgrimage to Ramdevra village to participate in a fair and pay homage at the village shrine. We turn off the highway, taking a road through the countryside to avoid the throng of pilgrims and traffic. We pass villages and towns holding tight clusters of huts and houses, built with earthen walls or brick and cement, topped with thatch or tin or concrete. Linked by trails and tree-lined roads, the settlements lie sparsely scattered across the open countryside. Nearby, small rectangular and geometric plots mark lands under cultivation or lying fallow, neatly set off from the wider landscape of pastures, dunes, and rocks where a multitude of trees and shrubs lie scattered. Dunes, ploughed fields, cultivated plots, savanna-like Israeli babool (*Acacia tortilis*) woodland, and stands of *bordi* whip past. Among the plants are dozens of native species still used by people to source foods, fibres, fuel wood, and traditional medicines.

Across the open terrain lies a smattering of *kair* and *thhor* bushes, flecks of green on sandy canvas, some topped with chattering flocks of itinerant rosy starlings, migrants on their own annual pilgrimage. By the road in a small village, we stop at a dhaba, a makeshift building and home with an old man and a little boy, a string cot placed out front and another within. Sparrows hop about in the dust, picking scraps and insects, carrying them to four nests tucked under the dhaba's ceiling, to feed cheeping chicks. Rufous-tailed larks merge into the fallows, while a rufous-fronted prinia and a small flock of blue-cheeked bee-eaters eye us from their perch on a wire running by the road. Doves coo and grey francolins cry out from the field beyond.

In a few minutes, we are served tasty rotis of bajra or pearl millet, with a *kadhi*, a gravy of chickpea flour, a spicy garlic chutney, and hot tea. Rajasthan, which languishes close to the bottom of the list of Indian states in human development, has the astonishing ability to transform arid lands into productive crops, to retain spaces for plants and animals in the villages, the fields, and the countryside, and to produce such rich flavours from the most modest ingredients.

After the meal, driving on towards Devikot and the Desert National Park, we are caught in a thundershower. In pouring rain, we stop where a river in spate, brown and churning with sediment, flows over the road. The land glistens with rain. Shining pools have formed all around while we wait a little for the rain and river to subside before driving through—the road, like the desert, rendered temporarily invisible under the waters.

Later as we reach the Desert National Park, around one village, or *dhani*, a man dressed in white with a bright red turban herds a flock of black-faced white sheep with a switch of *kheemp* in his hand. Splashes of water glisten in the bright afternoon sun as two men bathe and squealing boys jump into a shallow pool of ephemeral waters formed after the rare downpour. Some distance away, framed by a doorway, little girls stand watching. A couple of bottles and other litter lie scattered at the edge of the village, while a few of the ubiquitous plastic bags have snagged on the thorns of bushes. Across the open vista beyond, a train of five camels slowly undulates over the landscape. Even farther, two bright red-and-yellow specks turn out to be women carrying pots on their covered heads walking towards the *dhani* through amber evening light.

The arid Rajasthan countryside retains an aesthetic that, while it can scarcely be romanticized, appeals to one's eyes and spirit.

The countryside aesthetic emerges from a relationship between people and land that remains reverential and respectful, yet utilitarian.

## Desert Wilderness

A long cool wind blows with a low, melancholy voice carried over miles of desert and dune. Above, the dark cowl of night pinpricked with stars turns slowly. Scorpio crawls below a warm Mars and a cool Saturn; the Milky Way unfurls like a gossamer-thin scarf tossed to the skies. Below, amid thorn scrub and rolling dune, between slithering saw-scaled viper and skittering gerbils, I lie on the earth thinking about the morning's bustards.

In the bright morning at Sudasari in the Desert National Park, the knee-high grassland appears to stretch to the horizon, punctuated by the occasional thorny shrub or *Acacia* tree. Beyond the grasses, around 200 metres away, a long, low, rocky mound rises. And on its ridgeline, in stately silhouette, stand a pair of large birds, like small ostriches about a metre tall. Two long legs hold up each bird's dumpy body, shaped like the canopy of an *Acacia* tree—rounded at the top, flat as a sheet and parallel to the ground like a browse line below. Their long necks—white in the male in front, grey in the smaller female behind—rise like columns, holding proudly their black-capped heads. In unison and with grace, the birds preen, fanning tails upwards; the male fluffs out his throat plumes and sounds a deep booming call. They pause for a moment—a moment when the courting birds in profile seem inseparable from each other and the land.

A few minutes later, they turn together, float forwards in a purposeful walk, and launch themselves into the air. Even from afar, the five of us watching the birds thrill to the sound of the great whooshing wingbeats of the great Indian bustards in flight.

It is a sight we will remember: the silhouetted pair in stately pose on the distant mound against a backdrop of a vast expanse of desert and grassland. But even here, in the 3,162-square kilometre Desert National Park, this offers no vision of a wilderness untouched and unsullied by humans. Around Sudasari runs the Forest Department's fence of chain-link and barbed wire with grassland and antelope on the inside and camels and crop fields on the outside; the chinkara and the nilgai leaping over, between the two worlds, as the idea of conservation sits on a fence. And we will remember, too, the shape seen in the far distance, obscured by a low dust haze but unmistakable, between and behind the courting bustards, of a wind turbine spinning in the desert.

Driving back on the unsealed road outside the Sudasari fence in an SUV, we peer at the denizens of the Desert National Park through the lenses of our cameras and binoculars, the windscreen and windows of our vehicle. Wheatears in sand browns or smart black-and-whites flit from plants to earth and back hunting

insects, while shrikes with fierce hooked bills wait, scanning from low perches. A tawny eagle watches us pass from his perch atop a *kair* tree. A laggar falcon wings away in powerful flight low over the grassland. The open ground beside the unsealed road crawls with hundreds of spiny-tailed lizards: stocky brown lizards with stub-like heads, thick limbs, and flattened, almost crocodilian tails ringed with small conical spines. Many lie watchful, heads raised, stretched out near burrows in the earth, others merely poking their heads out. We have to only stick our lenses and heads out of the windows of our vehicle for the lizards to dash and duck into their burrows.

At night, on my back under the starlit sky I realize that the bustards from the morning were two of less than 200 individuals left in the wild. The Desert National Park remains the last stronghold of this critically endangered species. It has already disappeared from, or is on its last legs, in other wildlife reserves in the country. The National Park was set aside as a vast wildlife reserve and the undulating terrain and desert dunes stretching to the horizon does convey, momentarily, a sense of space, solitude, and natural beauty one imagines, or expects, a wilderness to contain.

The Park forms, however, a small part of the great Thar desert, the most densely populated desert in the world. Spanning 200,000 square kilometres along the boundary of India and Pakistan, the Thar holds a population of over 80 people per square kilometre. The human population density within the Desert National Park itself is lower, distributed across 73 villages or *dhanis*. And yet it is not the *dhanis* as much as the development of the region that now threaten the future of the bustards and the wilderness that remains. Roads snake through the park connecting to highway and city. Power lines and hundreds of wind power turbines stripe and dot the landscape that bustards, moving between the fields and fenced grasslands by day and night or on long flights of up to a hundred kilometres to feeding and breeding sites, have to navigate.

Over the last decade, seven bustards are known to have died after colliding into power lines. More deaths may have gone unnoticed and unreported.

Now, at night, below the vast firmament shimmering with stars, other lights burn along the horizon in the Desert National Park. The lights of villages and towns rimming the land cast a dull glow along the edge of earth and sky. And scattered all around, atop the innumerable wind turbines that sprout from the land, like mould on bread, red lights of warning pulse and pulse as the rotors spin and spin in the melancholy desert wind.

# Twinges of Longing, Passing Shadows

A primary concern in conservation is the extinction of species. Our work often leads us to ask: what should we do to save a species from extinction? The answer, or the search for answers, to this question spurs much of our research, our efforts. Yet, living as we are in the middle of an extinction spasm of the greatest import, we rarely ask the corollary: what should we do when a species does go extinct? In effect, when we fail to stave off an extinction? When a species passes on, should we just heave a collective gasp, drape a commiserative arm around our collective shoulders and move on to the next threatened species? Do we add another sample to the ever-growing database of extinct species for performing multi-dimensional analyses of extinction that incrementally develop our knowledge of why species go extinct? Or should there be something more to it? For with the passing of a species, we also lose any connection we once had with it.

Take a parallel from human life—when a friend passes away, when a close relationship is no more. What do we irrevocably lose, and how much? It is a kind of loss that defies quantification or commodification but, although difficult, it is not a loss that defies description or sentient perception. I realize that I am comparing the loss of species (non-human) with the loss of individuals (people). The loss of species presented this way conflates the loss of individuals within the species. Individuals that, in many animal species, have distinct identities and personalities and have come to occupy the imagination and affections of the people who have studied or got to know them. In any case, one presumes that the appreciation of individuals lost when a species goes extinct can, if anything, only heighten the magnitude of loss. And it is this loss of a species—including its individuals—along with a sense of our lost connections with them, which should not be overlooked when a species is no more. This may not be easy. In the timeless words by Peter Matthiessen in *The Snow Leopard*, which also inspired the title of this chapter:

> And only the enlightened can recall their former lives; for the rest of us, the memories of past existences are but glints of light, twinges of longing, passing shadows, disturbingly familiar, that are gone before they can be grasped, like the passage of that silver bird on Dhaulagiri.

And this we know, too, that the most enchanting of landscapes, to the discerning eye, may become bereaved and desolate with the passage of species. Expressions of this emerge from the best natural history writing and from poetry more often than from science or conservation writing. George Schaller conveys that deeper sense of loss with these words about the Himalaya in his *Stones of Silence*:

> For epochs to come the peaks will still pierce the lonely vistas, but when the last snow leopard has stalked among the crags and the last markhor has stood on a promontory, his ruff waving in the breeze, a spark of life will have gone, turning the mountains into stones of silence.

Ecologist Aldo Leopold, a leading figure in the wilderness and conservation movement, also wrote beautifully about landscapes and loss. In his most famous book, *A Sand County Almanac*, Leopold writes about the grouse in the American woods:

> Everyone knows, for example, that the autumn landscape in the north woods is the land, plus a red maple, plus a ruffed grouse. In terms of conventional physics, the grouse represents only a millionth of either the mass or the energy of an acre. Yet subtract the grouse and the whole thing is dead. An enormous amount of some kind of motive power has been lost.

That there is poetry in these words can hardly be denied. For themes of love, longing, and loss are, of course, within the domain also of great literature and poetry. In the poem, evocatively titled *Longing*, the poet Andrew Slattery conveys this, too.

> The mammoth and the dodo never saw it coming—
> in the end, there is only the idea of species, like a chair
> left swinging when the kids go in for lunch...

The extinction of a species may happen virtually unobserved. A species is there, or is declining, and, after a while, no trace is found of it in the wild. Often we see the causes—hunting or habitat loss or the crippling effects of an invasive species—that bring on the decline to the end. Only in exceptional cases do we know how the final blow was struck. This is probably true of the Great Auk, a flightless penguin-like alcid bird of the North Atlantic, where the last two known individuals, on the lonely island of Eldey in Iceland, were strangled to death and their egg smashed under a human boot. In the words of Peter Matthiessen, in his classic book *Wildlife in America*:

> One imagines with misgiving the last scene on desolate Eldey. Offshore, the longboat wallows in a surge of seas, then slides forward in the lull, its stern grinding hard on the rock ledge. The hunters hurl the two dead birds aboard and, cursing, tumble after, as the boat falls away into the wash.... The shell remnants lie at the edge of the tideline, and the last sea of the flood, perhaps, or a rain days later, washes the last piece into the water. Slowly it drifts down ... down at last to the deeps of the sea out of which, across slow eons of the Cenozoic era, the species first evolved.

We know, too, similarly of the extinction of the cheetah in India. After a long and sorry history of appropriation of habitat for agriculture, of hunting and capture, the cheetah's last days of roaming freely in Indian wilds arrived as the country gained its freedom in 1947. Although a handful of sight records are reported from a few scattered locations until the 1960s, the last definitive evidence is of three male cheetahs seen in Surguja district in Central India in 1947, by the gun-toting Maharajah Ramanuj Pratap Singh Deo who summarily shot them dead. The Private Secretary to the ruler wrote:

> All these three cheetahs were shot by the Durbar in our State (Korea—E. S. A.). He was driving at night and they were all seen sitting close to each other. They were all males.... The first bullet killed one and ... the second bullet after having gone through one struck the other, which was behind it, and killed it also. It is

> not known whether they were born in the State or had migrated from somewhere else. They were all of the same size, as you would see from the measurements and it is believed they were all from the same litter. There is no trace of their parents. They were in perfect condition.

The editors of the *Journal of the Bombay Natural History Society* published this record in 1948, highlighting with an editorial comment that the cheetah was a timid and harmless creature whose numbers had already declined precipitously. They also added scathingly:

> The editors were so nauseated by the account of this slaughter that their first impulse was to consign it to the waste-paper basket. Its publication here is intended in the nature of an impeachment rather than any desire on their part to condone or extol the deed. That anybody with the slightest claim to sportsmanship—and the general run of Indian princes justly prided themselves on that—should be so grossly ignorant of the present status of the Cheetah in India, or knowingly so wanton as to destroy such a rare and harmless animal when he has the phenomenal good fortune to run into not one but three together—probably the very last remnants of a dying race—is too depressing to contemplate. Further comment is needless.
>
> What adds to the heinousness of the episode is that the slaughter was done while motoring through the forest at night, presumably with the aid of powerful headlights or a spotlight. This, it will be recognised, is not only against all ethics of sport but it is a statutory offence deserving of drastic action by those whose business it should be to enforce the law.—Eds.

That we have ultimately lost so magnificent a species to so pathetic a demise leaves me distraught. Decades have passed since, with little effort to sustain the cheetah's memory in India or understand the effects of its absence on our landscapes, on our sentience, and on our lost connections. Like the Yangtze river dolphin or *baiji*, whose recent extinction in China has already relegated it to a fading memory as people slip-slide away in their ever-shifting baseline of awareness, the cheetah, too, vanished from India in more ways than one. Generations have since grown up in a cheetah-less nation, in landscapes bereft of its presence and its spirit. Now, an effort is proposed to bring back the cheetah to India, and a debate has ensued about the hows and whys of it. It strikes me that if one truly fathoms the sense of loss, what exactly we need to bring back will become clear. That one aspect confers utility, if utility be desired, to this process of appreciation of a species that has gone from an area, but not yet from everywhere.

On a wider canvas, many reasons to save species that are still extant have been articulated: there's economics (the money), there's utility (the products), there's ethics (the right to existence), there's aesthetics (the beauty), and there's ecosystem function (the web of life). In the market-driven, utilitarian world of today, ecologists and conservation biologists are going full tilt at the first couple of these reasons, and entire fields of work in environmental and ecological economics have been spawned speaking of valuation of and payments for biodiversity and ecosystem services, with talk of ecosystem marketplaces, of cap-and-trade systems, and sustainable use. This may be applauded as prudent or timely, as innovative or inevitable, and one can, with a little effort and a temporary suspension of a more fundamental awareness, even conjure some agreement over its immediate conservation value. Yet, if we do not take the right lessons from the extinction of species, if we forget the connections we have lost, the palpable, irreplaceable voids that have been created, we risk making a deep error. An error that only dulls the mind and hardens the heart to reconcile ourselves to a more impoverished existence in a more inconsiderate, inhuman world.

# Being with Dolphins

There is a dark sea above and a dark sea below. With one I am transfixed, with the other forever moving. Above, the arched firmament is smeared with galactic grey and sprinkled with silver brilliance of stars uncounted. Below, a fathomless depth hides under a smooth lustre, crested with white ribbons of surf and the luminescent wake of our passage.

And there is, with the wind, the gentle wind, tugging at my t-shirt, sifting through my hair, my eyes, eyelashes, over my hands and my legs, sighing in my ears, a light swell on which the boat rises, and a moment poised on a vertex of consciousness, filled with being.

In boundless seas, I am transfixed, I am moving, I am.

The moon is yet to rise. Behind me stretches the boat, its throbbing engine now silenced with a switch. The mizzen sail billows with mainsail and foresail, and the boat leans into the darkness. There is a lull and a surge of air as if the ocean has held its breath briefly: the sails slacken and then fill with a pop, like a slap on the rump of a horse that gets it going again.

There is no other boat or ship around. Except for the faded glow of an instrument panel astern, there is no other light not of the seas. There is just us, in a boat pointed towards an unseen island. People of a purpose sailing on the undefined and relentless purpose of the seas.

Dawn flenses the cowling of night off the waters, revealing clear blue unmarked by cloud. The world opens before us and the bow parts the brightened waters. Flying fish break forth, like a fountain of grasshoppers flushed in a meadow. They arch through the air gleaming and flashing in sunlight. They skitter the surface, rise briefly, and plunge. The water is glassy smooth and secretive again.

Suddenly the sea is alive with spinner dolphins. Their sleek and shining shapes course through the waters in a sibilant rush. In energetic waves they rise and breathe and curve and dip, sinuous undulations that scarcely mar the waters. In the distance others breach the surface into the air in exuberant bursts, spinning and twisting and falling in founts of spray.

The water is cobalt and clear and I watch a dolphin near me swimming his sea as he watches me sail through mine. His curved fin and flippers and flukes, the snout and streamlined body are all crafted to perfection in the waters. The dolphin effortlessly keeps pace, now scouting ahead, now falling back. And then with a surge he is gone. Gone, like the rest of them are gone. Barely ten minutes of being with dolphins and yet there is a pang of loss at their passing.

The boat cruises on and the sun rises into brilliant day. Did we come upon the dolphins or they come to us? The dolphins have the answer, and I wish I could ask them. I feel a strange kinship with them: is it because I know that they know?

The biologists have figured this much. Dolphins and their kin, porpoises and whales, are among the most intelligent mammals. Their large, intricate brains (in relation to their body mass) place them somewhere between humans and the great apes. Faced with a mirror, a bottlenose dolphin can recognize herself, a self-recognition that bespeaks a self-awareness and earns a membership in a small but growing club of animal species, one that includes the human being. Dolphins are social and empathic, intelligent, and emotive. They can be affectionate, enchanting, aggressive, playful, endearing. Their life is in the open sea. The life of the sea is in the dolphins.

The sun sears its way west. As dusk settles, a pod of pilot whale makes its way through the darkening waters. A brown haze hangs over the water, like an airy smog, the breath of a sea monster. Through the haze, the sun drops quickly from blood-red sky to bloodied sea. Our journey is not over.

The intelligence and sentience of dolphin and whale carries consequences, as do ours. Dolphins and whales such as orcas can be driven from delight and vitality to depression and debilitation when held captive in the artificial sea 'worlds' that are tanks and puddles. They can become extremely distressed when people drive them for slaughter or separate a mother and her calf for capture and trade. Then the dolphins and whales must pay for their lives, their existence of sorts, by succumbing to perform and amuse other people to the chimes of artificial music and the ringing of the cash registers. We know now for sure, the biologists say. They can feel pain. They can suffer. They are sentient beings, too.

Darkness returns and we are enveloped by the seas, with dolphins on our minds. What does it mean to be a human being in a world with other sentient beings? And what the moral imperative of our ability to bring far greater harm and pain to a dolphin than he or she can ever bring to us? Will our search for new worlds and other intelligent life bring us great discovery from the starry sea above, or from the yielding sea below? Or will it come instead from the sea within us, in surprise and joy and revelation? 'When it is dark enough,' wrote Ralph Waldo Emerson, 'you can see the stars.'

It is early yet in our quest into the lives and languages, the cultures and personalities, of dolphins and whales. The interpreters are still busy: marine scientists and other philosophers, the writers and the poets. Every day they probe the seas, fish out a nugget of knowledge or ravel out the skein of connections. It is an expansive, artful, expanding world.

Meanwhile, I am on the bow of the boat, cruising the dark seas. I sense an impending arrival at a place ordained but of my own choosing, too. And a sense of place impels me through waves of thought into a consciousness of what it means to be.

# Sentience for Conservation

What would our life be like if we could see, but not discern? If we could hear, but not listen, and if we could touch, but not feel? How would we experience life if we could taste and smell, but not savour? What would we be like, as a species and as individuals, if we could sense everything, yet make sense of nothing? Would our life be the same? Would we be the same? Would we even be human?

Biologists and philosophers have many lofty answers to the deeply fundamental questions of human existence. Ask Richard Dawkins and he will, delving into the firmaments of the science of evolutionary biology, essay answers to the question he posed in the opening of his famous book *The Selfish Gene*: why are people? The answers provide one view of our existence. Ask the philosophers and they will thread you through the arguments as to what sets apart *us* from *them*, and how we know we are who we are. The religions and the prophets have their own answers, too, some deep, many dubious. For me, as yet, the glimmerings of an understanding hover at the periphery of my vision, but it is clouded by an intellectual cataract that needs to be lifted.

We are a species named *Homo sapiens*, that is, the human that knows or the human who is wise. Sometimes it seems strange that *sapiens*, a Latin word meaning 'wise', is applied to our species. Behind and beyond our intellectual and cultural achievements is a litany of apparently senseless acts—war and plunder, environmental destruction and pillage, racism and genocide, crimes and violence—which questions the assumption that we are the wise ones. Are we truly sapient? I, for one, am not so sure.

We are also called *human beings*. I am not a trained philosopher, yet it seems to me this is a term of firmer substance. It suggests a species that has something above a mere functional existence, it hints at the possession of a *mind* of non-trivial cognitive capacity, and of certain existential qualities of perception and self-awareness. To me, it suggests and in some ways is inseparable from, a refined quality of sentience.

The dictionaries define *sentience* as 'the state of having or feeling sensation, or our faculty or readiness to perceive sensations'. We may perceive our own sentience and those of others in many ways. A neurologist may see it in the firing of neurons in the brain as clearly as a behaviourist may see it in the turn of a head. It may be in the dilation of the pupils, in a lump in the throat, or, during the aftermath of an emotive moment, in an averted glance or in words said or left unsaid. We *feel* it; it *affects* us.

Are we a sentient species? Sure, we are.

If we take sentience to refer to the form of perception or awareness of sensations emanating from our sense organs, we are clearly not alone, as a species on this planet, in being sentient. Yet, sentience has also been defined as 'an example of harmonious action between the intelligence and the sentiency of the mind.' Applied to us, this view of sentience suggests the need to strike a harmony between our intelligent understanding of the world and our mind influenced by sentient perception. It suggests a marriage between reason and affect. A marriage that, if performed, may justify our claim, as a species and as individuals, to being exceptional.

I think of human sentience often in the context of conservation. I think of it when a burst oil well a mile under the sea spews—*spews*, not spills—millions of litres of oil into the open ocean. When equatorial rainforest of exhilarating diversity is cut and burnt to make way for a vast plantation of one species. When the furrows of old roads and mines are still raw on the hills and the metal claws of heavy vehicles gouge for more. And when the rail track sings to the passing of an express train—sings a ringing requiem for the four elephants left behind, their life ebbing away in stunned and bloody repose. I think of it, even, when the man, by the side of the road, raises his crowbar to bring it down on the head of a small, harmless, and nearly-blind burrowing snake, just because it is a snake.

Aren't these, and many other human-nature interactions, matters that not only concern us, but *affect* us? Should we then approach solutions for a reconciliation purely through reason and science, as is a common refrain, or include in our ambit human emotion and feeling? Can we build a popular movement, devoted to a cause as to a nation, if we were to use only logic and dry fact, ignoring sentiment and disposition, music and arts, poetry and passion? Should we always seek answers in our intellect rather than in our humanity? In today's world, where credible science is called for to inform debate and decisions, human emotion and feeling is treated as an errant child to be kept in rein—side-lined, side-stepped—or as an unwanted churl who would confuse rather than clarify. In the process, a great and material part of human existence is brusquely overlooked.

I think an approach built on science, alone, cannot help conservation. We must include human sentience. Both reason and affect must be brought to bear on conservation problems. The idea is not new, yet it is seems worth articulating, reiterating. Fortunately, threads of support for this approach are emerging from diverse sources.

First, an over-reliance on science alone may turn out to be counterproductive (or at least insufficient) as seen in climate change campaigns. The environmental journalist George Monbiot writing about 'The Unpersuadables' says, 'The battle over climate change suggests that the more clearly you spell the problem out, the more you turn people away.' He sounds lost 'that there is no simple solution to public disbelief in science'. I cannot help wondering if an approach that did not rely only on science would help more.

Understanding human emotions and incorporating that into how we deal with human–wildlife interactions, conflicts, and conservation issues is now being suggested as an important direction to take. The discipline of 'conservation psychology' is also taking shape, hoping to link the understanding of human behaviour with conservation. Writing in the book *Who Cares about Wildlife?* Michael Manfredo presents emerging ideas and results of research on the effects of emotions on memory, decision processes, norms, values, attitudinal changes, and health. His tentative conclusion is that

> Emotions act with cognition to direct human behaviour. They play an important role in memory, decision making, and attitude change; they clarify roles and social structure.... Wildlife professionals should re-examine the widely held view that emotional response issues are trivial, unimportant, or

> non-informative. Emotions merit careful consideration and thoughtful response.

He also quotes Jon Elster, who in his book *Alchemies of the Mind: Rationality and the Emotions,* says more pithily, 'Emotions matter because if we did not have them nothing else would matter.'

Another line of argument comes from the work and ideas of the renowned primatologist Frans de Waal in his *The Age of Empathy: Nature's Lessons for a Kinder Society*. De Waal opens his book with these questions:

> Are we our brothers' keepers? Should we be? Or would this role only interfere with why we are on earth, which according to economists is to consume and produce and according to biologists is to survive and reproduce?

Linking both these ideas of competition-is-good-for-you to their origins around the time of the Industrial Revolution, de Waal presents a survey of modern research in animal behaviour, primatology, and anthropology, where there is compelling evidence for the importance of empathy in moulding social relationships. He examines social animals from dogs to dolphins, monkeys and apes, wolves and elephants. From this, he points out, 'If man is wolf to man, he is so in every sense, not just the negative one.'

He also does not shy away from talking about emotions and moods, greed and gratitude, attachments and morality. *The Age of Empathy* is an important book, from one of the world's leading primatologists, and what it addresses is tantalisingly pertinent. 'What is it that makes us care about the behaviour of others, or about others, period?' And empathy is central, too, as the philosopher Martha Nussbaum suggests in her book *Frontiers of Justice: Disability, Nationality, Species Membership,* for a world where peaceful coexistence among nations, justice for the less privileged, and concern for other species can flourish. Can we probe the hidden wells of human empathy for a more benign and graceful citizenry on this planet?

The foundations of a conservation ethic must be built on human sentience, tempered by empathy. And for this to work it may need to sincerely garner the support, not only of conservation scientists, but of painters and musicians, poets and songwriters, playwrights and psychologists, humourists and social workers. It needs, as is often said, to rebuild burnt bridges across the arts, humanities, and the sciences. It needs to bring back into serious discourse our motivations, emotions, passions, sensitivity, and humanism. Then, perhaps, in the years ahead, we will tread our path on planet Earth as *Homo sentiens*.

# Epilogue

## Tinker, Tailor, Soldier, Spy

### *A Personal Journey into the Wild Heart of India*

The door of my home in the Anamalai Hills frames a view of the distant dome of Anaimudi, 2,695 metres, the highest mountain in the Western Ghats. The rounded peak, like the head of a giant elephant, hides in haze, nestles in clouds, or rises in austere grace into gin-clear skies, depending on the moods of earth, sky, and wind. Living here at the edge of a hill town in a peopled landscape of forests and plantations, I have seen from my doorstep wild elephants, great hornbills, bright mongooses, sounders of wild pig and murmurations of starlings, stealthy leopards, shy sambar, tranquil snakes, skies flickering with swifts and buzzing with migrating dragonflies, little pied bushchats singing from the wires, the Malabar whistling-thrush serenading rain and river, and so much more. From the shola forests and grasslands in the distance, clear streams flow into the River Sholayar, named after the rainforests the waters emerge from, the dark, wet evergreen forests that covered this landscape until about 125 years ago. Now, in the surrounding landscape of the Valparai plateau lie the rainforest fragments and the tea and coffee plantations where I have been working with Divya Mudappa, Ananda Kumar, and many colleagues, friends, students, partners, and volunteers over the last two decades, carrying out research on wildlife, ecologically restoring degraded rainforests, and trying to enable coexistence between people and species such as leopards and elephants. It is from here that I travelled to places across India, visiting busy cities, savouring India's rich and beautiful countryside, exploring wild forests and grasslands and deserts, only to return again, as one always must, to *oikos*, to home. My travels, my time in nature, my research studies, have carried me through defining moments and enriching chapters of my life that have formed a particular and personal journey into the wild heart of India.

Moments on the journey are imprinted in memory: the last banyan that stood by the road, for instance; the great peak that rose above its range seeming to grow taller as I moved away; the wise elephant that paused and looked me in the eye as I passed by. But the journey itself has taken an arbitrary course and can take an arbitrary name. To loosely thread the narrative through the chapters, a title borrowed from a John le Carré novel, *Tinker Tailor Soldier Spy*, will do nicely, but set in reverse.

## Spy

> Is it really worth while to spend our time, the time which escapes us so swiftly, this stuff of life, as Montaigne calls it, in

> gleaning facts of indifferent moment and highly contestable utility? Is it not childish to enquire so minutely into an insect's actions? Too many interests of a graver kind hold us in their grasp to leave us any leisure for these amusements. That is how the harsh experience of age impels us to speak; that is how I should conclude, as I bring my investigations to a close, if I did not perceive, amid the chaos of my observations, a few gleams of light touching the loftiest problems which we are privileged to discuss.
>
> —*The Mason Wasps*, Jean-Henri Fabre

If a spy is taken to be an observer, someone who sources information to analyse and report factually on it, doing this in a sort of self-effacing manner, remaining detached, possibly hidden, inconspicuous, then perhaps naturalists and field biologists are spies, too.

My initiation as a spy-naturalist began the day I, as an eleven-year old, took my parents' binoculars from the bedroom cupboard, walked up to the terrace of our Chennai home, in Mylapore, and watched for the first time the rose-ringed parakeets careening into the trees in our neighbour's yard.

The Mylapore house was built in 1972, the year I was born, the year India passed its Wildlife Protection Act, 10 years after *Silent Spring*, in the decade when the country and the wider world was awakening to environmental consciousness, legislation, and activism. The decade witnessed the Chipko movement, the creation of wildlife sanctuaries such as Guindy, Kalakad, and Anamalai that would later entrance me for years, and culminated in the passage of the Forest Conservation Act in 1980. Sheltered at home and school, I remained oblivious to these developments. It was the simple, joyous act of birdwatching that first pulled me out of home and the confines of classrooms into contact with the daily wonders outside the walls.

Birding in my backyard began my ecological education. Bird research in forest and field continues my learning as a wildlife scientist. It has taught me that an astonishing diversity of birds thrives in the city, the country, and the wild—from parakeets to tropicbirds and partridges in Chennai; peacock-pheasants to tesias and wren-babblers in the jhum landscapes of Mizoram; warblers to flycatchers and hornbills in the Western Ghats rainforests. Opening your eyes, ears, and imagination to the birds around you brings new perception and wonder. The migratory warbler arrives in your garden marking the passage of another year, linking your home to distant lands. The whistling-thrush and river tern and fish-eagle, through their voices and arresting presence, relay the health of streams and rivers. The imperial-pigeons and hornbills and flowerpeckers, locked in an intimate ecological embrace with the fruits of trees and mistletoes, signify connectedness and renewal. Every bird, like every other animal or plant, has a story waiting to be discovered by the spies who care to open their senses and sensitivity to nature.

I have been fortunate to gain the company of many others who share such curiosity and concerns over nature that helped expand my ecological education in this world of wounds and impel me into conservation. My friends and colleagues at my institution, the Nature Conservation Foundation, work in diverse landscapes and waterscapes across India. They dive in Lakshadweep and the Andaman and Nicobar islands to study the ecology and trajectory of coral reefs and marine life; they roam the cold desert steppes of the high Himalaya in the land of the snow leopard to protect wildlife with people; they strive in the dense forests of Arunachal Pradesh and northeast hill states to conserve hornbills and other endangered species; they work in the Western Ghats to alleviate human impacts on forests and learn to live with wildlife; they survey plains and wetlands from Gujarat to Bihar to conserve cranes and waterbirds; they engage local communities, citizens, and children across India in watching and appreciating nature; they try to fathom the behaviour of primates and elephants and birds and ungulates and fish and people. Out in the Anamalai Hills, as a team of biologists and conservationists working out of a field research station, we now gather daily the stories of plants and animals in rainforest fragments and plantations of tea, eucalyptus, coffee, and cardamom. We see more clearly the wonders and wounds in the landscape, the newer cuts and older scar-tissue, continue our efforts to understand what we can through research, and work to protect what remains.

## Soldier

> War must be, while we defend our lives against a destroyer who would devour all; but I do not love the bright sword for its sharpness, nor the arrow for its swiftness, nor the warrior for his glory. I love only that which they defend ...
>
> —*The Lord of the Rings: The Two Towers*, J.R.R. Tolkien

One feels compelled to rise to the defence of what one cherishes when it is threatened. Just as the villagers of Khejarli and Garhwal rose to protect the trees they valued in their landscape of desert or mountain in decades past, the conservation movement today has its foot soldiers everywhere. And as the wider world trundles on with an overwhelming focus on industrialization and economic growth, the conservation battles they fight are neither trivial nor easy.

There are city dwellers who rise again and again in the defence of street trees and urban wetlands and parks, from Bengaluru to Gurgaon, Mumbai to Guwahati. In rural areas, people rise to defend their own lands from acquisition for destructive projects, highways, and dams, or to safeguard the vitality of the countryside itself: the great trees along country roads, the diversity of crops, livestock species, and ways of life evident from paddy fields to jhum fields and fished waters to pastures across India. Even as the government went about creating a slew of over 600 wildlife reserves across India, other areas, too, had their protectors. Communities protect hundreds of sacred groves and sites across India's forest landscapes, although many of these have declined or disappeared in recent years. In Niyamgiri in Odisha, the Dongria Kondh tribal people saved their forests and sacred mountain from mining. In Nagaland, the people of Khonoma set aside a community forest reserve. In Bitra island in the Lakshadweep, fishers created a no-fishing zone in the ocean to protect a rare fish-spawning aggregation. In these and other examples, the motivations and approaches that sparked the soldiers into action differed, but there is little doubt that they all helped contribute to conservation.

And so we soldiered on, too, in the Anamalai Hills. There were forest fragments to protect, lying in private lands outside designated wildlife reserves. The remnants lay in an ecological hinterland, a paradoxical one, which one sees with fresh eyes leaving behind rainforests deep inside reserves and entering an apparently human domain. In this landscape with over 70,000 people thrive troops of lion-tailed macaques, great flocks of hornbills and pigeons, butterflies and babblers, elephants, deer, fungi, mosses, incredible insects ... life, everywhere. The forest remnants lay ignored by the state Forest Department as they were outside the boundaries of reserves. They were overlooked by private plantations that focused on their businesses and crop production. There were magnificent trees—rainforest giants and hornbill homes—to save from girdling and felling. There were fences to erect to fight abuse and overuse of land, to stave off the insidious attrition of forest remnants due to road expansion and degradation. The fragments and the wildlife they sheltered seemed to need soldiers, not so much as protectors, but as champions who could turn other appreciative eyes towards their conservation.

Protecting land for conservation also creates conflicts. Conflicts between the state that sets aside wildlife reserves and the local people who feel dispossessed and excluded. Conflicts between tribal ways of life or conservation needs and industrial expansion. Conflicts between those in a community out to conserve a resource for the long-term and those out to make a quick buck of it. Conflicts outside designated wildlife reserves between people and other species using the landscape: the elephants and leopards and nilgai and wild pig that seem suddenly out of place.

As we went about working to reduce conflicts in the landscape, we came to realize that the arena of conflict over land is defined by dimensions of possession, protection, and punishment. It is a cold and brutal and lonely space. The possession of land creates owners, not guardians; protection also builds walls between people and nature; and punishment singes the psyche, provokes retaliation and resentment.

## Tailor

Before I built a wall I'd ask to know
What I was walling in or walling out,
And to whom I was like to give offence.
Something there is that doesn't love a wall,
That wants it down ...

— 'Mending Wall', Robert Frost

The questions then arose: how can we redraw the connections eroded and lost as chasms grew or barriers rose between protected wildlife reserves and surrounding landscapes? How does one quilt the city, the country, and the wild into one ecologically viable landscape? How does one stitch the cuts and tears on the land to heal the world of wounds?

Across India, efforts by a variety of people point the way. Some work to include in the ambit of conservation areas outside designated wildlife reserves such as wetlands, pastures, agroforests, grasslands, and forest remnants, where many species, including threatened ones, still survive. They foster partnerships between various people who own, use, or manage land, or stake claims based on culture and belonging and livelihoods, working to moderate land uses, sustain harvests, or mitigate conflicts with wildlife. Others work with forest dwellers to secure their rights to land and to conserve their resources, under the Forest Rights Act, although its implementation by many states remains tardy and incomplete. As consumer awareness grows, citizens pulling a product off a market shelf increasingly sense the thread that connects their purchase to distant places and modes of production and grab the opportunity which it creates to vote with their wallets to influence business and land-use practices for the better.

In the Anamalais, our research revealed, too, that plantations were not devoid of conservation values. A graded intensity of production where agriculture becomes industry is matched by a gradation in the diversity of life. Plantations could act as refuges for many species, when they hold habitats such as forest and grassland remnants or use native trees of the region as shade for their crops. Plantations with their embedded habitat remnants could also function as stepping stones and corridors for the movement of animals from hornbills to dhole. With land-use insensitive to ecology, elephants and leopards come into conflict with people, causing loss of life or damage to property, and rivers and streams turn to trickles and drains, with water becoming a scarce resource even in areas of highest rainfall.

We try to bridge the gaps between the Forest Department and private land owners to better appreciate landscapes outside protected reserves and liaise with each other for conservation. We work to rekindle the interest of local communities to appreciate and marvel at what has always been around them, but subtly hidden, through conservation education and awareness initiatives. We work on the belief that planters can be influenced to take a larger landscape perspective, consider spans of time well beyond their annual balance sheets. Perhaps then, plantation companies would begin to not merely exercise their rights as owners of land but act on responsibilities as guardians of landscapes, respond not merely to demands of company shareholders and employees but to concerns of community stakeholders and citizens. Perhaps we would, by working with all of them, come to understand them and our own place in the scheme of things.

Aldo Leopold wrote,

> We abuse land because we regard it as a commodity belonging to us. When we see land as a community to which we belong, we may begin to use it with love and respect.

For us, the work to build such bridges, tailor relationships, and stitch broken connections to regain that sense of community is a work in progress.

## Tinker

> To keep every cog and wheel is the first precaution of intelligent tinkering.
>
> — *A Sand County Almanac*, Aldo Leopold

If conservation entails a looking back to prevent what may be lost, restoration envisages a looking forward to what may be gained. When deserts and wetlands and savanna grasslands and rainforests are damaged or destroyed, one may never be able to recover everything they had in their undiminished states. Can ecological restoration then help attain a new flourishing and regain complexity, diversity, and beauty through intelligent tinkering by people who care for the land? Can we transform landscapes of conflict between people and wildlife, wrought by relationships gone awry, into landscapes of coexistence? Tantalizing hints from the efforts of local people, field researchers, and conservation practitioners indicate the possibility.

The first step, as Aldo Leopold wrote, is to keep all the pieces: people, wildlife, forests, plantations, rivers. No country offers greater opportunities for this, perhaps, than India. In India, one can still see people living in close contact with nature and myriad wildlife species in the city, the country, and the wild. City dwellers living alongside leopards in Mumbai or alongside thousands of waders and waterbirds; farmers and shepherds sharing spaces with sarus cranes and wolves and foxes in the vibrant countryside of central and northern India; forest dwellers sharing space with elephants and tigers in the forests of the Western Ghats—all testify both to what remains and what is possible to conserve in the world's second-most populous nation. But we cannot be too complacent. Many species remain on tenterhooks, for instance, the endangered greater adjutant storks that shuttle between Guwahati's garbage dump and their tree nests in people's backyards and home gardens in Assam's villages, or the last few great Indian bustards roaming the deserts and grasslands of India navigating landscapes sliced by power lines, or the small populations of hangul and sangai living precariously in wildlife reserves.

The second step is to rekindle relationships between people and place, between people and the rest of nature. This requires nurturing a land ethic, perhaps one constituted as Leopold described:

> All ethics so far evolved rest upon a single premise: that the individual is a member of a community of interdependent parts. The land ethic simply enlarges the boundaries of the community to include soils, waters, plants and animals, or collectively the land.

Or it could entail reviving and cherishing a land ethic that already exists but has waned, as among the Bishnoi or myriad other human communities from hunter-gatherers and pastoralists to fishers and farmers across India. Inspiring examples are coming forth across India: from communities working to conserve their forests and fisheries and pastures, from conservationists carrying out ecological restoration in the deserts of Rajasthan or in the forests of the Himalaya and Western Ghats, from citizens participating in movements to change the manner in which land and nature are treated and respected.

And so in the Anamalai landscape, in our small way, we pursue our own tinkering. We plant saplings of more than one hundred and fifty species of native rainforest trees raised in our nursery in degraded rainforest remnants in an attempt at ecological restoration. We track the movement of wild elephants, conveying their presence to local people by text message, cable television, or alert beacons to ensure their safety and enable the animals to pass by in peace. We build bridges, literally with canvas, linking the canopies of trees on either side of highways to enable macaques and tree-dwelling animals to cross safely from one side to another. We conduct training and outreach and foster dialogues with estate workers, plantation managers, Forest Department officers and field staff, tribal people, tourists, and citizens, to renew their connections with land and alter their actions to better accommodate the wild species in their midst.

With slow, halting progress, sudden slip-slides of regress, we inch forward in our stick-in-the-mud form of conservation. Every year, in drenching monsoons, as we plant rainforest saplings in the earth, a palpable feeling emerges. As the saplings grow into spindly stems, and some into robust treelets, as canopies spread and rainforest birds grace the shrubbery, as leopards sharpen their claws on the bark of trees planted by us within the last two decades, as the young *Vernonia* and *Heynea* trees burst into flower for the first time and the fruits mature to send their seeds flying on the wind or in the bellies of visiting birds, we watch and wonder at nature resurgent, at rainforest revenant.

Time will carry us ahead, even as there is a longing to go back. Back to those rainforests, in the heart of Kalakad, which left their mark on us along our journey. To a field station surrounded by rainforests and rivers, remote from people and the chattels of modernity, without electricity or internet, to days before phones became mobiles, when letters were written by hand and books were only read by turning the paper over, when one could sit and gaze at mountains with meaning.

But it is here in the Anamalai Hills, among people living on mountains named for the elephants, drinking the waters of a river named for the rainforest, that we have come to feel that we are only a member of that larger community that Leopold called the land. It is here that we gained a sense that, in a world of wounds, to intelligently tinker with and restore land, one needs to nurture a land ethic. A land ethic and place in a community, open to all who care to participate, who will feel moved to act and make space for other species in their lives and in their hearts.

# Notes

## Six Seasons in the City

**p. 4** *Madras Naturalists' Society* See www.blackbuck.org.in.

**p. 5** *Ritusamhara* Kalidasa. 1947. *Ritusamhara or Pageant of the Seasons.* Translated from Sanskrit by R.S. Pandit. The National Information and Publications Ltd., Bombay.

**p. 6** *crowd the woodland* Raman, T.R.S., Menon, R.K.G., and Sukumar, R. 1996. Ecology and Management of Chital and Blackbuck in Guindy National Park, Madras. *Journal of the Bombay Natural History Society* 93: 178–92.

**p. 8** *larger herds form* Raman, T.R.S. 1997. Factors Influencing seasonal and Monthly Changes in the Group Size of Chital or Axis Deer in Southern India. *Journal of Biosciences* 22: 203–18.

**p. 9** *Juvenile males attaining hard antler* Raman, T.R.S. 1998. Antler Cycles and Breeding Seasonality of Chital (*Axis axis* Erxleben) in Southern India. *Journal of the Bombay Natural History Society* 95: 377–91.

**p. 11** *yearly turmoil of rut* See Selvakumar, R. 1979. On the Ecology and Ethology of Blackbuck *Antilope cervicapra* (Linnaeus) and the Chital *Axis axis* (Erxleben) at the Guindy Deer Sanctuary. MSc Dissertation, Madras Christian College, Chennai.

**p. 12** *botanists from the Madras Christian College* My thanks to P. Dayanandan, C. Livingstone, D. Narasimhan, Ravichandran, and A. Vasuki for identifying the plants.

**p. 17** *events of every month* See Rajasekhar, B. 1992. Observations on the Vegetation of Guindy National Park. *Blackbuck* 8(2): 38–42.

## Night Life in Chennai

**p. 20** *system of navigation of bats* Griffin, D.R. 1959. *Listening in the Dark: Acoustic Orientation of Bats and Men.* Yale University Press, New Haven, Connecticut.

**p. 22** *11 kilohertz to 212 kilohertz* Jones, G. and Holderied, M.W. 2007. Bat Echolocation Calls: Adaptation and Convergent Evolution. *Proceedings of the Royal Society B* 274: 905–12.

**p. 22** *evolutionary biologist, Richard Dawkins* Dawkins, R. 1986. *The Blind Watchmaker.* W.W. Norton and Co., New York.

**p. 24** *more insights have emerged* Jones, G. and Teeling, E.C. 2006. The Evolution of Echolocation in Bats. *Trends in Ecology and Evolution* 21: 149–56.

**p. 25** *flying foxes and other bats* see Bates, P.J.J. and Harrison, D.L. 1997. *Bats of the Indian Subcontinent.* Harrison Zoological Museum Publication, Kent, England.

**p. 31** *As Henry Beston wrote* see Beston, H. 1928. *The Outermost House: A Year of Life on the Great Beach of Cape Cod.* Ballantine Books, New York.

## Lone Palm Tree, Sir!

**p. 34** *The Mountain Gorilla* Schaller, G.B. 1963. *The Mountain Gorilla: Ecology and Behaviour.* University of Chicago Press, Chicago.

**p. 36** *Allee effect at work* Allee effect, in population biology, refers to a positive correlation between population density and measures of fitness, like breeding success.

**p. 36** *discussed in leading journals* Stephens, P.A. and Sutherland, W.J. 1999. Consequences of the Allee Effect for Behaviour, Ecology and Conservation. *Trends in Ecology and Evolution* 14: 401–5.

**p. 37** *1974 work on sampling methods* Altmann, J. 1974. Observational Study of Behavior: Sampling Methods. *Behaviour* 49: 227–67.

**p. 38** *Shingo Miura on social behaviour* Miura, S. 1981. Social Behaviour of the Axis Deer during the Dry Season in Guindy Sanctuary, Chennai. *Journal of the Bombay Natural History Society* 78: 125–38.

**p. 38** *publishing a paper in 1982* Menon, R.K. 1982. Observations on Cheetal at Guindy National Park, Chennai. *Cheetal* 24(1): 37–40.

**p. 38** *paper by Anderson and Pospahala* Anderson, D.R. and Pospahala, R.S. 1970. Correction of Bias in Belt Transect Studies of Immotile Objects. *Journal of Wildlife Management* 34: 141–6.

**p. 40** *bird counts ... at Vedanthangal* Santharam, V. and Menon, R.K.G. 1991. Some Observations on the Water-Bird Populations of the Vedanthangal Bird Sanctuary. *Newsletter for Birdwatchers* 31(11 and 12): 6–8.

## The Tropicbirds of Memory

**p. 44** *this excellent bird journal* The *Indian Birds* journal is now an online publication. URL: www.indianbirds.in.

**p. 44** *Notes on Indian rarities* Praveen, J., Jayapal, R., and Pittie, A. 2013. Notes on Indian Rarities-1: Seabirds. *Indian Birds* 8(5): 113–25.

**p. 45** *white-tailed tropicbird* Also known earlier as the White tropicbird.

## Fording the Flood

**p. 55** *establish the Nature Conservation Foundation* As more people joined over the years, the organization grew to work across India. See website: www.ncf-india.org.

## Answering the Call of the Hoolock Gibbon

**p. 57** *carry out a survey* Our key findings from the survey are recorded in the following publications: Raman, T.R.S., Mishra, C., and Johnsingh, A.J.T. 1995. Survey of Primates in Mizoram, Northeast India. *Primate Conservation* 16: 59–62; Mishra, C., Raman, T.R.S., and Johnsingh, A.J.T. 1998. Hunting, Habitat and Conservation of Rupicaprines in Mizoram, Northeast India. *Journal of the Bombay Natural History Society* 95: 215–20.

**p. 61** *knowledge about this species* Gupta, A.K. and Kumar, A. 1994. Feeding Ecology and Conservation of the Phayre's Leaf Monkey *Presbytis phayrei* in Northeast India. *Biological Conservation* 69: 301–6.

**p. 63** *available on hoolock gibbons* For an up-to-date account about the species, see Kakati, K., Alfred, J.R.B., and Sati, J.P. 2013. Hoolock Gibbon *Hoolock hoolock* and *Hoolock leuconedys,* 332–54 in A.J.T. Johnsingh and N. Manjrekar, Editors. *Mammals of South Asia.* Volume 1. Universities Press, Hyderabad.

## Bamboo Bonfires and Biodiversity

**p. 68** *changes in vegetation and wildlife* The results of the studies appeared in the following publications: Raman, T.R.S. 1996. Impact of Shifting Cultivation on Diurnal Squirrels and Primates in Mizoram, Northeast India—A Preliminary Study. *Current Science* 70: 747–50; Raman, T.R.S., Rawat, G.S., and Johnsingh, A.J.T. 1998. Recovery of Tropical Rainforest Avifauna in Relation to Vegetation Succession Following Shifting Cultivation in Mizoram, North-East India. *Journal of Applied Ecology* 35: 217–31; Raman, T.R.S. 2001. Effect of Slash-and-Burn Shifting Cultivation on Rainforest Birds in Mizoram, Northeast India. *Conservation Biology* 15: 685–98.

**p. 69** *after the bamboos* My recent research suggests that this was a simplistic expectation. Bamboos regenerated and persisted after the 2007 flowering at these sites.

**p. 72** *an organic system of multiple cropping* See Raman, T.R.S. 2000. Jhum: Shifting Opinions. *Seminar* 486: 15–18.

## In Clouded Leopard Country

**p. 76** *clouded leopard density in Dampa* See Singh, P. and Macdonald, D.W. 2017. Populations and Activity Patterns of Clouded Leopards and Marbled Cats in Dampa Tiger Reserve, India. *Journal of Mammalogy* 98: 1453–62.

**p. 77** *last vestiges of rainforests* Sharma, N., Madhusudan, M.D., and Sinha, A. 2012. Socio-Economic Drivers of Forest Cover Change in Assam: A Historical Perspective. *Economic and Political Weekly* 47: 64–72.

**p. 77** *Wildlife scientists have recorded* See Sharma, N., Madhusudan, M.D., and Sinha, A. 2014. Local and Landscape Correlates of Primate Distribution and Persistence in the Remnant Lowland Rainforests of the Upper Brahmaputra Valley, Northeastern India. *Conservation Biology* 28: 95–106; Kakati, K. 2004. Impact of Forest Fragmentation on the Hoolock Gibbon in Assam, India. Doctoral thesis, University of Cambridge, Cambridge, UK; Kakati, K., Raghavan, R., Chellam, R., Qureshi, R., and Chivers, D.J. 2009. Status of Western Hoolock Gibbon (*Hoolock hoolock*) Populations in Fragmented Forests of Eastern Assam. *Primate Conservation* 24: 127–37.

**p. 77** *dams, roads, mining and plantations* See Vagholikar, N. 2011. Dams and Environmental Governance in India, 360–9 in *India Infrastructure Report 2011—Water: Policy and Performance for Sustainable Development*, Infrastructure Development Finance Company and Oxford University

Press, New Delhi; and for a recent, readable account, Datta, A. 2017. When Dams Loom Large: Missing the Big Picture. *Current Conservation* 11(4): 8–13.

## The Dance of the Bamboos

**p. 82** *other bamboos may be found* I am grateful to David Vanlalfakawma for sharing his knowledge on the bamboos of Mizoram.

**p. 86** *the decks are stacked* See 'Bamboozled by Land-Use Policy'.

## Bird by Bird in the Rainforest

**p. 96** *two decades ago* See Chapter 'Bamboo Bonfires and Biodiversity'.

**p. 93** *book on writing by Anne Lamott* Lamott, A. 1994. *Bird by Bird: Some Instructions on Writing and Life*. Anchor Books, New York.

## Abode of Rainforest Rarities

**p. 98** *Sri Lanka bay-owl* See Raman, T.R.S. 2001. Observations on the Oriental Bay Owl (*Phodilus badius*) and Range Extensions in the Western Ghats, India. *Forktail* 17: 110–11. Split off recently from the Oriental bay-owl by taxonomists, the bird is now considered a separate species.

**p. 99** *35 regions as 'hotspots'* See www.cepf.net for resources on the world's biodiversity hotspots.

**p. 101** *orange-lipped forest lizard* The species was initially identified as *Calotes andamanensis*.

## Shadowing Civets

**p. 105** *daybeds used by brown palm civets* Mudappa, D. 2006. Day-Bed Choice by the Brown Palm Civet (*Paradoxurus jerdoni*) in the Western Ghats, India. *Mammalian Biology* 71: 238–43.

**p. 106** *Around 60 species* Mudappa, D., Kumar, A., and Chellam, R. 2010. Diet and Fruit Choice of the Brown Palm Civet *Paradoxurus jerdoni*, a Viverrid Endemic to the Western Ghats Rainforest, India. *Tropical Conservation Science* 3: 282–300.

**p. 107** *larger than 300 hectares* Rabinowitz, A.R. 1991. Behaviour and Movements of Sympatric Civet Species in Huai Kha Khaeng Wildlife Sanctuary, Thailand. *Journal of Zoology* 223: 281–98.

**p. 107** *home ranges ... Malay civet* Colón, C.P. 2002. Ranging Behaviour and Activity of the Malay civet (*Viverra tangalunga*) in a Logged and an Unlogged Forest in Danum Valley, East Malaysia. *Journal of Zoology* 257: 473–85.

**p. 108** *survey and later work* Rajamani, N., Mudappa, D., and Van Rompaey, H. 2002. Distribution and Status of the Brown Palm Civet in the Western Ghats, South India. *Small Carnivore Conservation* 27: 6–11.

**p. 108** *the restoration programme* See Mudappa, D. and Raman, T.R.S. 2007. Rainforest Restoration and Wildlife Conservation on Private Lands in the Western Ghats, 210–40 in G. Shahabuddin and M. Rangarajan, Editors. *Making Conservation Work*. Permanent Black, Ranikhet.

## Kalakad: Three Years in the Rainforest

**p. 112** *small mammals and birds* For an account of Divya's work, see 'Shadowing Civets,' and chapter notes; for my work on birds, see Raman, T.R.S., Joshi, N.V., and Sukumar, R. 2005. Tropical Rainforest Bird Community Structure in Relation to Altitude, Tree Species Composition, and Null Models in the Western Ghats, India. *Journal of the Bombay Natural History Society* 102: 145–57; Raman, T.R.S. and Sukumar, R. 2002. Responses of Tropical Rainforest Birds to Abandoned Plantations, Edges, and Logged Forest in the Western Ghats, India. *Animal Conservation* 5: 201–16.

**p. 115** *restore degraded rainforest fragments* See Mudappa, D. and Raman, T.R.S. 2007. Rainforest Restoration and Wildlife Conservation on Private Lands in the Western Ghats, 210–40 in G. Shahabuddin and M. Rangarajan, Editors. *Making Conservation Work*. Permanent Black, Ranikhet.

## Feathered Foresters

**p. 118** *threat of a proposed hydroelectric project* See Muringatheri, M. 2011. Athirapilly Project Threatens Hornbills. *The Hindu*, 19 February, available at: www.thehindu.com/sci-tech/energy-and-environment/Athirappilly-project-threatens-hornbills/article15454755.ece, accessed on 20 January 2019; Bhaduri, A. 2013. Athirapilly Falls under Threat. India Water Portal, 23 November, available at: www.indiawaterportal.org/articles/athirapilly-falls-under-threat, accessed on 20 January 2019.

**p. 118** *a quest along the Western Ghats* Our survey findings are detailed in Mudappa, D. and Raman, T.R.S. 2009. A Conservation Status Survey of Hornbills (Bucerotidae) in the Western Ghats, India. *Indian Birds* 5: 90–102.

**p. 121** *brings up a bigger question* For a study, set in a different context, on the effects of hornbill decline on forest regeneration see Naniwadekar, R.,

Shukla, U., Isvaran, K., and Datta, A. 2015. Reduced Hornbill Abundance Associated with Low Seed Arrival and Altered Recruitment in a Hunted and Logged Tropical Forest. *PLoS ONE* 10(3): e0120062.

## Deep Forest: Namdapha

**p. 123** *her research and conservation work* For a detailed account about Namdapha and the conservation issues there, see Datta, A. 2007. Threatened Forests, Forgotten People, 166–209 in G. Shahabuddin and M. Rangarajan, Editors. *Making Conservation Work*. Permanent Black, Ranikhet.

## The Beleaguered Blackbuck

**p. 132** *Guindy ... owes its existence* Menon, R.K.G. 1986. The Guindy National Park: Its History and Physiogeography. *Blackbuck* 2(1): 14–21.

**p. 132** *occurence of chital in Guindy* M. Krishnan mentions the introduction of chital into Guindy around 1947 (See Krishnan, M. 2013. Whitebuck. *Blackbuck* 29 (1–4): 40–2; and Krishnan, M. 1972. An Ecological Survey of the Larger Mammals of Peninsular India. *Journal of the Bombay Natural History Society* 69: 469–501). But in a book published in 1900, Isabel Savory describes deer already present in Guindy Park: 'Six miles in circumference, well wooded, and rather overstocked, if anything, with spotted deer and antelope, there are plenty of hares, snipe, and quail as well.' (Savory, I. 1900. *A Sportswoman in India: Personal Adventures and Experiences of Travel in Known and Unknown India*, Hutchinson and Co., London.) I thank A. Rajaram, Madras Naturalists' Society, for bringing this to my notice.

**p. 133** *population of blackbuck ... had plummeted* Raman, T.R.S., Menon, R.K.G., and Sukumar, R. 1995. 'Decline of Blackbuck (*Antilope cervicapra* L.) in an Insular Nature Reserve: The Guindy National Park, Madras.' *Current Science* 68: 578–80.

**p. 136** *biomass into the park ecosystem* Raman, T.R.S., Menon, R.K.G., and Sukumar, R. 1996. Ecology and Management of Chital and Blackbuck in Guindy National Park, Chennai. *Journal of the Bombay Natural History Society* 93: 178–92.

## A Bounty of Deer

**p. 141** *68 extant deer species* Based on Wilson, D.E. and Reeder, D.M. 2005. *Mammal Species of the World: A Taxonomic and Geographic Reference.*

3rd Edition. Johns Hopkins University Press, Baltimore; considering three species of chevrotains in India and Sri Lanka.

**p. 142** *striking pair of branching antlers* See Lincoln, G.A. 1992. Biology of Antlers. *Journal of Zoology* 226: 517–28.

**p. 142** *first pair of antlers* See Raman, T.R.S. 1998. Antler Cycles and Breeding Seasonality of Chital (*Axis axis* Erxleben) in Southern India. *Journal of the Bombay Natural History Society* 95: 377–91.

**p. 146** *aspects of deer biology* A relevant—now classic—study from India: Schaller, G.B. 1967. *The Deer and the Tiger*. University of Chicago Press, Chicago.

**p. 147** *The hangul population thrives* For recent estimates of hangul population and conservation status, including the effects of the sheep farm (since removed from Dachigam) and conflicts with herders, see Ahmad, K., Sathyakumar, S., and Qureshi, Q. 2009. Conservation Status of the Last Surviving Wild Population of Hangul or Kashmir deer *Cervus elaphus hanglu* in Kashmir, India. *Journal of the Bombay Natural History Society* 106: 245–55; Parvaiz, A. 2018. As Forces Block Nomads' Access to Pastures, Grazing Poses Major Challenge to Hangul Conservation. *Kashmir Observer* 28 March, available at: kashmirobserver.net/2018/features/forces-block-nomads-access-pasturesgrazing-poses-major-challenge-hangul-conservation, accessed on 21 January 2019.

**p. 148** *threats continue to affect* See Laithangbam, I. 2017. Can Hope Float Loktak's Phumdis? *The Hindu*, 15 July, available at: www.thehindu.com/society/can-hope-float-loktaks-phumdis/article19276922.ece, accessed on 21 January 2019.

## Hornbills: Giants Among Forest Birds

**p. 151** *57 species of hornbills* Poonswad, P., Kemp, A., and Strange, M. 2013. *Hornbills of the World: A Photographic Guide*. Draco Publishing, Singapore, and Hornbill Research Foundation, Bangkok.

**p. 151** *The Oriental region* For a good overview of Asian hornbills, see Kinnaird, M. and O'Brien, T. 2008. *The Ecology and Conservation of Asian Hornbills*. University of Chicago Press, Chicago.

**p. 152** *peculiar breeding habits* Kemp, A. 1995. *The Hornbills, Bucerotiformes*. Oxford University Press, Oxford.

**p. 152** *Ornithologists in the Western Ghats* Mudappa, D. and Kannan, R. 1997. Nest-Site Characteristics and Nesting Success of the Malabar Gray Hornbill in the Southern Western Ghats, India. *Wilson Bulletin* 109: 102–11; James, D.A. and Kannan, R. 2007. Wild Great Hornbills (*Buceros bicornis*)

Do Not Use Mud to Seal Nest Cavities. *The Wilson Journal of Ornithology* 119: 118–21.

**p. 152** *Indian grey hornbill uses* Santhoshkumar, E. and Balasubramanian, P. 2010. Breeding Behaviour and Nest Tree Use by Indian Grey Hornbill *Ocyceros birostris* in the Eastern Ghats, India. *Forktail* 26: 82–5.

**p. 152** *Eastern Himalaya in northeast India* Datta, A. and Rawat, G.S. 2004. Nest-Site Selection and Nesting Success of Three Hornbill Species in Arunachal Pradesh, North-east India: *Buceros bicornis, Aceros undulatus* and *Anthracoceros albirostris. Bird Conservation International* 14: S249–S262.

**p. 153** *In southwest India* Kannan, R. and James, D.A. 1997. Breeding Biology of the Great Pied Hornbill (*Buceros bicornis*) in the Anaimalai Hills of Southern India. *Journal of the Bombay Natural History Society* 86: 448–9; Mudappa, D. 2000. Breeding Biology of the Malabar Grey Hornbill (*Ocyceros birostris*) in Southern Western Ghats. *Journal of Bombay Natural History Society* 97: 15–24.

**p. 153** *as early as December* See Pawar, P.Y., Naniwadekar, R., Raman, T.R.S., and Mudappa, D. 2018. Breeding biology of Great Hornbill *Buceros bicornis* in Tropical Rainforest and Human-Modified Plantation Landscape in Western Ghats, India. *Ornithological Science* 17: 205–16.

**p. 154** *fruit crops on fig trees* O'Brien, T.G., Kinnaird, M.F.,. Dierendfeld, E.S., Conklin, N.L., Wrangham, R.W., and Silver, S.C. 1998. What's so Special about Figs? *Nature* 392: 668.

**p. 155** *tree species with large fruits* Kitamura, S. and Poonswad, P. 2013. Nutmeg-Vertebrate Interactions in the Asia-Pacific Region: Importance of Frugivores for Seed Dispersal in Myristicaceae. *Tropical Conservation Science* 6: 608–36.

**p. 156** *goats from the island* Khan, T., Lenin, J., Mistry, U., Mudappa, D., Raman, T.R.S., Varma, K., and Whitaker, R. 2010. Goats and the Narcondam Hornbill *Aceros narcondami. Indian Birds* 6: 28.

**p. 158** *lopping of fig trees* Kannan, R. and James, D.A. 2008. Fig Trees, Captive Elephants and Conservation of Hornbills and Other Frugivores in an Indian Wildlife Sanctuary. *Journal of the Bombay Natural History Society* 105: 238–42.

**p. 158** *co-operation of local tribals* Bachan, K.H.A., Kannan, R., Muraleedharan, S., and Kumar, S. 2011. Participatory Conservation and Monitoring of Great Hornbills and Malabar Pied Hornbills with the Involvement of Endemic Kadar Tribe in the Anamalai Hills of Southern Western Ghats, India. *The Raffles Bulletin of Zoology Suppl.* 24: 37–43.

## A Life of Courage and Conviction

**p. 160** *hitherto amounted to nothing* Sankaran, R. 2001. The Status and Conservation of the Edible-Nest Swiftlet (*Collocalia fuciphaga*) in the Andaman and Nicobar Islands. *Biological Conservation* 97: 283–294; Sekhsaria, P. 2009. Edible-Nest Swiftlet *Collocalia fuciphaga*: Extinction by Protection. *Indian Birds* 5: 126–7.

**p. 162** *numerous substantive publications* Rahmani, A.R. 2009. Ravi Sankaran's Ornithological Contribution. *Indian Birds* 5: 121–5.

## Cavities, Caves, and a Caveat

**p. 166** *swiftlet populations in the islands* See Sankaran, R. 2001. The Status and Conservation of the Edible-Nest Swiftlet (*Collocalia fuciphaga*) in the Andaman and Nicobar Islands. *Biological Conservation* 97: 283–294.

**p. 166** *Centre's scientist Dr Shirish Manchi* See for example, Manchi, S.S. and Sankaran, R. 2010. Foraging Habits and Habitat Use by Edible-Nest and Glossy Swiftlets in the Andaman Islands, India. *The Wilson Journal of Ornithology* 122: 259–72.

**p. 167** *Nests may be harvested* See Sankaran, R. 2001. The Status and Conservation of the Edible-Nest Swiftlet (*Collocalia fuciphaga*) in the Andaman and Nicobar Islands. *Biological Conservation* 97: 283–294; Sekhsaria, P. 2009. Edible-Nest Swiftlet Collocalia fuciphaga: Extinction by Protection. *Indian Birds* 5: 126–7.

## Life in the Garbage Heap

**p. 171** *population trending downhill* BirdLife International. 2016. *Leptoptilos dubius. The IUCN [International Union for Conservation of Nature] Red List of Threatened Species 2016*: e.T22697721A93633471, available at: dx.doi.org/10.2305/IUCN.UK.2016-3.RLTS.T22697721A93633471.en, accessed on 31 March 2018.

**p. 172** *wetland of international importance* See Ramsar Sites Information Service, available at: rsis.ramsar.org/ris/1207, accessed on 21 January 2019.

**p. 172** *polluted by pesticide and fertilizer* Bhattacharyya, K.G. and Kapil, N. 2010. Impact of Urbanization on the Quality of Water in a Natural Reservoir: A Case Study with the Deepor Beel in Guwahati City, India. *Water and Environment Journal* 24: 83–96.

**p. 173** *Deepor Beel has lost 41%* Deka, A.J., Tripathi, O.P., Khan. M.L. 2011. A Multi-Temporal Remote Sensing Approach for Monitoring Changes in Spatial Extent of Freshwater Lake of Deepor Beel Ramsar Site, a Major Wetland of Assam. *Journal of Wetlands Ecology* 5: 40–7.

**p. 173** *appear to be in decline* PTI. 2014. Population of Migratory Water Birds Declines in India. *Zee News*, 17 February, available at: www.zeenews.india.com/news/eco-news/population-of-migratory-water-birds-declines-in-india_912166.html, accessed on 6 February 2019.

**p. 173** *the trains will not stop* Mitra, S., and Bezbaruah, A.N. 2014. Railroad Impacts on Wetland Habitat: GIS and Modeling Approach. *Journal of Transport and Land Use* 7: 15–28; see also Dasgupta, S. and Ghosh, A.K. 2015. Elephant–Railway Conflict in a Biodiversity Hotspot: Determinants and Perceptions of the Conflict in Northern West Bengal, India. *Human Dimensions of Wildlife* 20: 81–94.

## Death on the Highway

**p. 176** *5.6 million kilometres Basic Road Statistics of India 2015–16*, Ministry of Road Transport and Highways, Government of India, available at: www.morth.nic.in/showfile.asp?lid=3100; also see An Overview: Road Network of India, available at: www.morth.nic.in/showfile.asp?lid=364, accessed on 28 January 2019.

**p. 176** *registered motor vehicles* Figures obtained from the Open Government Data (OGD) Platform India, available at: www.community.data.gov.in/registered-motor-vehicles-in-india-as-on-31-03-2015/ and www.visualize.data.gov.in/?inst=7a82b08b-d7fe-4edb-82f7-216a1f1bed07, accessed on 28 January 2019.

**p. 176** *total passenger movement by road India in Figures 2018*, available at: www.mospi.gov.in/sites/default/files/publication_reports/India_in_figures-2018_rev.pdf, Ministry of Statistics and Programme Implementation, India, accessed on 28 January 2019; and The Working Group Report on Road Transport in India for the Eleventh Five Year Plan, Planning Commission, India, 2007, available at: planningcommission.nic.in/aboutus/committee/wrkgrp11/wg11_roadtpt.pdf, accessed on 28 January 2019.

**p. 177** *studies from ... Indian forests* For a review and references, see Raman, T.R.S. 2011. Framing Ecologically Sound Policy on Linear Intrusions Affecting Wildlife Habitats. Background paper for the National Board for Wildlife, India. For the 2017 study, see Jeganathan, P., Mudappa, D., Kumar, M.A., and Raman, T.R.S. 2017. Seasonal Variation in Wildlife Roadkills in Plantations and Tropical Rainforest in the Anamalai Hills, Western Ghats, India. *Current Science* 114: 619–26.

**p. 177** *Studies from elsewhere* Eigenbroda, F., Hecnarb, S.J., and Fahrig, L. 2008. The Relative Effects of Road Traffic and Forest Cover on Anuran Populations. *Biological Conservation* 141: 35–46.

**p. 177** *Another study estimates* Gibbs, J.P. and Shriver, W.G. 2005. Can Road Mortality Limit Populations of Pool-Breeding Amphibians? *Wetlands Ecology and Management* 13: 281–89.

**p. 178** *an ecological death-trap* Horváth, G., Kriska, G., Malik, P., and Robertson, B. 2009. Polarized Light Pollution: A New Kind of Ecological Photopollution. *Frontiers in Ecology and the Environment* 7: 317–325; Robertson, B.A. and Hutto, R.L. 2006. A Framework for Understanding Ecological Traps and an Evaluation of Existing Evidence. *Ecology* 87: 1075–85.

**p. 179** *such highly intelligent species* Hockings, K.J., Anderson, J.R., and Matsuzawa, T. 2006. Road Crossing in Chimpanzees: A Risky Business. *Current Biology* 16: R668–70; Blake, S., Deem, S.L., Strindberg, S., Maisels, F., Momont, L., Isia, I., Douglas-Hamilton, I., Karesh, W.B., and Kock, M.D. 2008. Roadless Wilderness Area Determines Forest Elephant Movements in the Congo Basin. *PLoS ONE* 3(10): e3546.

**p. 179** *As roads become wider* Seiler, A. 2003. The Toll of the Automobile: Wildlife and Roads in Sweden. PhD thesis. Swedish University of Agricultural Sciences, Up-psala; Laurance, S.G. and Gomez, M.S. 2005. Clearing Width and Movements of Understorey Rainforest Birds. *Biotropica* 37: 149–52; Laurance, S.G., Stouffer, P.C. and Laurance, WF. 2004. Effects of Road Clearings on Movement Patterns of Understorey Rainforest Birds in Central Amazonia. *Conservation Biology* 18: 1099–109; Goosem, M. 2007. Fragmentation Impacts Caused by Roads through Rainforests. *Current Science* 93: 1587–95.

**p. 180** *various ecological changes* Forman, R.T.T. and Alexander, L.E. 1998. Roads and Their Major Ecological Effects. *Annual Review of Ecology and Systematics* 29: 207–31; Trombulak, S.C. and Frissell, C.A. 2000. Review of Ecological Effects of Roads on Terrestrial and Aquatic Communities. *Conservation Biology* 14: 18–30; Fahrig, L., and Rytwinski, T. 2009. Effects of Roads on Animal Abundance: An Empirical Review and Synthesis. *Ecology and Society* 14(1): 21.

**p. 180** *weeds spread along roads* Gelbard, J.L. and Belnap, J. 2003. Roads as Conduits for Exotic Plant Invasions in a Semiarid Landscape. *Conservation Biology* 17: 420–32.

**p. 181** *No study has yet* For a recent attempt, see Jeganathan, P., Mudappa, D., Kumar, M.A., and Raman, T.R.S. 2017. Seasonal Variation in Wildlife Roadkills in Plantations and Tropical Rainforest in the Anamalai Hills, Western Ghats, India. *Current Science* 114: 619–26.

**p. 181** *Garo Hills in Meghalaya* Bera, S.K., Basumatary, S.K., Agarwal, A. and Ahmed, M. 2006. Conversion of Forest Land in Garo Hills,

Meghalaya for Construction of Roads: A Threat to the Environment and Biodiversity. *Current Science* 91: 281–84. For the 2018 study in the Western Ghats, see Krishnadas, M., Agarwala, M., Sridhara, S., and Eastwood, E. 2018. Parks protect forest cover in a tropical biodiversity hotspot, but high human population densities can limit success. *Biological Conservation* 223:147–155.

**p. 181** *A range of measures* Laurance, W.F., Goosem, M. and Laurance, S.G., 2009. Impacts of Roads and Linear Clearings on Tropical Forests. *Trends in Ecology and Evolution*, 24: 659–69. 3. Forman, R.T.T., Sperling, D., Bissonette, J.A., Clevenger, A.P., Cutshall, C.D., Dale, V.H., Fahrig, L., France, R.L., Heanue, K., Goldman, C.R., Jones, J., Swanson, F., Turrentine, T., and Winter, T.C. 2003. *Road Ecology: Science and Solutions*. Island Press, Washington, DC; van der Ree, R., Smith, D.J. and Grilo, C. 2015. *Handbook of Road Ecology*. John Wiley, Chichester, UK.

**p. 182** *the first nine months* The Zanzibar Red Colobus Monkey: Behavior, Ecology, and Conservation. DVD documentary, T.T. Struhsaker, Department of Biological Anthropology and Anatomy, Duke University, USA.

**p. 183** *including road planning* Morgan, D. and Sanz, C. 2007. Best Practice Guidelines for Reducing the Impact of Commercial Logging on Great Apes in Western Equatorial Africa. IUCN [International Union for Conservation of Nature] SSC Primate Specialist Group (PSG), Gland, Switzerland; for a comprehensive recent publication see Rainer, H., White, A., and Lanjouw, A. Editors. 2018. *State of the Apes: Infrastructure Development and Ape Conservation*. Cambridge University Press and Arcus Foundation, Cambridge, UK.

**p. 183** *guidelines in India* Rajvanshi, A., Mathur, V.B., Teleki, G.C., Mukherjee, S.K. 2001. Roads, Sensitive Habitats and Wildlife: Environmental Guidelines for India and South Asia. Wildlife Institute of India, Dehradun; Raman, T.R.S. 2011. Framing Ecologically Sound Policy on Linear Intrusions Affecting Wildlife Habitats. Background paper for the National Board for Wildlife, India; WII 2016. *Eco-friendly Measures to Mitigate Impacts on Linear Infrastructures on Wildlife*. Wildlife Institute of India, Dehradun.

**p. 184** *roads by actively removing them* Switalski, T.A., Bissonette, J.A., DeLuca, T.H., Luce, C.H., and Madej, M.A. 2004. Benefits and Impacts of Road Removal. *Frontiers in Ecology and the Environment* 2: 21–8.

## Natural Engineering: India's Green Infrastructure

**p. 185** *Economic Outlook for 2009–10* Website archive of the Economic Advisory Council, available at: eac.gov.in, accessed on 21 January 2019.

## The Long Road to Growth

**p. 190** *recently permitting Central agencies* Jain, A. 2015. NGT Stays Govt. Order on Felling of Trees to Push Linear Projects. *The Hindu*, 21 February, available at: www.thehindu.com/news/cities/Delhi/ngt-stays-govt-order-on-felling-of-trees-to-push-linear-projects/article6917758.ece, accessed on 21 January 2019.

**p. 190** *litany of associated problems* See Raman, T.R.S. and Madhusudan, M.D. 2013. Development Minus Green Shoots. *The Hindu*, 13 February, available at: www.thehindu.com/opinion/op-ed/Development-minus-green-shoots/article12343002.ece, accessed on 7 February 2019.

**p. 191** *A 2006 study noted* Sidle, R.C., Ziegler, A.D., Negishi, J.N., Nik, A.R., Siew, R., and Turkelboom, F. 2006. Erosion Processes in Steep Terrain—Truths, Myths, and Uncertainties related to Forest Management in Southeast Asia. *Forest Ecology and Management* 224: 199–225.

**p. 191** *1-2 million large animals* Gaskill, M. 2013. Rise in Roadkill Requires New Solutions. *Scientific American*, 16 May, available at: www.scientificamerican.com/article/roadkill-endangers-endangered-wildlife/, accessed on 21 January 2019.

**p. 191** *89–340 million birds every year* Loss, S.R., Will, T. and Marra, P.P. 2014. Estimation of Bird-Vehicle Collision Mortality on U.S. Roads. *Journal of Wildlife Management* 78: 763–71.

**p. 192** *In Bandipur Tiger Reserve* Prasad, A.E. 2009. Tree Community Change in a Tropical Dry Forest: The Role of Roads and Exotic Plant Invasion. *Environmental Con-servation* 36: 201–7.

**p. 192** *a comprehensive scientific review* Fahrig, L., and Rytwinski, T. 2009. Effects of Roads on Animal Abundance: An Empirical Review and Synthesis. *Ecology and Society* 14(1): 21.

**p. 193** *detailed background paper* See Raman, T.R.S. 2011. Framing Ecologically Sound Policy on Linear Intrusions Affecting Wildlife Habitats. Background paper for the National Board for Wildlife, India; For more on the NBWL's work, see Bindra, P.S. 2017. *The Vanishing: India's Wildlife Crisis*. Penguin Random House, Gurgaon.

## Watering Down Forest Protection

**p. 198** *overlooks the fact* Rajshekar, M. 2014. New Forest Diversion Norms Too Simplistic. *Economic Times*, 11 June, available at economictimes.indiatimes.com/news/economy/policy/new-forest-diversion-norms-too- simplistic/articleshow/36397578.cms, accessed on 21 January 2019.

**p. 198** *biologically significant non-forest areas* Veldman, J.W., Overbeck, G.E., Negreiros, D., Mahy, G. Stradic, S.L., Fernandes, G.W., Durigan G., Buisson, E., Putz, F. E., and Bond, W. J. 2015. Where Tree Planting and Forest Expansion are Bad for Biodiversity and Ecosystem Services. *BioScience* 65: 1011–18.

**p. 199** *Water Resources Information System* See www.india- wris.nrsc.gov.in/wris.html, accessed on 21 January 2019.

**p. 199** *less than 35 blocks* Chakraborty, S. and Sethi, N. 2015. Final Number of Inviolate Coal Blocks Down from 206 to Less Than 35. *Business Standard,* 2 September, available at: www.business-standard.com/article/economy-policy/final-number-of-inviolate-coal-blocks-down-from-206-to-less-than-35-115090200004_1.html, accessed on 6 February 2019.

**p. 199** *iron ore mines within Kudremukh* See Impact of Iron Ore Mining in Kudremukh on Bhadra River Ecosystem, available at admin.indiaenvironmentportal.org.in/files/KUDREMUKH, accessed on 21 January 2019.

**p. 199** *A 2015 study from Meghalaya* Mallik, S.K., Sarma, D., Sarma, D., and Shahi, N. 2015. Acid Mine Drainage, a Potential Threat to Fish Fauna of Simsang River, Meghalaya. *Current Science* 109: 687–90.

## Protecting the Wildlife Protection Act

**p. 202** *NBWL's credibility and effectiveness* Available at: www.conservationindia.org/campaigns/campaign-for-a-credible-national-board-for- wildlife-board-nbwl, accessed on 21 January 2019; Menon, M. 2014. NGOs Protest Dilution of National Board for Wildlife, *The Hindu*, 5 August; Sinha, N. 2014. Protecting the Constituency of Nature. *The Hindu,* 4 August.

**p. 205** *Postscript* Statistics on the number and rate of project clearances are from *State of India's Environment 2019,* Centre for Science and Environment, New Delhi.

## Living with Leopards in Countryside and City

**p. 209** *his 2015 Masters dissertation* Surve, N. 2015. Ecology of Leopard (*Panthera pardus*) in Sanjay Gandhi National Park, Maharashtra with Special Reference to its Abundance, Prey Selection and Food Habits. MSc Dissertation, Wildlife Institute of India and Saurashtra University, Rajkot.

**p. 209** *An early study in SGNP* Edgaonkar, A., and Chellam, R. 1998. A Preliminary Study on the Ecology of the Leopard, *Panthera pardus fusca* in the Sanjay Gandhi National Park, Maharashtra. Wildlife Institute of India, Dehradun.

**p. 209** *encroachment has continued* Soumya, E. 2014. The Leopards of Mumbai: Life and Death among the City's 'Living Ghosts'. *The Guardian*, 26 November, available at www.theguardian.com/cities/2014/nov/26/leopards-mumbai-life-death-living-ghosts-sgnp, accessed on 21 January 2019.

**p. 210** *targeting the wrong individuals* Lenin, J. 2013. Spotting the Killer. *The Hindu* 30 August, available at www.thehindu.com/news/cities/chennai/chen-columns/spotting-the-killer/article5075349.ece, accessed on 21 January 2018.

**p. 210** *in a 2007 publication* Athreya, V.R., Thakur, S.S., Chaudhuri, S., and Belsare, A. 2007. Leopards in Human-Dominated Areas: A Spillover from Sustained Translocations into Nearby Forests? *Journal of the Bombay Natural History Society* 104: 13–18.

**p. 211** *translocation programme in the area* Athreya, V., Odden, M., Linnell, J.D.C., and Karanth, K.U. 2011. Translocation as a Tool for Mitigating Conflict with Leopards in Human-Dominated Landscapes of India. *Conservation Biology* 25: 131–41.

**p. 211** *increased conflict because of the translocation* Athreya, V. 2006. Is Relocation A Viable Management Option for Unwanted Animals? The Case of the Leopard in India. *Conservation and Society* 4: 419–23; Athreya, V.R., Thakur, S.S., Chaudhuri, S., and Belsare, A. 2007. Leopards in Human-Dominated Areas: A Spillover from Sustained Translocations into Nearby Forests? *Journal of the Bombay Natural History Society* 104: 13–18; Athreya, V., Odden, M., Linnell, J.D.C., and Karanth, K.U. 2011. Translocation as a Tool for Mitigating Conflict with Leopards in Human-Dominated Landscapes of India. *Conservation Biology* 25: 131–41.

**p. 211** *A 2014 study in Maharashtra* Odden, M., Athreya, V., Rattan, S. and Linnell, J.D.C. 2014. Adaptable Neighbours: Movement Patterns of GPS-Collared Leopards in Human Dominated Landscapes in India. *PLoS ONE* 9(11): e112044.

**p. 212** *movie by the same name* Ajoba, trailer available at: www.youtube.com/watch?v=pZNLbBP6Q0M, accessed 6February 2019.

**p. 212** *slipping back to the method* Lenin, J. 2013. Why Do Leopards Kill Humans? *The Hindu*, 13 September, available at www.thehindu.com/news/cities/chennai/chen-columns/why-do-leopards-kill-humans/article5124009.ece, accessed on 18 January 2019.

**p. 212** *A spate of 6 incidents in 2017* Chatterjee, B. 2017. Sudden Rise in Leopard Attacks in Mumbai's Aarey Colony in 2017: What's the Reason? *Hindustan Times*, 23 October, available at: www.hindustantimes.com/mumbai-news/how-some-mumbai-hamlets-live-dangerously-close-to-leopards/story-Q2C4REMAF4fk4MXq93YrAI.html, accessed on 30 January 2019.

**p. 212** *narratives of 'killer leopards'* Hayden, M.E. 2013. Cat's Cradle. *The Caravan*, September, available at www.caravanmagazine.in/reportage/cats-cradle, accessed on 20 January 2019.

**p. 212** *Mumbaikars for SGNP* See www.mumbaikarsforsgnp.com, accessed on 20 January 2019.

**p. 213** *adapted quite well* Lenin, J. 2014. Leopards in the Neighbourhood. *BBC Wildlife* December: 36–44.

**p. 213** *agriculture in Maharashtra* Athreya, V., Odden, M., Linnell, J.D.C., Krishnaswamy, J., and Karanth, K.U. 2013. Big Cats in our Backyards: Persistence of Large Carnivores in a Human Dominated Landscape in India. *PLoS ONE* 8(3): e57872.

**p. 213** *Western Ghats mountains* Navya, R., Athreya, V., Mudappa, D. and Raman, T.R.S. 2014. Assessing Leopard Occurrence in the Plantation Landscape of Valparai, Anamalai Hills. *Current Science* 107: 1381–5.

**p. 213** *multiple-use landscapes in Karnataka* See ncf- india.org/projects/the-secret-lives-of-leopards, accessed on 20 December 2019.

**p. 213** *available prey species* Sidhu, S., Raman, T.R.S., and Mudappa, D. 2015. Prey Abundance and Leopard Diet in a Plantation and Rainforest Landscape, Anamalai Hills, Western Ghats. *Current Science* 109: 323–30; Athreya, V., Odden, M., Linnell, J.D.C., Krisnaswamy, J., and Karanth, K.U. 2016. A Cat among the Dogs: Leopard Panthera pardus Diet in a Human-Dominated Landscape in Western Maharashtra, India. *Oryx* 50: 56–162.

**p. 213** *navigate human-dominated terrain* See Odden, M., Athreya, V., Rattan, S. and Linnell, J.D.C. 2014. Adaptable Neighbours: Movement Patterns of GPS-Collared Leopards in Human Dominated Landscapes in India. *PLoS ONE* 9(11): e112044.

**p. 213** *domain of neutral interaction* For a recent paper that discusses this, see Sidhu, S., Raghunathan, G., Mudappa, D., and Raman, T.R.S. 2017. Conflict to Coexistence: Human–Leopard Interactions in a Plantation Landscape in Anamalai Hills, India. *Conservation and Society* 15: 474–82.

**p. 213** *At the national level* See Guidelines for Human–Leopard Conflict Management, Ministry of Environment and Forests, India, available at envfor.nic.in/sites/default/files/moef-guidelines-2011-human-leopard-conflict-management.pdf, accessed on 7 February 2019.

**p. 214** *Africa, Europe, and North America* See Chapron, G., Kaczensky, P., Linnell, J.D.C., von Arx, M., Huber, D., Andrén, H. and others. 2014. Recovery of Large Carnivores in Europe's Modern Human-Dominated Landscapes. *Science* 346: 1517–19; Morrell, V. 2013. Predators in the 'Hood. *Science* 341: 1332–5; Nijhuis, M. 2014. The Carnivores Next Door. *New Yorker*, 18 December, available at: www.newyorker.com/tech/elements/carnivores-next-door, accessed on 20 December 2019; Woodroffe, R., Thirgood, S., and

Rabinowitz, A. Editors. 2005. *People and Wildlife: Conflict or Coexistence?* Cambridge University Press, Cambridge, UK.

**p. 214** *people and wildlife can coexist* Lenin, J. 2014. Leopards in the Neighbourhood. *BBC Wildlife,* December: 36–44; Rangarajan, M., Madhusudan, M.D., and Shahabuddin, G. 2014. *Nature without Borders.* Orient Blackswan, Hyderabad.

## The Culling Fields

**p. 215** *the difference of views The Hindu.* 2016. Maneka Slams Javadekar over Culling of Animals. *The Hindu,* 9 June, available at www.thehindu.com/news/national/Maneka-slams- Javadekar-over-culling-of-animals/article14413150.ece, accessed on 21 January 2019.

**p. 215** *kill ... with few safeguards* Raman, T.R.S. 2016. The Silence of India's Wildlife Scientists, Including Myself, Rings Louder Than Gunshots. *Scroll.in* 20 September, available at scroll.in/article/816752/the-silence-of-indias-wildlife-scientists-including-myself-rings-louder-than-gunshots, accessed on 20 January 2019.

**p. 216** *may even exacerbate them* Athreya, V. 2016. Culling Wild Animals Isn't Part of the Indian Ethos—We Can Do Better to Avoid Conflict. *Scroll.in,* 6 December, available at scroll.in/article/819662/culling-the-conflict-killing-animals-does-not-work-prevention-must-focus-on-people, accessed on 16 December 2018.

**p. 216** *Himachal Pradesh ... killed hundreds* Radhakrishna, S. and Raman, T.R.S. 2016. Get the Monkey Off the Back. *The Tribune,* 23 August, available at www.tribuneindia.com/news/comment/get-the-monkey-off-the-back/284023.html, accessed on 16 December 2018; Makhaik, R. 2019. Efforts Fail to Contain Man-Monkey Conflict. *The Tribune,* 5 January, available at: www.tribuneindia.com/news/weekly-pullouts/himachal-tribune/efforts-fail-to-contain-man-monkey-conflict/709020.html, accessed on 7 February 2019; Singh, M., Kumara, H.N., and Velankar, A.D. 2015. Population status of rhesus macaque (*Macaca mulatta*) in Himachal Pradesh, India. Technical Report (PR-150), Sálim Ali Centre for Ornithology and Natural History, India, available at: www.sacon.in/wp-content/uploads/2015/06/FT-2016-PR150-Himachal_rhesus_report.pdf, accessed on 7 February 2019.

**p. 219** *turned out to be a non-starter* Makhaik, R. 2019. Efforts Fail to Contain Man-Monkey Conflict. *The Tribune,* 5 January, available at: www.tribuneindia.com/news/weekly-pullouts/himachal-tribune/efforts-fail-to-contain-man-monkey-conflict/709020.html, accessed on 7 February 2019.

## Bamboozled by Land-Use Policy: Jhum and Oil Palm in Mizoram

**p. 221** *Two spectacular bamboo dances* See 'The Dance of the Bamboos' for a description of the Cheraw and jhum.

**p. 222** *Mizoram ... first in India* Sethi, N. 2004. Organic States. *Down to Earth* 31 August, available at: www.downtoearth.org.in/news/organic-states-11703, accessed 22 December 2018.

**p. 222** *New Land Use Policy* See nlup.mizoram.gov.in, accessed on 7 February 2019.

**p. 222** *claiming they are a better* See the *Mizoram Economic Survey 2012–13,* available at: www.planning.mizoram.gov.in, accessed 16 January 2019.

**p. 222** *contained fires in March* See also Darlong, V.T. 2002. Traditional Community-Based Fire Management among the Mizo Shifting Cultivators of Mizoram in Northeast India. In Moore, P., Ganz, D., Tan, L.C., Enters, T. and Durst, P. B., Editors. Communities in Flames: Proceedings of an International Conference on Community Involvement in Fire Management, RAP Publications 2002/25, FAO, Bangkok.

**p. 223** *led by P. S. Ramakrishnan* Ramakrishnan, P.S. 1992. *Shifting Agriculture and Sustainable Development: An Interdisciplinary Study from North-Eastern India.* UNESCO, Parthenon, Paris.

**p. 222** *2011 ... Forest Report* See www.fsi.org.in, accessed on 8 January 2019.

**p. 224** *over Rs 2,800 crores* See nlup.mizoram.gov.in/page/concise-summary-of-nlup.html, accessed on 9 January 2019.

**p. 224** *NLUP's primary objective ... misdirected* Leblhuber, S.K. and Vanlalhruaia, H. 2012. Jhum Cultivation versus the New Land Use Policy: Agrarian Change and Transformation in Mizoram. In Münster, D., Münster, U., and Dorondel, S., Editors. *Fields and Forests: Ethnographic Perspectives on Environmental Globalization. RCC Perspectives* 5: 83–9.

**p. 225** *My earlier research* See 'Bamboo Bonfires and Biodiversity.'

**p. 225** *three ... oil palm companies* Goswami, R. 2011. Mizoram Lures Edible Oil Makers. *The Telegraph,* 12 December, available at: www.telegraphindia.com/states/north-east/mizoram-lures-edible-oil-makers/cid/474734, accessed on 7 February 2019.

**p. 226** *Daman Singh notes* Singh, D. 1996. *The Last Frontier: People and Forests in Mizoram.* Tata Energy Research Institute, Delhi.

**p. 227** *expansion and impacts of oil palm* Obidzinski, K., Andriani, R., Komarudin, H., and Andrianto, A. 2012. Environmental and Social Impacts of Oil Palm Plantations and Their Implications for Biofuel

Production in Indonesia. *Ecology and Society* 17(1): 25; Carlson, K.M., Curran, L.M., Asner, G.P., Pittman, A.M., Trigg, S.N., and Adeney, J.M. 2013. Carbon Emissions from Forest Conversion by Kalimantan Oil Palm Plantations. *Nature Climate Change* 3: 283–7; Fitzherbert, E.B., Struebig, M.J., Morel, A., Danielsen, F., Brühl, C.A., Donald, P.F., and Phalan, B. 2008. How Will Oil Palm Expansion Affect Biodiversity? *Trends in Ecology and Evolution* 23: 538–45; Area under oil palm production based on FAO data, available at: www.fao.org/faostat/en/#data/, accessed on 28 January 2019.

**p. 229** *A study from Thailand* Aratrakorn, S., Thunhikorn, S., and Donald, P. 2006. Changes in Bird Communities Following Conversion of Lowland Forest to Oil Palm and Rubber Plantations in Southern Thailand. *Bird Conservation International* 16: 71–82.

**p. 229** *quality in hill streams* Carlson, K. M., Curran, L.M., Ponette-González, A.G., Ratnasari, D., Ruspita, Lisnawati, N., Purwanto, Y., Brauman, K.A., and Raymond, P. A. 2014. Influence of Watershed-Climate Interactions on Stream Temperature, Sediment Yield, and Metabolism along a Land Use Intensity Gradient in Indonesian Borneo. *Journal of Geophysical Research Biogeosciences* 119: 1110–28.

**p. 229** *My own research* See 'The march of the triffids.'

**p. 231** *A 2019 study by Purabi Bose* Bose, P. 2019. Oil Palm Plantations vs. Shifting Cultivation for Indigenous Peoples: Analyzing Mizoram's New Land Use Policy. *Land Use Policy* 81: 115–123.

**p. 232** *science and sustainability of jhum* Grogan, P., Lalnunmawia, F., and Tripathi, S.K. 2012. Shifting Cultivation in Steeply Sloped Regions: A Review Of management Options and Research Priorities for Mizoram State, Northeast India. *Agroforestry Systems* 84:163–77.

**p. 233** *view oil palm expansion* Mazoomdar, J. 2013. Reading the Palm Right. *Tehelka*, 13 December. Available at www.tehelka.com/2013/12/reading-the-palm-right, accessed on 3 January 2019; Srinivasan, U. 2014. Oil Palm Expansion: Ecological Threat to North-East India. *Economic and Political Weekly* 49, available at www.epw.in/journal/2014/36/reports-states-web-exclusives/oil-palm-expansion.html, accessed on 3 January 2019.

## The March of the Triffids

**p. 235** *Oil palm expansion in India* See Srinivasan, U. 2014. Oil Palm Expansion: Ecological Threat to North-East India. *Economic and Political Weekly* 49, available at www.epw.in/journal/2014/36/reports-states-web-exclusives/oil-palm-expansion.html, accessed on 3 January 2019; Dasgupta, S. 2014. India Plans Huge Palm Oil Expansion, Puts Forests at Risk. *Mongabay*,

14 October, available at news.mongabay.com/2014/10/india-plans-huge-palm-oil-expansion-puts-forests-at-risk/, accessed on 3 January 2019.

**p. 236** *In a 2001 publication* See Raman, T.R.S. 2001. Effect of Slash-and-Burn Shifting Cultivation on Rainforest Birds in Mizoram, Northeast India. *Conservation Biology* 15: 685–98.

**p. 236** *To find answers* Mandal, J. and Raman, T.R.S. 2016. Shifting Agriculture Supports More Tropical Forest Birds Than Oil Palm or Teak Plantations in Mizoram, Northeast India. *The Condor: Ornithological Applications* 18: 345–59.

**p. 238** *land-use planning and practices* Azhar, B., Saadun, N., Puan, C.L., Kamarudin, N., Aziz, N., Nurhidayu, S., and Fischer, J. 2015. Promoting Landscape Heterogeneity to Improve the Biodiversity Benefits of Certified Palm Oil Production: Evidence from Peninsular Malaysia. *Global Ecology and Conservation* 3: 553–561; Koh, L. P. 2008. Can Oil Palm Plantations be Made More Hospitable for Forest Butterflies and Birds? *Journal of Applied Ecology* 45: 1002–9.

## How Green is Your Tea?

**p. 240** *as 'green deserts'* Lenin, J. 2005. Rainforest Revival. *The Hindu Sunday Magazine,* 17 July. Although the word 'desert' is often used disparagingly for areas poor in plant or animal life, real deserts—hot deserts such as the Thar and cold deserts as in Ladakh—are valuable natural ecosystems in their own right with their own unique species.

**p. 240** *role in wildlife conservation* Mudappa, D., Kumar, M.A., and Raman, T.R.S. 2014. Restoring Nature: Wildlife Conservation in Landscapes Fragmented by Plantation Crops in India, 178–214 in M. Rangarajan, M.D. Madhusudan, and G. Shahabuddin, Editors. *Nature without Borders.* Orient Blackswan, New Delhi.

## Rhythms of Renewal

**p. 246** *a thoughtful report* See: Rangarajan, M., Desai, A., Sukumar, R., Easa, P.S., Menon, V., Vincent, S., Ganguly, S., Talukdar, B.K., Singh, B., Mudappa, D., Chowdhury, S. Prasad, A.N. 2010. *Gajah: Securing the Future for Elephants in India.* The Report of the Elephant Task Force, Ministry of Environment and Forests, Government of India, New Delhi.

**p. 248** *January 2019* For dilutions of environmental regulations, see: *State of India's Environment 2019,* Centre for Science and Environment, New Delhi; Raman, T.R.S. and Madhusudan, M.D. 2013. Development

Minus Green Shoots. *The Hindu*, 13 February, available at: www.thehindu.com/opinion/op-ed/Development-minus-green-shoots/article12343002.ece, accessed on 7 February 2019; On elephant population, see: Ganesan, R. 2018. As Living Room Runs Out in India, the Slaughter of its Elephants Escalates, *IndiaSpend*, 3 November, available at: www.indiaspend.com/as-living-room-runs-out-in-india-the-slaughter-of-its-elephants-escalates/, accessed on 3 February 2019; On climate change, $CO_2$, and extreme weather events see: Ramachandran, T.R. and Raman, T.R.S. 2013. Living in an Extreme World. *The Hindu*, 13 July, available at: www.thehindu.com/opinion/lead/living-in-an-extreme-world/article4909364.ece, accessed on 7 February 2019; Jones, N. 2017. How the World Passed a Carbon Threshold and Why it Matters, Yale Environment 360, January 26, available at: e360.yale.edu/features/how-the-world-passed-a-carbon-threshold-400ppm-and-why-it-matters, accessed on 3 February 2019; Kahn, B. 2017. The Climate Could Hit a State Unseen in 50 Million Years, *Climate Central*, 4 April, available at: www.climatecentral.org/news/climate-change-unseen-50-million-years-21312, accessed on 3 February 2019.

## Conserving a Connected World

**p. 252** *in a famous speech* See Rangarajan M. 2009. Striving for a Balance: Nature, Power, Science and India's Indira Gandhi, 1917–1984. *Conservation and Society* 7:299–312; Gandhi, I. 1972. Man and His World, Speech at the UN Conference on the Human Environment, Stockholm, 14 June 1972, 60–7 in Gandhi, I. 1983. *On Peoples and Problems*, Second edition. Hodder and Stoughton, London.

## Integrating Ecology and Economy

**p. 256** *Barack Obama had said* Friedman, T.L. 2014. Obama on Obama on climate. *New York Times*, 7 June, available at www.nytimes.com/2014/06/08/opinion/sunday/friedman-obama-on-obama-on-climate.html, accessed on 3 January 2019.

**p. 257** *redefine the criteria for inviolate forests* Goswami, U. 2014. Environment Ministry Reworks Yardsticks for Giving Mining Nod. *Economic Times*, 6 June, available at: economictimes.indiatimes.com/news/economy/policy/environment-ministry-reworks-yardsticks-for-giving-mining-nod/articleshow/36119511.cms, accessed on 7 February 2019.

**p. 257** *diluting environmental norms State of India's Environment 2019*, Centre for Science and Environment, New Delhi.

**p. 257** *700,000 hectares of forest* Sethi, N. 2014. Modi Faces a Jungle of Pending Forest Clearances. *Business Standard,* 23 May, available at: www.business-standard.com/article/economy-policy/modi-faces-a-jungle-of-pending-forest-clearances-114052201673_1.html, accessed on 7 February 2019.

**p. 258** *NPV—at 2014 rates* Verma, M., Negandhi, D., Wahal, A.K. and Kumar, R. 2013. Revision of Rates of NPV Applicable for Different Class/Category of Forests. Indian Institute of Forest Management, Bhopal; following a revised report in 2014 the NPV rates were revised in 2016 to a range of Rs 5.54 lakh to Rs 50.72 lakh per hectare, see Shrivastava, K.S. 2016. India's Forests Valued at Rs 115 Trillion, but Tribals Unlikely to Get a Share, *Hindustan Times,* 10 August, available at: www.hindustantimes.com/india-news/india-s-forests-valued-at-rs-115-trillion-but-tribals-unlikely-to-get-a-share/story-L06QqOb6PGrkcxZ7l1ACDI.html, accessed on 3 February 2019.

**p. 258** *artificial forests are no substitute* Gibson, L., Lee, T.M., Koh, L.P., Brook, B.W., Gardner, T.A., Barlow, J., Peres, C.A., Bradshaw, C.J. A., Laurance, W.F., Lovejoy, T.E., and Sodhi, N.S. 2011. Primary Forests are Irreplaceable for Sustaining Tropical Biodiversity. *Nature* 478: 378–81.

## The Health of Nations: The Other Invisible Hand

**p. 262** *its predicted destiny* Smialek, J. 2015. These Will be the World's 20 Largest Economies in 2030. *Bloomberg* 10 April, available at: www.bloomberg.com/news/articles/2015-04-10/the-world-s-20-largest-economies-in-2030, accessed on 7 February 2019.

**p. 262** *lower priority ... health and environment* Kalra, A. 2014. India Slashes Health Budget, Already One of the World's Lowest. *Reuters,* 23 December, available at: in.reuters.com/article/india-health-budget-idINKBN0K10Y020141223, accessed on 7 February 2019; Chakraborty, S. 2015. Modi's Budget Slashes Environmental Funding for India. *Reuters,* March 12, available at: in.reuters.com/article/india-budget-energy-idINKBN0M812D20150312, accessed on 7 February 2019.

**p. 263** *increased public revenues* Dreze, J. and Sen, A. 2011. Putting Growth in Its Place. *Outlook,* 14 November, available at www.outlookindia.com/magazine/story/putting-growth-in-its-place/278843, accessed on 5 January 2019.

**p. 263** *air pollution crisis* Death by Breath: What's Happening to Delhi Air Will Have You Gasping for Breath. *Indian Express,* 4 April, available at indianexpress.com/article/india/india-others/death-by-breath-an-exhaustive-series-on-delhis-air-pollution/, accessed on 9 December 2018; Ghosal, A. and Chatterjeee, P. 2015. Landmark Study Lies Buried: How Delhi's Poisonous Air is Damaging Its Children for Life. *Indian Express*, 2 April, available at indianexpress.com/article/india/india-others/landmark-study-lies-buried-how-delhis-poisonous-air-is-damaging-its-children-for-life/, accessed on 12 December 2018.

**p. 263** *Over 660 million people* Greenstone, M., Nilekani, J., Pande, R., Ryan, N., Sudarshan, A., and Sugathan, A. 2015. Lower Pollution, Longer Lives: Life Expectancy gains if India Reduced Particulate Matter Pollution. *Economic and Political Weekly* 50(8): 40–6.

**p. 263** *air quality database* WHO Global Ambient Air Quality Database (update 2018), available at: www.who.int/airpollution/data/cities/en/, accessed on 3 February 2019.

**p. 263** *20 million asthma cases* Goenka, D., Jawahar, P., and Guttikunda, S. 2015. Regulating Air Pollution from Coal-Fired Power Plants in India. *Economic and Political Weekly* 50(1): 62–7.

**p. 264** *yield loss up to* Burney, J. and Ramanathan, V. 2014. Air Pollution Impacts on Indian Agriculture. *Proceedings of the National Academy of Sciences* 111: 16319–24; see also Auffhammer, M., Ramanathan, V., and Vincent, J.R. 2006. Integrated Model Shows that Atmospheric Brown Clouds and Greenhouse Gases have Reduced Rice Harvests in India. *Proceedings of the National Academy of Sciences* 103(52): 19668–72.

**p. 264** *over 400 million people* See cleancookstoves.org/country-profiles/focus-countries/5-india.html, accessed on 9 January 2019.

**p. 264** *A study published in 2019* India State-Level Disease Burden Initiative Air Pollution Collaborators. 2019. The Impact of Air Pollution on Deaths, Disease Burden, and Life Expectancy Across the States of India: the Global Burden of Disease Study 2017, *The Lancet Planetary Health* 3(1): e26–e39.

**p. 264** *2018 report ... Central Pollution Control Board* See *State of India's Environment 2019*, Centre for Science and Environment, New Delhi.

**p. 264** *2013 World Bank study* World Bank. 2013. *India: Diagnostic Assessment of Select Environmental Challenges an Analysis of Physical and Monetary Losses of Environmental Health and Natural Resources* (in Three Volumes). World Bank Report No. 70004-IN, World Bank, Washington, DC.

**p. 265** *World Health Organisation (WHO) study* Prüss-Üstün, A., and Corvalán, C. 2006. Preventing Disease through Healthy Environments: Towards an Estimate of the Environmental Burden of Disease. World Health Organisation, Geneva.

**p. 265** *country profile for India* See www.who.int/nmh/countries, accessed on 9 January 2019.

**p. 265** *a global problem* See www.who.int/gho/ncd, accessed on 9 January 2019.

**p. 265** *health situation will worsen* Friel, S., Bowen, K., Campbell-Lendrum, D., Frumkin, H., McMichael, A.J., and Rasanathan, K. 2011. Climate Change, Noncommunicable Diseases, and Development: The Relationships and Common Policy Opportunities. *Annual Review of Public Health* 32: 133–47; Haines, A., Kovats, R.S., Campbell-Lendrum, D. and Corvalan, C. 2006. Climate Change and Human Health: Impacts, Vulnerability and Public Health. *Public Health* 120: 585–96.

**p. 265** *the next few decades* Smith, K.R., Woodward, A., Campbell-Lendrum, D., Chadee, D.D., Honda, Y., Liu, Q., Olwoch, J.M., Revich, B., and Sauerborn, R. 2014. Human Health: Impacts, Adaptation, and Co-benefits, 709–54 in C.B. Field, V.R. Barros, D.J. Dokken, K.J. Mach, M.D. Mastrandrea, T.E. Bilir, M. Chatterjee, K.L. Ebi, Y.O. Estrada, R.C. Genova, B. Girma, E.S. Kissel, A.N. Levy, S. MacCracken, P.R. Mastrandrea, and L.L. White, Editors. *Climate Change 2014: Impacts, Adaptation, and Vulnerability. Part A: Global and Sectoral Aspects.* Contribution of Working Group II to the Fifth Assessment Report of the Intergovernmental Panel on Climate Change, Cambridge University Press, Cambridge, UK.

**p. 265** *One study ... 2014 report* Mahal, A., Karan, A., and Michael, E. 2010. The Economic Implications of Non-Communicable Disease for India. Health, Nutrition and Population (HNP) discussion paper. World Bank, Washington, DC; Bloom, D.E., Cafiero-Fonseca, E.T., Candeias, V., Adashi, E., Bloom, L., Gurfein, L., Jané-Llopis, E., Lubet, A., Mitgang, E., O'Brien J, C., and Saxena, A. 2014. Economics of Non-Communicable Diseases in India: The Costs and Returns on Investment of Interventions to Promote Healthy Living and Prevent, Treat, and Manage NCDs. Harvard School of Public Health, World Economic Forum, Geneva.

**p. 266** *in recent debates* Choudhury, C. 2015. Not Just a Coal Block. *People's Archive of Rural India*, 2 January, available at ruralindiaonline.org/articles/not-just-a-coal-block- hasdeo-arand/, accessed on 9 January 2019; Bhandari, A. 2015. Coal Kills Indians. Can the Sun Power the Country? *IndiaSpend*, 30 May, available at: archive.indiaspend.com/cover-story/coal-kills-indians-can-the-sun-power-india-32906, accessed on 7 February 2019.

**p. 266** *health impact assessments* Pradyumna, A. 2015. Health Aspects of the Environmental Impact Assessment Process in India. *Economic and Political Weekly* 50(8): 57–64.

**p. 266** *higher financial outlays* See blogs.bmj.com/bmj/2015/03/19/a-public-health-commentary-on-indias-draft-national-health-policy-2015/, accessed on 9 January 2019.

## The Wild Heart of India

**p. 269** *The Wild Heart of India* On the term 'heart of India,' historian Nayanjot Lahiri writes in her book *Time Pieces: A Whistle-stop Tour of Ancient India* (2018, Hachette, Gurugram): 'The earliest expression of an artistic sense in our region is very likely to have been the one visible on rocks in the "heart of India"–an expression coined by a geographer to describe the rocky area of hills and plateaus between the Gangetic plains and the Deccan. This hilly landscape is spectacular. It comprises long finger-like ridges and jaw-dropping rock formations, and masses of scarped plateaus dotted with rock shelters, often to be found in the midst of scattered jungles and grasslands. This stretch was once home to prehistoric inhabitants living remarkably basic lives.'

**p. 270** *effusion of fragrant flowers* See Jha, M. 2018. The mahua story, *Fountain Ink*, May, available at: fountainink.in/interactive/the-mahua-story/, accessed on 7 February 2019.

**p. 272** *landmark book* Schaller, G.B. 1967. *The Deer and the Tiger*. University of Chicago Press, Chicago.

## Close Encounters of the Third Kind

**p. 277** *Greater adjutant stork landed* I thank Jaydev Mandal for the opportunity to watch storks from his bamboo platform.

## Who Gives a Fig?

**p. 285** *a stately banyan* The banyan *Ficus benghalensis* is the National Tree of India.

## Welcome Back, Warblers

**p. 289** *leaf warblers are back* This essay is partly based on the research of Madhusudan Katti: Katti, M. and Price, T. 1996. Effects of Climate on Palaearctic Warblers Over-Wintering in India. *Journal of the Bombay Natural History Society* 93: 411–27; Katti, M. and Price, T. 1999. Annual Variation in Fat Storage by a Migrant Warbler Overwintering in the Indian Tropics.

*Journal of Animal Ecology* 68: 815–23; Katti, M. 2001. Vocal Communication and Territoriality during the Non-Breeding Season in a Migrant Warbler. *Current Science* 80: 419–23.

## Musician of the Monsoon

**p. 294** *blue whistling-thrush* This species was also known earlier as the Himalayan whistling-thrush.

## The Caricature Monkey

**p. 299** *family of the macaques* I am indebted to Anindya Sinha for sharing his knowledge of bonnet macaques and research publications, including: Ram, S., Venkatachalam, S., and Sinha, A., 2003. Changing Social Strategies of Wild Female Bonnet Macaques During Natural Foraging and on Provisioning. *Current Science* 84: 780–90; Sinha, A. 2001. *The Monkey in the Towns Commons: A Natural History of the Indian Bonnet Macaque.* NIAS Report No. R2-2001, National Institute of Advanced Studies, Bengaluru.

## Turning the Turtle

**p. 302** *turning to the wrong horizon* Karnad, D., Isvaran, K., Kar, C.S., and Shanker, K. 2009. Lighting the Way: Towards Reducing Misorientation of Olive Ridley Hatchlings due to Artificial Lighting at Rushikulya, India. *Biological Conservation*, 142: 2083–8.

## The Deaths of Osama

**p. 304** *killed twice in India* Bignell, P. 2006. 'Osama' the Serial Killer Elephant is Shot Dead—Or Is He? *The Independent*, 18 December; Reuters. 2006. Serial Killer Elephant Shot Dead in India. *NBC News,* 17 December, available at: www.nbcnews.com/id/16248233/ns/world_news-south_and_central_asia/t/serial-killer-elephant-shot-dead-india/, accessed on 7 February 2019; Reuters. 2008. Killer Elephant 'Osama' Shot Dead in Jharkhand. *Reuters*, 31 May, in.reuters.com/article/idINIndia-33839920080531, accessed on 7 February 2019.

**p. 305** *serial killers and raging bulls* Loudon, B. 2006. Raging Bull Elephant Osama to Be Shot Dead. *The Australian*, 16 December, available at: www.theaustralian.com.au/news/world/raging-bull-elephant-osama-to-be-shot-dead/news-story/9d51d5c684078c8afa0c0f170526713a, accessed on 7 February 2019.

**p. 305** *of rogues and raiders* Sukumar, R. 1992. *The Asian Elephant: Ecology and Management.* Cambridge University Press, Cambridge; Sukumar, R. 1990. Ecology of the Asian Elephant in Southern India. II. Feeding Habits and Crop Raiding Patterns. *Journal of Tropical Ecology* 6: 33–53; Sukumar, R. 1995. Elephant Raiders and Rogues, *Natural History* 104(7): 52–61.

**p. 306** *coming up with their answers* Lenin, J. and Sukumar, R. 2011. Action Plan for the Mitigation of Elephant-Human Conflict In India, Unpublished Report, Asian Nature Conservation Foundation, Bengaluru; Fernando, P., Kumar, M.A., Williams, A.C., Wikramanayake, E., Aziz, T., and Singh, S.M. 2008. Review of Human-Elephant Conflict Mitigation Measures Practiced in South Asia. WWF AREAS Technical Support Document, WWF, Geneva.

**p. 307** *J.M. Coetzee writes* Coetzee, J.M. 1999. *The Lives of Animals.* Princeton University Press, Princeton, NJ.

**p. 308** *As Bradshaw notes* Bradshaw, G.A. 2009. *Elephants on the Edge: What Animals Teach us about Humanity.* Yale University Press, New Haven.

## An Apology to the Iyerpadi Gentleman

**p. 312** *gentleman of Iyerpadi* The Iyerpadi Gentlemen, or IG, as he is called by elephant scientists working in the landscape, still roams peaceably over these mountains.

## An Enduring Relevance

**p. 314** *The book, Silent Spring* Carson, R. 1962. *Silent Spring.* Houghton Mifflin, Boston.

**p. 314** *sparks that lit the fire* Griswold, E. 2012. How 'Silent Spring' Ignited the Environmental Movement. *New York Times,* 21 September, available at: www.nytimes.com/2012/09/23/magazine/how-silent-spring-ignited-the-environmental-movement.html, accessed on 7 February 2019.

**p. 314** *peregrine falcons and bald eagles* How Important was Rachel Carson's Silent Spring in the Recovery of Bald Eagles and Other Bird Species? *Scientific American,* available at www.scientificamerican.com/article/rachel-carson-silent-spring-1972-ddt-ban-birds-thrive, accessed on 9 January 2019.

**p. 315** *so the accusation goes* See correspondence of journalist George Monbiot and Stewart Brand, available at www.monbiot.com/2010/11/06/correspondence-with-stewart-brand, accessed on 9 January 2019.

**p. 315** *allowing the use of DDT* List of Pesticides which are Banned, Refused Registration and Restricted in Use (As on 20th October 2015), Central Insecticides Board and Registration Committee, India, available at: www.cibrc.nic.in, accessed on 9 January 2019.

**p. 316** *issues to be discussed* Sharma, V.P. 2003. DDT: The Fallen Angel. *Current Science* 85: 1532–7; Lalmalsawmzauva, K.C. 2012. Reproductive Health in Mizoram with Special Reference to Champhai District. PhD thesis, North-eastern Hill University, Shillong, available at: hdl.handle.net/10603/5373, accessed on 9 January 2019; Prakash, A., Bhattacharyya, D.R., Mohapatra, P.K., Goswami, B. K., and Mahanta, J. 2008. Community Practices of Using Bed Nets and Acceptance and Prospects of Scaling Up Insecticide Treated Nets in North-East India. *Indian Journal of Medical Research* 128: 623–29.

**p. 316** *Rainforest Alliance certification* See www.rainforest-alliance.org, accessed on 9 January 2019.

**p. 318** *historian J.R. McNeill recounts* McNeill, J.R. 2001. *Something New Under the Sun: An Environmental History of the Twentieth-Century World.* W.W. Norton and Company, New York.

**p. 319** *in this context* For Carson's biography, see Lear, L. 2009. [1997] *Rachel Carson: Witness for Nature.* Mariner Books, Houghton Miffling Harcourt, Boston.

## Behind the Onstreaming

**p. 332** *the Cauvery basin* See india-wris.nrsc.gov.in, accessed on 9 January 2019.

**p. 336** *65 dams already, plus* See india-wris.nrsc.gov.in, accessed on 9 January 2019.

**p. 336** *River Basin Atlas of India* India-WRIS. 2012. *River Basin Atlas of India*, RRSC-West, National Remote Sensing Centre, ISRO, Jodhpur, available at: www.india-wris.nrsc.gov.in/Publications/RiverBasinAtlasChapters/RiverBasinAtlas_Full.pdf, accessed on 7 February 2018.

**p. 338** *T. Janakiraman and Chitti* Janakiraman, T., and Chitti. 1993. *Eternal Kaveri: The Story of a River.* Translation. Penguin, Delhi.

**p. 339** *a delightful travel book* Seshadri, P. and Sundararaghavan, P.M. 2012. *It Happened Along the Kaveri – A Journey Through Space and Time.* Niyogi Books, New Delhi.

**p. 339** *book he carries with him* Borges, J.L. 2000. *Fictions.* Translated from Spanish by Andrew Hurley, Penguin, London.

## Earth-Scar Evening

**p. 343** *future forest is a robusta* On robusta as an invasive plant, see Joshi, A., Mudappa, D. and Raman, T.R.S. 2009. Brewing Trouble: Coffee Invasion in Relation to Edges and Forest Structure in Tropical Rainforest Fragments of the Western Ghats, India. *Biological Invasions* 11: 2387–400.

## The Butchery of the Banyans

**p. 348** *Research has so far* Shanahan, M., So, S., Gompton, S.G., and Corlett, R. 2001. Fig-Eating by Vertebrate Frugivores: A Global Review. *Biological Reviews* 76: 529–72.

## Forest of Aliens

**p. 354** *the Jarawa forest* UNESCO. 2010. The Jarawa Tribal Reserve Dossier: Cultural and Biological Diversities in the Andaman Islands. Pankaj Sekhsaria and Vishvajit Pandya, Editors. UNESCO, Paris.

**p. 355** *weeds in the undergrowth* Rasingam, L. and Parthasarathy, N. 2009. Diversity of Understorey Plants in Undisturbed and Disturbed Tropical Lowland Forests of Little Andaman Island, India. *Biodiversity and Conservation* 18: 1045–65.

**p. 356** *many animals have been brought* Ali, R., 2006. Issues Relating to Invasives in the Andaman Islands. *Journal of the Bombay Natural History Society* 103: 349–66.

**p. 357** *a world apart* For more context on the Jarawa and the islands, see Sekhsaria, P. 2017. *Islands in Flux: the Andaman and Nicobar Story.* HarperCollins, Noida.

## The Pigeon's Passengers

**p. 357** *passenger pigeons were decimated* For a debate on one contested aspect of their extinction, see 1491 and Passenger Pigeons, available at: brushwoodcenter.wordpress.com/2012/09/19/1491-and-passenger-pigeons, accessed on 18 December 2018.

## The Mistletoe Bird

**p. 368** *pollinate its flowers and disperse its seeds* See in particular the research of Priya Davidar in south India: Davidar, P. 1983. Similarity between Flowers and Fruits in Some Flowerpecker Pollinated Mistletoes. *Biotropica* 15: 32–7; Davidar, P. 1985. Ecological Interactions between Mistletoes and Their Avian Pollinators in South India. *Journal of the Bombay Natural History Society* 82: 45–60.

**p. 369** *Recent research suggests* Watson, D.M. 2009. Parasitic Plants as Facilitators: More Dryad than Dracula? *Journal of Ecology* 97: 1151–9; Watson, D.M., and Herring, M. 2012. Mistletoe as a Keystone Resource: An Experimental Test. *Proceedings of the Royal Society Series B* 279: 3853–60.

## The Walk that Spun the World

**p. 373** *forests once again cover* Recent trends, however, indicate ongoing deforestation again: Olofsson, P., Holden, C.E., Bullock, E.L., and Woodcock, C.E. 2016. Time Series Analysis of Satellite Data Reveals Continuous Deforestation of New England Since the 1980s. *Environmental Research Letters* 11(6): p.064002.

**p. 374** *special place in India's environmental history* See Guha, R. 2000. *The Unquiet Woods: Ecological Change and Peasant Resistance in the Himalaya*. University of California Press, Berkeley; Tucker, R.P. 2012. *A Forest History of India*. SAGE Publications, New Delhi; Pandit, M.K. 2017. *Life in the Himalaya: An Ecosystem at Risk*. Harvard University Press, Cambridge, Massachusetts.

## Aesthetics in the Desert

**p. 377** *Mehrangarh Museum Trust* See mehrangarh.org and raojodhapark.com, both acccessed on 6 January 2019.

**p. 381** *Machia Biological Park* See www.machiabiopark.com/Facilities.aspx, accessed on 16 January 2019.

**p. 384** Near Bishnoi villages Hall, J.C. and Chhangani, A.K. 2015. Cultural Tradition and Wildlife Conservation in the Human-Dominated Landscape of Rural Western Rajasthan, India. *Indian Forester* 141: 1011–19.

**p. 387** *The countryside aesthetic emerges* Jha, M., Vardhan, H., Chatterjee, S., Kumar, K., and Sastry, A.R.K. 1998. Status of Orans

(Sacred groves) in Peepasar and Khejarli Villages in Rajasthan, 263–75 in P.S. Ramakrishnan, U.M. Chandrashekhara, and K.G. Saxena, Editors. *Conserving the Sacred for Biodiversity Management*. UNESCO and Oxford-IBH Publishing, New Delhi; Hall, J.C. and Hamilton, I.M. 2014. Religious Tradition of Conservation Associated with Greater Abundance of a Keystone Tree Species in Rural Western Rajasthan, India. *Journal of Arid Environments* 103: 11–16.

**p. 390** *bustards ... have died* See Bindra, P.S. 2017. *The Vanishing: India's Wildlife Crisis.* Penguin Random House, Gurgaon.

## Twinges of Longing, Passing Shadows

**p. 391** *an extinction spasm* Ellis, R. 2004. *No Turning Back: The Life and Death of Animal Species*. HarperCollins, New York; Kolbert, E. 2014. *The Sixth Extinction: An Unnatural History*. Henry Holt and Company, New York.

**p. 392** *Matthiessen, The Snow Leopard* Matthiessen, P. 1989. [1979]. *The Snow Leopard*. Harvill Press, London.

**p. 395** *the cheetah in India* Divyabhanusinh. 1995. *The End of a Trail—The Cheetah in India*. Banyan Books, New Delhi.

**p. 395** *published this record in 1948* van Ingen and van Ingen. 1948. Interesting Shikar Trophies: Hunting Cheetah *Acinonyx jubatus* (Schreber). *Journal of the Bombay Natural History Society* 47: 718–20.

**p. 396** *baiji, whose recent extinction* Turvey, S.T., Pitman, R.L., Taylor, B.L., Barlow, J., Akamatsu, T., Barrett, L.A., Zhao, X., Reeves, R.R., Stewart, B.S., Wang, K. Wei, Z., Zhang, X., Pusser, L.T., Richlen, M. Brandon, J.R., and Wang, D. 2007. First Human-Caused Extinction of a Cetacean Species? *Biology Letters* 3: 537–40.

## Sentience for Conservation

**p. 404** *'The Unpersuadables'* See www.monbiot.com/2010/03/08/the-unpersuadables, accessed on 6 January 2019.

**p. 404** *discipline of 'conservation psychology'* Clayton, S. and Myers, G. 2015. *Conservation Psychology: Understanding and Promoting Human Care for Nature.* John Wiley and Sons, Chichester.

**p. 405** *primatologist Frans de Waal* De Waal, F. 2009. *The Age of Empathy: Nature's Lessons for a Kinder Society.* Three Rivers Press, New York.

## Epilogue: Tinker, Tailor, Soldier, Spy: A Personal Journey into the Wild Heart of India

**p. 410** *at my institution* Visit the Nature Conservation Foundation (NCF) website to know more about NCF, download our publications, or support our research and conservation work across India: www.ncf-india.org.

# Common and Scientific Names of Species

Here's an alphabetical list of common and scientific names of species mentioned in the text. Local names are in bold italics. See the Select Bibliography for taxonomic sources for the names.

**African elephant** *Loxodonta africana*

**American marten** *Martes americana*

**Amur falcon** *Falco amurensis*

**Andaman drongo** *Dicrurus andamanensis*

**Andaman treepie** *Dendrocitta bayleii*

**Ash** *Fraxinus sp.*; American white ash *Fraxinus americana*; American black ash *Fraxinus nigra*; Himalayan ash *Fraxinus micrantha, F. floribunda*

**Ashy-headed green-pigeon** *Treron phayrei*

**Asian brown flycatcher** *Muscicapa dauurica*

**Asian elephant** *Elephas maximus*

**Asian fairy-bluebird** *Irena puella*

**Asian openbill** *Anastomus oscitans*

**Assamese macaque** *Macaca assamensis*

**Baiji** Yangtze river dolphin *Lipotes vexillifer*

***Bajra*** Pearl millet *Pennisetum glaucum*
**Bald eagle** *Haliaeetus leucocephalus*
**Bamboo rat** Lesser bamboo rat *Cannomys badius*, Hoary bamboo rat *Rhizomys pruinosus*
**Banyan** *Ficus benghalensis*
**Baobab** *Adansonia sp.*
**Barasingha** Swamp deer *Rucervus duvaucelii*
**Barn owl** *Tyto alba*
**Basswood** American basswood *Tilia americana*
**Beautiful nuthatch** *Sitta formosa*
**Beaver** North American beaver *Castor canadensis*
**Beech** *Fagus sp.*, American beech *Fagus grandifolia*
**Bengal florican** *Houbaropsis bengalensis*
**Binturong** *Arctictis binturong*
**Bird's-nest fern** *Asplenium nidus*
**Black dammar** *Canarium strictum*
**Black eagle** *Ictinaetus malaiensis*
**Black kite** *Milvus migrans*
**Black redstart** *Phoenicurus ochruros*
**Black stork** *Ciconia nigra*
**Black-naped monarch** *Hypothymis azurea*
**Blackbuck** *Antilope cervicapra*
**Blue pitta** *Hydrornis cyaneus*
**Blue whistling-thrush** *Myophonus caeruleus*
**Blue-cheeked bee-eater** *Merops persicus*
**Blue-faced malkoha** *Phaenicophaeus viridirostris*
**Blue-throated flycatcher** *Cyornis rubeculoides*
**Bonnet macaque** *Macaca radiata*
***Bordi*** Jujube tree *Ziziphus nummularia*
**Bougainvillea** *Bougainvillea sp.*
**Broad-tailed grassbird** *Schoenicola platyurus*
**Bronzeback tree snake** Common bronzeback tree snake *Dendrelaphis tristis*
**Brown hawk-owl** *Ninox scutulata*
**Brown hornbill** *Anorrhinus austeni*
**Brown palm civet** *Paradoxurus jerdoni*
**Brown-cheeked fulvetta** *Alcippe poioicephala*
**Buff-breasted babbler** *Pellorneum tickelli*
**Camel** Dromedary *Camelus dromedarius*
***Canarium*** see black dammar
**Cane** Rattan *Calamus sp.*
**Capped langur** *Trachypithecus pileatus*
**Cardamom** *Elettaria cardamomum*
**Cedar** Northern white cedar *Thuja occidentalis*
***Chalthe*** *Pseudostachyum polymorphum*
**Cheetah** *Acinonyx jubatus*
**Chestnut-shouldered petronia** *Petronia xanthocollis*
**Chestnut-tailed starling** *Sturnia malabarica*
**Chevron-breasted babbler** Cachar wedge-billed babbler *Stachyris roberti*
**Chickpea** *Cicer arietinum*
**Chimpanzee** *Pan troglodytes*
**Chinkara** Indian gazelle *Gazella bennettii*
**Chital** *Axis axis*

**Chousingha** Four-horned antelope *Tetracerus quadricornis*
***Chromolaena*** *Chromolaena odorata*
**Cinnamom** *Cinnamomum sp.*
**Clouded leopard** *Neofelis nebulosa*
**Coffee** Arabica *Coffea arabica,* Robusta *Coffea canephora*
**Colocolo opossum** *Dromiciops gliroides*
**Common green-magpie** *Cissa chinensis*
**Common krait** *Bungarus caeruleus*
**Common palm civet** *Paradoxurus hermaphroditus*
**Common sandpiper** *Actitis hypoleucos*
**Common tailorbird** *Orthotomus sutorius*
**Common woodshrike** *Tephrodornis pondicerianus*
**Cougar** Mountain lion or puma *Puma concolor*
**Coyote** *Canis latrans*
**Creeping lily** *Gloriosa superba*
**Crested serpent eagle** *Spilornis cheela*
**Cup-and-saucer plant** *Holmskioldia sanguinea*
**Darter** Oriental darter *Anhinga melanogaster*
**Deodar** Himalayan cedar *Cedrus deodara*
**Desert cat** Asiatic wildcat *Felis silvestris*
**Desert fox** *Vulpes vulpes grifitthi*
**Desert monitor** *Varanus griseus*
**Dhole** Asiatic wild dog *Cuon alpinus*
**Dipper** *Cinclus sp.*
**Dodo** *Raphus cucullatus*
**Dog** Domestic dog *Canis familiaris*
**Dollarbird** *Eurystomus orientalis*
**Earpod wattle** *Acacia auriculiformis*
**Edible-nest swiftlet** now called white-nest swiftlet *Aerodramus fuciphagus*
**Eld's deer** See Sangai
**Emerald dove** Asian emerald dove *Chalcophaps indica*
**Eurasian collared-dove** *Streptopelia decaocto*
**Flame-of-the-forest** *Butea monosperma*
**Flamingo** Greater flamingo *Phoenicopterus roseus,* Lesser flamingo *Phoeniconaias minor*
**Flying fish** Tropical two-wing flyingfish *Exocoetus volitans*
**Flying fox** Indian flying fox *Pteropus giganteus*
**Flying lizard** South India flying lizard *Draco dussumieri*
**Forest wagtail** *Dendronanthus indicus*
**Forktail** *Enicurus sp.*
**Four-horned antelope** see chousingha
**Garlic** *Allium sativum*
**Gaur** *Bos gaurus*
**Gerbil** Rodents in Family Muridae, Subfamily Gerbillinae
**Gharial** *Gavialis gangeticus*
***Ghatbor*** *Fleuggia leucopyrus*
**Giant salvinia** *Salvinia molesta,* now *S. adnata*
**Giant wood spider** *Nephila pilipes*
**Goat** Domestic goat *Capra aegagrus hircus*
**Golden jackal** *Canis aureus*

***Goondi*** *Cordia sinensis*
**Gorilla** Western gorilla *Gorilla gorilla*, eastern gorilla *G. beringei*
**Great Indian bustard** *Ardeotis nigriceps*
**Great argus** *Argusianus argus*
**Great auk** *Pinguinus impennis*
**Great hornbill** *Buceros bicornis*
**Great slaty woodpecker** *Mulleripicus pulverulentus*
**Greater adjutant** *Leptoptilos dubius*
**Green imperial-pigeon** *Ducula aenea*
**Green pigeons** *Treron sp.*
**Greenish warbler** *Phylloscopus trochiloides* (the similar green warbler *P. nitidus*, now recognized as a distinct species, is also common in southern India)
**Grey bushchat** *Saxicola ferreus*
**Grey francolin** *Francolinus pondicerianus*
**Grey heron** *Ardea cinerea*
**Grey peacock-pheasant** *Polyplectron bicalcaratum*
**Grey-bellied tesia** *Tesia cyaniventer*
**Grey-headed canary-flycatcher** *Culicicapa ceylonensis*
**Grouse** Ruffed grouse *Bonasa umbellus*, Spruce grouse *Falcipennis canadensis*
**Hangul** Kashmir stag, red deer *Cervus elaphus hanglu*, Sikkim stag or shou *C. e. wallichi*
**Helmeted hornbill** *Buceros vigil*
**Hemlock** Eastern hemlock *Tsuga canadensis*
***Heynea*** *Heynea trijuga*
**Hill myna** Common hill myna *Gracula religiosa*
**Hill partridges** *Arborophila sp.*
**Himalayan goral** *Naemorhedus goral*
**Himalayan striped squirrel** *Tamiops mcclellandii*
***Hingoto*** Desert date *Balanites roxburghii*
**Hoary-bellied squirrel** *Callosciurus pygerythrus*
**Hog badger** *Arctonyx collaris*
**Hog deer** *Hyelaphus porcinus*
***Holematthi*** *Terminalia arjuna*
**Horseshoe bats** *Rhinolophus* species
**Hyacinth** Water hyacinth *Eichhornia crassipes*
**Ibisbill** *Ibidorhyncha struthersii*
**Indian ash tree** *Lannea coromandelica*
**Indian crested porcupine** *Hystrix indica*
**Indian false vampire** *Megaderma lyra*
**Indian giant flying squirrel** *Petaurista phillipensis*
**Indian giant squirrel** *Ratufa indica*
**Indian grey hornbill** *Ocyceros birostris*
**Indian hare** *Lepus nigricollis*
**Indian laburnum** *Cassia fistula*
**Indian muntjac** *Muntiacus muntjak*
**Indian nightjar** *Caprimulgus asiaticus*
**Indian rhinoceros** *Rhinoceros unicornis*
**Indian scops-owl** *Otus bakkamoena*
**Indian spotted chevrotain** *Moschiola indica*
**Israeli babool** *Acacia tortilis*

**Jackal** see golden jackal
***Jamun*** *Syzygium cumini*
**Jerdon's courser** *Rhinoptilus bitorquatus*
**Indian vultures** Indian long-billed vultures *Gyps indicus*
***Kair*** *Capparis decidua*
***Kanakambara*** *Crossandra infundibuliformis*
***Kheemp*** *Leptadenia pyrotechnica*
***Kheer kheemp*** *Sarcostemma acidum*
***Khejri*** *Prosopis cineraria*
***Khuangthli*** *Bischofia javanica*
**King cobra** *Ophiophagus hannah*
**Knobbed hornbill** *Rhyticeros cassidix*
**Koel** Asian koel *Eudynamys scolopaceus*
***Kumatiyo*** Gum arabic *Acacia senegal*
**Ladybug** Ladybird beetle, Family Coccinellidae; species in Genus *Stethorus* are known predators of spider mites
**Laggar falcon** *Falco jugger*
**Langur** Grey langur *Semnopithecus sp.*
***Lantana*** *Lantana camara*
**Laughing dove** *Streptopelia senegalensis*
**Leaf muntjac** Leaf deer, *Muntiacus putaoensis*
**Leaf-nosed bats** Family Hipposideridae
**Leopard** *Panthera pardus*
**Leopard cat** *Prionailurus bengalensis*
**Lesser fish-eagle** *Haliaeetus humilis*
**Lesser florican** *Sypheotides indicus*
**Lion** *Panthera leo*
**Lion-tailed macaque** *Macaca silenus*
**Little spiderhunter** *Arachnothera longistra*
**Little swift** *Apus affinis*
**Long-tailed broadbill** *Psarisomus dalhousiae*
**Lynx spider** *Peucetia viridana*
**Magpie-robin** Oriental magpie-robin *Copsychus saularis*
***Mahua*** *Madhuca longifolia*
**Maize** *Zea mays*
**Malabar grey hornbill** *Ocyceros griseus*
**Malabar pied-hornbill** *Anthracoceros coronatus*
**Malabar spiny dormouse** Spiny tree mouse *Platacanthomys lasiurus*
**Malabar trogon** *Harpactes fasciatus*
**Malabar whistling-thrush** *Myophonus horsfieldii*
**Malay civet** *Viverra tangalunga*
**Malayan giant squirrel** *Ratufa bicolor*
***Mallige*** *Jasminum sp.*
**Mammoth** *Mammuthus sp.*
**Mango** *Mangifera indica*
***Manilkara*** *Manilkara littoralis*
**Marbled cat** *Pardofelis marmorata*
**Marshweed** *Limnophila heterophylla*
***Mautak*** *Melocanna baccifera*
**Mesquite** *Acacia auriculiformis*
***Mikania*** *Mikania micrantha*
***Missi*** Cowpea witchweed *Striga gesnerioides*
**Mistle thrush** *Turdus viscivorus*

**Mistletoebird** *Dicaeum hirundinaceum*
**Mongoose** *Herpestes sp.*
**Moose** *Alces alces*
**Mountain imperial-pigeon** *Ducula badia*
**Mugger crocodile** Marsh crocodile *Crocodylus palustris*
**Musk deer** Kashmir musk deer *Moschus cupreus,* alpine musk deer *M. chrysogaster,* Himalayan musk deer *M, leucogaster,* and black musk deer *M. fuscus*
**Narcondam hornbill** *Rhyticeros narcondami*
**Necklaced laughingthrush** Greater necklaced laughingthrush *Ianthocincla pectoralis*
**Neem** *Azadirachta indica*
**Nicobar scrubfowl** Nicobar megapode *Megapodius nicobariensis*
**Night heron** Black-crowned night heron Nycticorax *nycticorax*
***Nila palai*** *Wrightia tinctoria*
**Nilgai** *Boselaphus tragocamelus*
**Nilgiri flowerpecker** *Dicaeum concolor*
**Nilgiri langur** *Semnopithecus johnii*
**Nilgiri marten** *Martes gwatkinsii*
**Oil palm** *Elaeis guineensis, E. oleifera*
**Olive ridley sea turtle** *Lepidochelys olivacea*
**Olive-backed pipit** *Anthus hodgsoni*
**Orange-bellied leafbird** *Chloropsis hardwickii*
**Orange-headed ground-thrush** Orange-headed thrush *Geokichla citrina*
**Orangutan** *Pongo pygmaeus*
**Oriental pied-hornbill** *Anthrococeros albirostris*
***Padauk*** *Pterocarpus dalbergioides*
**Paddy** see rice
**Pale-headed woodpecker** *Gecinulus grantia*
**Pallas's squirrel** *Callosciurus erythraeus*
**Palm swift** Asian palm-swift *Cypsiurus balasiensis*
**Palmyra** *Borassus flabellifer*
**Para grass** *Brachiaria mutica*
***Parthenium*** *Parthenium hysterophorus*
**Passenger pigeon** *Ectopistes migratorius*
**Peafowl** Indian peafowl *Pavo cristatus*
***Peeloo*** *Salvadora persica*
***Peepal*** *Ficus religiosa*
**Peregrine falcon** *Falco peregrinus*
**Phayre's langur** *Trachypithecus phayrei*
***Phulrua*** *Dendrocalamus hamiltonii*
**Pied bushchat** *Saxicola caprata*
**Pin-tailed pigeon** *Treron apicauda*
**Pineapple** *Ananas comosus*
**Pipistrelle** Indian pipistrelle *Pipistrellus coromandra*
**Plain flowerpecker** *Dicaeum minullum*
**Plumbeous redstart** *Phoenicurus fuliginosus*
***Pongam*** *Pongamia pinnata*
**Pratincole** *Glareola sp.*
**Pygmy hog** *Porcula salvania*
**Racket-tailed drongo** Greater racket-tailed drongo *Dicrurus paradiseus*

***Raga*** Himalayan spruce *Picea smithiana*
**Rain tree** *Albizia saman*
**Rat snake** Indian rat snake *Ptyas mucosa*
**Raven** Common raven *Corvus corax*
***Rawnal*** *Dendrocalamus longispathus*
***Rawthing*** *Bambusa tulda*
**Red deer** *Cervus elaphus*
**Red maple** *Acer rubrum*
**Red spider mite** *Oligonychus coffeae*
**Red velvet mite** *Trombidium sp.*
**Red-headed trogon** *Harpactes erythrocephalus*
**Red-vented bulbul** *Pycnonotus cafer*
**Redstart** *Phoenicurus sp.*
**Reindeer** Caribou *Rangifer tarandus*
**Rhesus macaque** *Macaca mulatta*
**Rhino, rhinoceros** see Indian rhinoceros
**Rhinoceros hornbill** *Buceros rhinoceros*
**Rice** *Oryza sativa*
**River lapwing** *Vanellus duvaucelii*
**River tern** *Sterna aurantia*
**Rosy starling** *Pastor roseus*
**Rubber** *Hevea brasiliensis*
***Rudraksh*** *Elaeocarpus serratus*
**Ruffed grouse** see grouse
**Rufous-fronted prinia** *Prinia buchanani*
**Rufous-tailed lark** *Ammomanes phoenicura*
***Sairil*** *Melocalamus compactiflorus*
***Sal*** *Shorea robusta*
**Sambar** *Rusa unicolor*
**Sangai** Eld's deer or Brow-antlered deer or Thamin *Rucervus eldii*
**Sarus crane** *Antigone antigone*
**Saw-scaled viper** *Echis carinatus*
**Sea sparkle** *Noctiluca scintillans*
**Serow** Red serow *Capricornis rubidus*
**Sheep** *Ovis aries*
**Sheildtail snake** Family Uropeltidae
**Short-nosed fruit-bat** *Cynopterus sphinx*
**Shou** see hangul
**Shrike** *Lanius sp.*
**Silver oak** *Grevillea robusta*
**Slaty-bellied tesia** *Tesia olivea*
**Slaty-headed babbler** Indian scimitar-babbler *Pomatorhinus horsfieldii*
**Sloth bear** *Melursus ursinus*
**Small Indian civet** *Viverricula indica*
**Small-clawed otter** Oriental small-clawed otter *Aonyx cinereus*
**Smooth-coated otter** *Lutrogale perspicillata*
**Snow leopard** *Panthera uncia*
**Snowy-browed flycatcher** *Ficedula hyperythra*
**Snowy-throated babbler** *Stachyris oglei*
**Sparrow** House sparrow *Passer domesticus*
**Spider mite** mites of Family Tetranychidae, see also red spider mite
**Spinner dolphin** *Stenella longirostris*
**Spiny-tailed lizard** *Saara hardwickii*
**Spittle bug** Froghopper
**Spot-bellied eagle-owl** *Bubo nipalensis*

**Spotted dove** *Streptopelia chinensis*
**Spotted linsang** *Prionodon pardicolor*
**Spotted owlet** *Athene brama*
**Spruce** Red spruce *Picea rubens*
**Sri Lanka bay-owl** *Phodilus assimilis*
**Stripe-necked mongoose** *Herpestes vitticollis*
**Stump-tailed macaque** *Macaca arctoides*
**Sugar maple** *Acer saccharum*
**Sumba hornbill** *Rhyticeros everetti*
**Swamp deer** see barasingha
***Symplocos*** *Symplocos sp.*
**Tahr** Nilgiri tahr *Nilgiritragus hylocrius*
**Takin** *Budorcas taxicolor*
**Tamarind** *Tamarindus indica*
**Tea** *Camellia sinensis*
**Teak** *Tectona grandis*
***Tetrameles*** *Tetrameles nudiflora*
***Thhor*** Leafless spurge *Euphorbia caducifolia*
**Thick-billed pigeon** *Treron curvirostra*
**Tiger** *Panthera tigris*
**Tokay gecko** *Gekko gecko*
**Tricoloured munia** *Lonchura malacca*
**Tufted deer** *Elaphodus cephalophus*
**Umbrella thorn** *Acacia planifrons*
***Vernonia*** *Vernonia arborea*
**Wagtail** *Motacilla sp.*
**Ward's trogon** *Harpactes wardi*
**Water buffalo** *Bubalus arnee*
**Water deer** *Hydropotes inermis*
**Western hoolock gibbon** *Hoolock hoolock*
**Wheatear** *Oenanthe sp.*
**Whistling-duck** *Lesser whistling-duck Dendrocygna javanica,* Fulvous whistling-duck *D. bicolor*
**White dammar** *Vateria indica*
**White-bellied blue flycatcher** *Cyornis pallipes*
**White-browed bulbuls** *Pycnonotus luteolus*
**White-browed piculet** *Sasia ochracea*
**White-crested laughingthrush** *Garrulax leucolophus*
**White-eared bulbul** *Pycnonotus leucotis*
**White-rumped munia** *Lonchura striata*
**White-tailed tropicbird** *Phaethon lepturus*
**White-throated bulbul** *Alophoixus flaveolus*
**Wild balsam** *Impatiens sp.*
**Wild dog** see Dhole
**Wild nutmeg** *Myristica dactyloides*
**Wild pig** *Sus scrofa*
**Wolf** *Canis lupus*
**Wood apple** *Limonia acidissima*
**Wreathed hornbill** *Rhyticeros undulatus*
**Yellow oleander** *Cascabela thevetia*
**Yellow-backed sunbird** Crimson sunbird *Aethopyga siparaja*
**Yellow-bellied warbler** *Abroscopus superciliaris*
**Zanzibar red colobus** *Piliocolobus kirkii*

# Select Bibliography

*Readings that influenced my thinking on nature and conservation*

Carson, R. 1962. *Silent Spring.* Houghton Mifflin, Boston.
Carson, R. 1998. *Lost Woods: The Discovered Writing of Rachel Carson.* Beacon Press, Massachusetts.
Coetzee, J.M. 1999. *The Lives of Animals.* Princeton University Press, Princeton, NJ.
De Waal, F. 2009. *The Age of Empathy: Nature's Lessons for a Kinder Society.* Three Rivers Press, New York.
Flader, S.L., and Callicott, J.B. 1991. *The River of the Mother of God and Other Essays by Aldo Leopold.* The University of Wisconsin Press, Wisconsin.
Guha, R. 2014. *Environmentalism: A Global History.* Allen Lane, Gurgaon.
Haskell, D.G. 2017. *The Songs of Trees: Stories from Nature's Great Connectors.* Viking, New York.
Leopold, A. 1949. *A Sand County Almanac and Sketches Here and There.* Oxford University Press, New York. Also see the 1966 edition: *A Sand County Almanac with Essays on Conservation from Round River.* Ballantine Books, New York.)
McNeill, J.R. 2001. *Something New Under the Sun: An Environmental History of the Twentieth-Century World.* W.W. Norton and Company, New York.

Monbiot, G. 2013. *Feral: Searching for Enchantment on the Frontiers of Rewilding.* Allen Lane, London.

Nussbaum, M. 2007. *Frontiers of Justice: Disability, Nationality, Species Membership. Belknap Press, Cambridge, Massachusetts.*

Worster, D. 1994. *Nature's Economy: a History of Ecological Ideas.* Second edition. Cambridge University Press, New York.

*Readings on contemporary nature and conservation in India*

Ali, R. 2018. *Running Away from Elephants: The Adventures of a Wildlife Biologist.* Speaking Tiger, New Delhi.

Bindra, P.S. 2017. *The Vanishing: India's Wildlife Crisis.* Penguin Random House, Gurgaon.

Chundawat, R. 2018. *The Rise and Fall of the Emerald Tigers.* Speaking Tiger, New Delhi.

Dutt, B. 2014. *Green Wars: Dispatches from a Vanishing World.* HarperCollins, Noida.

Gadgil, M., and Guha, R. 2000. *The Use and Abuse of Nature.* Oxford University Press. New Delhi.

Ghosh, A. 2016. *The Great Derangement: Climate Change and the Unthinkable.* Penguin Random House, Gurgaon.

Gubbi, S. 2018. *Second Nature: Saving Tiger Landscapes in the Twenty-First Century.* Rainfed Books, Chennai.

Iyer, R.R. (Editor). 2015. *Living Rivers, Dying Rivers.* Oxford University Press, New Delhi.

Johnsingh, A.J.T. 2018. *On Jim Corbett's Trail and other Tales from the Jungle.* Natraj Publishers, Dehradun.

Karanth, K.U. 2007. *A View from the Machan: How Science Can Save the Fragile Predator.* Permanent Black, Ranikhet.

Lenin, J. 2012. *My Husband and Other Animals.* Westland, Chennai.

Lenin, J. 2018. *My Husband and Other Animals 2: The Wildlife Adventure Continues.* Westland, Chennai.

Mazoomdar, J. 2016. *The Age of Endlings.* HarperCollins, Noida.

Nagendra, H. 2016. *Nature in the City: Bengaluru in the Past, Present, and the Future.* Oxford University Press, New Delhi.

Narain, S. 2017. *Conflicts of Interest: My Journey through India's Green Movement.* Penguin Random House, Gurgaon.

Pandit, M.K. 2017. *Life in the Himalaya: an Ecosystem at Risk.* Harvard University Press, Cambridge, Massachusetts.

Rangarajan, M., Madhusudan, M.D., and Shahabuddin, G. (Editors). 2014. *Nature Without Borders.* Orient Blackswan, Hyderabad.

Sekhsaria, P. 2017. *Islands in Flux: the Andaman and Nicobar Story.* HarperCollins, Noida.

Shahabuddin, G. 2010. *Conservation at the Crossroads: Science, Society, and the Future of India's Wildlife.* Orient Blackswan, Hyderabad.

Shahabuddin, G., and Rangarajan, M. (Editors). 2007. *Making Conservation Work.* Permanent Black, Ranikhet.

Shanker, K. 2016. *Soup to Superstar: the Story of Sea Turtle Conservation along the Indian Coast.* HarperCollins, Noida.

Shrivastava, A., and Kothari, A. 2012. *Churning the Earth: the Making of Global India.* Penguin, New Delhi.

Srinivasan, U. and Velho, N. 2018. Editors. *Conservation from the Margins.* Orient Blackswan, Hyderabad.

Subramanian, M. 2015. *Elemental India: The Natural World at a Time of Crisis and Opportunity.* HarperCollins, Noida.

*Field guides and sources for common and scientific names*

Aengals, R., Kumar, V.S. and Palot, M.J., 2018. A checklist of reptiles of India. Version 3.0. Zoological Survey of India, Kolkata. Available at zsi.gov.in/WriteReadData/userfiles/file/Checklist/Reptile%20Checklist%20(May%202018).pdf. Accessed on 4 February 2019.

Clements, J.F., Schulenberg, T.S., Iliff, M.J., Roberson, D., Fredericks, T.A., Sullivan, B.L., and Wood, C. L. 2018. *The eBird/Clements checklist of birds of the world: v2018.* Available at www.birds.cornell.edu/clementschecklist/download/. Accessed on 4 February 2019.

Gamble, J.S. and Fischer, C.E.C. 1935. *Flora of the Presidency of Madras,* Parts I to XI. Secretary of State for India, London.

Grimmett, R., Inskipp, C., and Inskipp, T. 2011. *Birds of the Indian Subcontinent,* Second Edition. Oxford University Press, London.

IUCN [International Union for Conservation of Nature] 2017. *The IUCN Red List of Threatened Species.* Version 2017-3. Available at www.iucnredlist.org. Accessed on 31 March 2018.

Johnsingh, A.J.T. and Manjrekar, N. Editors. 2013. *Mammals of South Asia.* Volume 1, Universities Press. Hyderabad.

Johnsingh, A.J.T. and Manjrekar, N. Editors. 2015. *Mammals of South Asia.* Volume 2, Universities Press. Hyderabad.

The Plant List. 2013. Version 1.1. Available at www.theplantlist.org. Accessed on 31 March 2018.

Whitaker, R., Captain, A. and Ahmed, F. 2004. *Snakes of India.* Draco Books, Chennai.

# Publication Credits

Prologue Appeared earlier on my blog *View from Elephant Hills* on the Coyotes Network (coyot.es/elephanthills), 15 February 2013.

## Field Days: An Ecological Education

1. Six Seasons in the City: This essay was written specially for this volume.
2. Night Life in Chennai: A previous version of this article appeared in *Focus*, the now-defunct magazine of the Indian Institute of Technology, Chennai [Sridhar, T.R., 1992, Night Life, *Focus*, 14(3), 32–36].
3. Lone Palm Tree, Sir!: Reproduced by permission of the Madras Naturalists' Society [Raman, T.R.S., 2009, 'Lone Palm Tree, Sir!' *Blackbuck*, 25(1 and 2), 1–14].
4. The Tropicbirds of Memory: Appeared on *EcoLogic*, blog of the Nature Conservation Foundation, Mysore, *EcoLogic*, 21 September 2013.
5. Fording the Flood: An earlier version appeared in the newsletter of the Wildlife Institute of India [Sridhar, T.R., 1994, 'Let's Get Back to Base Camp,' *WII Newsletter*, 9(3), 18–20].

6. Answering the Call of the Hoolock Gibbon: Edited version, reproduced by permission of the Madras Naturalists' Society [Sridhar, T.R., 1994, 'Answering the Call of the Hoolock Gibbon,' *Blackbuck*, 10(1), 1–8].
7. Bamboo Bonfires and Biodiversity: Edited version, reproduced by permission of the Wildlife Conservation Society [Raman, T.R.S., 2007, 'On Fire,' *Wildlife Conservation*, May–June, 46–51].
8. In Clouded Leopard Country: Reproduced by permission of *The Hindu* [Raman, T.R.S., 2016, 'In Clouded Leopard Country,' *The Hindu Sunday Magazine*, 8 October, 1–2].
9. The Dance of the Bamboos: Edited and combined version of the following essays reproduced by permission of the Anandabazar Patrika (ABP) Group, People's Archive of Rural India, and *The Frontier Despatch* [Raman, T.R.S., 'Field and Fallow, Farm and Forest,' *The Telegraph*, 12 April 2014; 'Crop Cycles: Fire and Renewal in Mizoram,' *People's Archive of Rural India*, 21 April 2015; and 'Why Mizoram Must Revive, Not Eradicate, Jhum,' *The Frontier Despatch*, March 4–10, 2016, 6].
10. Bird by Bird in the Rainforest: Appeared earlier on my blog *View from Elephant Hills*, 11 February 2014.
11. Abode of Rainforest Rarities: Reproduced by permission from *The Hindu* [Raman, T.R.S., 'Abode of Rainforest Rarities,' *The Hindu Sunday Magazine*, 1 November, 2, 1998].
12. Shadowing Civets: This essay was first published in *GEO*. Reproduced by permission of *The Outlook* group of publications [Raman, T.R.S., 2009, 'Shadowing Civets: On the Trail of Asia's Most Elusive Small Carnivores,' *GEO*, 26–29 July].
13. Kalakad: Written with Divya Mudappa; appeared earlier on my blog *View from Elephant Hills*, 9 December 2014.
14. Feathered Foresters: Written with Divya Mudappa; reproduced with permission from *Saevus* [Mudappa, D. and Raman, T.R.S., 2012, 'The Feathered Foresters,' *Saevus*, 1(4, Sep/Oct), 28–33].
15. Deep Forest: This essay was first published in *Outlook Traveller*. Edited and reproduced by permission of *The Outlook* group of publications [Raman, T.R.S., 2009, 'Deep Forest: Miao-Vijayanagar Road,' *Outlook Traveller*, September, 126–40].

## Conservation: A World of Wounds

The Rachel Carson quote is from the documentary 'The Silent Spring of Rachel Carson' aired on *CBS Reports*, USA, on 3 April 1963.

The Ravi Sankaran quote is from a video of his lecture to students at a workshop for ornithologists at Gurukula Kangri University, Haridwar, Uttarakhand, 27–30 November 2008.

16. The Beleagured Blackbuck: This essay was written specially for this volume.
17. A Bounty of Deer: Edited version, reproduced by permission of Indian Academy of Sciences [Raman, T.R.S., 1996, 'A Horde of Indian Deer,' *Resonance—Journal of Science Education*, July, 52–61].
18. Hornbills: Written with Divya Mudappa, edited version; reproduced by permission of Indian Academy of Sciences [Raman, T.R.S. and Mudappa, D., 1998, 'Hornbills—Giants among Forest Birds,' *Resonance—Journal of Science Education*, 8, 56–65].
19. A Life of Courage and Conviction: Written with Divya Mudappa, edited version; reproduced by permission of Oriental Bird Club [Raman, T.R.S. and Mudappa, D., 2009, 'Ravi Sankaran—A Life of Courage and Conviction,' *BirdingASIA*, 11, 126–7].
20. Cavities, Caves, and a Caveat: Written with Divya Mudappa; reproduced by permission from *The Hindu* [Raman, T.R.S. and Mudappa, D., 2012, 'Conservation Caveats,' *The Hindu Sunday Magazine*, 26 February, 4].
21. Life in the Garbage Heap: Appeared earlier on my blog *View from Elephant Hills*, 'Blowin' in the Wind—II,' 18 February 2015.
22. Death on the Highway: Reproduced by permission from *The Hindu* [Raman, T.R.S., 2009, 'Death on the Highway,' *The Hindu Survey of the Environment 2009*, 113–18].
23. Natural Engineering: Reproduced by permission from *Deccan Herald* [Raman, T.R.S., 2010, 'Natural Engineering: India's Green Infrastructure,' *Deccan Herald*, 15 February, 9].
24. The Long Road to Growth: Reproduced by permission from *The Hindu* [Raman, T.R.S., 'The Long Road to Growth,' *The Hindu*, 19 March, 9, 2015].
25. Watering Down Forest Protection: Reproduced by permission from *Scroll.in* [Raman, T.R.S., 2015 'India is Diluting Forest Protection by Ignoring this One Vital Thing,' *Scroll.in*, 30 October].
26. Protecting the Wildlife Protection Act: This article was first published in *India Together* (indiatogether.org), with the support of Oorvani Foundation, community-funded media for the new India; reproduced with permission [Raman, T.R.S., 2014, 'An Uncertain Future for Our Fauna,' *India Together*, 9 September].
27. Living with Leopards in Countryside and City: Reproduced by permission from *Economic and Political Weekly* [Raman, T.R.S., 2015,

'Leopard Landscapes: Coexisting with Carnivores in Countryside and City,' *Economic and Political Weekly*, Web Exclusive, 3 January].

28. The Culling Fields: Reproduced by permission from *The Hindu* [Raman, T.R.S., 2016, 'The Culling Fields,' *The Hindu*, 17 June, 9].
29. Bamboozled by Land-use Policy: Edited and combined version of the following essays reproduced by permission of *The Hindu*, *Newslink*, and *The Frontier Despatch* [Raman, T.R.S., 'Mizoram: Bamboozled by Land Use Policy,' *The Hindu*, 14 May, 9, 2014; 'Perils of Oil Palm,' *Newslink*, 20 August, 16(189), 2, 2014; 'Is Oil Palm Expansion Good for Mizoram?' *The Frontier Despatch*, March 18–24, 6–7, 2016].
30. The March of the Triffids: Reproduced by permission of the British Ornithologists' Union, BOU (bou.org.uk) [Raman, T.R.S., 2016, 'The March of the Triffids,' *#TheBOUblog*, 8 August].
31. How Green is your Tea?: Written with Divya Mudappa; reproduced by permission from *The Hindu* [Raman, T.R.S. and Mudappa, D., 2014, 'How Green is Your Tea?' *Blink: The Hindu Business Line*, 27 September, 10–11].
32. Rhythms of Renewal: Written with Divya Mudappa; reproduced by permission of *The Hindu* [Raman, T.R.S. and Mudappa, D., 'Rhythms of Renewal,' *The Hindu Sunday Magazine*, 2 January, 5, 2011].
33. Conserving a Connected World: Reproduced by permission of *The Hindu* [Raman, T.R.S., 'One Earth, One Chance: Conserving a Connected World,' *The Hindu Sunday Magazine*, 5 June, 1 and 4, 2011].
34. Integrating Ecology and Economy: Reproduced by permission of *The Hindu* [Raman, T.R.S., 'Integrating Ecology and Economy,' *The Hindu*, 3 July, 9, 2014].
35. The Health of Nations: Reproduced by permission of the *International Health Policies Blog* (internationalhealthpolicies.org) [Raman, T.R.S., 'The Other Invisible Hand,' *International Health Policies Blog*, 4 June 2015].

## Reflections: Our Place in Nature

36. The Wild Heart of India: Published in *Fountain Ink* magazine, reproduced by permission [Raman, T.R.S., 2014, 'The Wild Heart of India,' *Fountain Ink*, March 3(5), 103–10].
37. Close Encounters of the Third Kind: One section of this essay appeared earlier in *Orion*; reproduced with permission [Raman, T.R.S., 2016, 'Elephant Crossing,' *Orion*, May–June, 35(3), 6].

38. Who Gives a Fig?: Reproduced by permission from *The Hindu* [Raman, T.R.S., 'Who Gives a Fig?' *The Hindu Sunday Magazine*, 26 July, 5, 2009].
39. Welcome Back, Warblers: Reproduced by permission from *The Hindu* [Raman, T.R.S., 2009, 'Welcome Back, Warblers,' *The Hindu Sunday Magazine*, 1 November, 5].
40. Musician of the Monsoon: Reproduced by permission from *The Hindu* [Raman, T.R.S., 2009, 'Musician of the Monsoon,' *The Hindu Sunday Magazine*, 6 September, 5].
41. The Caricature Monkey: Reproduced by permission from *The Hindu* [Raman, T.R.S., 2011, 'The Caricature Monkey,' *The Hindu Sunday Magazine*, 10 July, 8].
42. Turning the Turtle: Appeared on *EcoLogic* blog, 17 August 2012.
43. The Deaths of Osama: Reproduced by permission from *Deccan Herald* [Raman, T.R.S., 2011, 'Death of Two Osamas,' *Deccan Herald Spectrum*, 24 May, 4].
44. An Apology to the Iyerpadi Gentleman: Reproduced by permission from *The Hindu* [Raman, T.R.S., 2013, 'An Apology to the Iyerpadi Gentleman,' *The Hindu Sunday Magazine*, 14 April, 4].
45. An Enduring Relevance: Appeared earlier on my blog *View from Elephant Hills*, 'The Enduring Relevance of Rachel Carson,' 16 June 2014.
46. River Reverie: Reproduced by permission from *The Hindu* [Raman, T.R.S., 2010, 'River Reverie,' *The Hindu Sunday Magazine*, 7 March, 7].
47. Behind the Onstreaming: Appeared earlier on my blog *View from Elephant Hills*, 11 December 2013.
48. Earth-scar Evening: Appeared on *EcoLogic* blog, 4 May 2009.
49. The Butchery of the Banyans: Written with Divya Mudappa; reproduced by permission from *Deccan Herald* [Raman, T.R.S. and Mudappa, D., 2009, 'Requiem for Hacked Banyans,' *Deccan Herald Spectrum*, 28 July, 1].
50. Of Tamarind and Tolerance: Reproduced by permission from *The Hindu* [Raman, T.R.S., 2012, 'Of Tamarind and Tolerance,' *The Hindu Sunday Magazine*, 17 June, 4].
51. Forest of Aliens: Reproduced by permission from *The Hindu* [Raman, T.R.S., 2012, 'Forest of the Aliens,' *The Hindu Sunday Magazine*, 1 January, 4].
52. The Tall Tree: Appeared in the *Rainforest Revival* blog (presently defunct) of the Nature Conservation Foundation, Mysore, 20 December 2010.

53. The Pigeon's Passengers: Reproduced by permission from *The Hindu* [Raman, T.R.S., 2012, 'The Pigeon's Passengers,' *The Hindu Sunday Magazine*, 6 May, 4].
54. The Mistletoe Bird: Edited version, reproduced by permission from *Scroll.in* [Raman, T.R.S., 2017, 'Why the Evolutionary Link between Flowerpeckers and Mistletoes is Crucial to the Forests,' *Scroll.in*, 19 August].
55. The Walk that Spun the World: Appeared earlier on my blog *View from Elephant Hills*, 'The Walk that Spun the World,' 8 October 2014.
56. Aesthetics in the Desert: A section of this essay appeared earlier in *The Hindu*; reproduced with permission [Raman, T.R.S., 2017, 'How a Desert was Saved from Trees,' *The Hindu Sunday Magazine*, 26 November, 6].
57. Twinges of Longing, Passing Shadows: Appeared on *EcoLogic* blog, 19 September 2010.
58. Being with Dolphins: Reproduced by permission from *The Hindu* [Raman, T.R.S., 2012, 'Dancing with Dolphins,' *The Hindu Sunday Magazine*, 18 March, 4].
59. Sentience for Conservation: Published in *Eternal Bhoomi*, reproduced with permission [Raman, T.R.S., 2012, 'Sentience for Conservation,' *Eternal Bhoomi*, July–September, 3(3), 36–37].

Epilogue An earlier version, co-created with Divya Mudappa, appeared on the presently defunct *Rainforest Revival* blog, 'Tinker, Tailor, Soldier, Spy: A Personal Decade from 9/11,' 11 September 2011.

# Acknowledgements

When I set out on an attempt to pull together this set of essays written over a period spanning more than two decades, I knew that there would be a large list of people I must acknowledge for their unstinting help, guidance, words of encouragement, support, and helpful criticism. Still, when I made the list, it was humbling to realize that my time in the field, my wildlife research career, and my writing itself, would have been impossible without the help of so many people. I would not have gained the experiences I did, I would not have been able to collect it in these words. So even as I acknowledge more specific debts below, it is a deeper, heartfelt thanks for this wider and more formative influence that I wish to express.

Early on, as a student and naturalist, I was fortunate to fall into the company of friends like V. Santharam, Ragupathy Kannan, R.K.G. Menon, and other members of the Madras Naturalists'

Society (MNS) including K.V. Sudhakar, K.V. Prabhakar, B. Rajasekhar, R. Selvakumar, and A. Rajaram among many others. They suffered my company on birding and field trips; let me regularly borrow their books; read and commented on my early manuscripts and natural history notes; and tolerated my pestering without complaint, sharing freely their time, suggestions, and knowledge. For all this, I am ever grateful. A special note of thanks to Theodore Baskaran who introduced me to MNS and for his later encouragement of our work in the Western Ghats over the years. My debt to R.K.G. Menon is recounted in 'Lone Palm Tree, Sir!'

At the Wildlife Institute of India (WII), I thank A.J.T. Johnsingh and G.S. Rawat for guiding me through my Masters field research in Mizoram, the subject of several essays in this volume. Other faculty at WII shared their knowledge and advice during my studies, particularly Ravi Chellam, S.P. Goyal, Y.V. Jhala, Qamar Qureshi, and K. Sankar. It was at WII that I first met Mahesh Rangarajan. It is his constant support and advice through the years of my field research and writing, and his energetic encouragement to work on a book that finally resulted in this volume. I am deeply grateful for his guidance and his belief that I could and should write this book.

I thank Raman Sukumar of the Centre for Ecological Sciences (CES), Indian Institute of Science (IISc), for his encouragement, support, and mentoring during my first forays into ecological field research in Guindy and during my doctoral research at CES under his guidance. He gave me the freedom to pursue different ideas and gain field experiences, some of which I have recounted in some chapters in this book, for which I shall remain ever grateful. A special thanks, too, to Niranjan V. Joshi, for his insightful and unstinting guidance and advice. From him I came to understand that mathematical ecology and numbers and patterns in nature can be both revelatory and beautiful. A number of friends and colleagues made my time at IISc memorable, helped me through my research, commented on my papers and the drafts of some of the essays collected in this volume. I thank Sonia Kaushik, K. Vedham, Kartik Shanker, Savitha Moorthy, Robert John, Sumana Rao, Anuradha Bhat, J. Santosh, and, especially, Anindya 'Rana' Sinha. Rana, friend, colleague, (tor)mentor: thank you for those countless walks to Coffee Board, the endless discussions and poring over draft manuscripts, the insane argument over hyphens and dashes that we are yet to resolve to *your* satisfaction.

It has also been a privilege to work in the company of stellar colleagues, friends, and associates, to together build the Nature Conservation Foundation (NCF), Mysore, as a non-profit conservation research institution of repute. Few students anywhere in the world could have been as fortunate as I was to gain as classmates, friends, and colleagues a truly extraordinary batch of people with whom I have stood and walked and worked alongside over the years, and who remain a pervasive influence on my thinking and writing. To Teresa Alcoverro, Rohan Arthur, Yashveer Bhatnagar, Aparajita Datta, Advait Edgaonkar, Kavita Isvaran, M.D. Madhusudan, Charudutt Mishra, Divya Mudappa, Kakoli Mukhopadhyay, Suhel Quader, Pavithra Sankaran, S.U. Saravanakumar, Anindya Sinha, K.S. Gopi Sundar, and Kulbhushansingh Suryawanshi, I extend my thanks for the company, comments, critiques, friendship, support, and affection. My experiences in the Anamalais were deeply influenced and continue to be moulded by shared experiences with Divya Mudappa, M. Ananda Kumar, P. Jeganathan, Swati Sidhu, Ganesh Raghunathan, Anand Osuri, Manish Chandi, Nisarg Prakash, Sreedhar Vijayakrishnan, K. Srinivasan, and colleagues, students, and visitors at our Rainforest Research Station in the Anamalais: I am deeply grateful to all of them. I also thank Kalyan Varma,

N.S. Prashanth, Tanya Seshadri, Harsha Jayaramaiah, and Payal Mehta for their company on many field trips; for help with several chapters; for freely sharing their knowledge and exceptional photographs; and for a level of daily tolerance of my bad puns which remains unsurpassed.

In the course of travel, learning, and research for writing many of the essays collected here, many people shared their knowledge, ideas, and experience, or joined field trips and collaborative efforts. For this, I extend my heartfelt thanks to Joke Aerts, Vidya Athreya, Praveen Bhargav, Gunnel Cederlöf, Ravi Chellam, Shekar Dattatri, Eben Goodale, Uromi Goodale, Nandita Hazarika, James Lalnunzira Hrahsel, Bhagyashree Ingle, Kashmira Kakati, Kaberi Kar-Gupta, Madhusudan Katti, Tasneem Khan, Swati Kittur, Jagdish Krishnaswamy, Ajith Kumar, Janaki Lenin, Shirish Manchi, Jaydev Mandal, Nima Manjrekar, Uttara Mendiratta, Umeed Mistry, Shomita Mukherjee, Ranjini Murali, Goutam Narayan, Bivash Pandav, Milind Pariwakam, Aasheesh Pittie, Asad Rahmani, Jayashree Ratnam, Mahesh Sankaran, Pankaj Sekhsaria, Narayan Sharma, Mrinalini K. Siddhartha, Mewa Singh, Neha Sinha, Ramki Sreenivasan, Krithika Srinivasan, Nikit Surve, S.K. Tripathi, Neeraj Vagholikar, Abi Tamim Vanak, David Vanlalfakawma, M.M. Venkatachalam (Venky Muthiah), Romulus Whitaker, A. Christy Williams, and Claire F.R. Wordley. A special thanks to Smita Prabhakar and Rakhee Karumbaya at NCF for their friendship and support over the years. My NCF colleagues and friends—too many to list—helped in so many ways and I can only record my gratitude here to all of them.

I thank Paul Devanesan and Mary Sugantha for being wonderful neighbours and for enabling us to make our home and establish our field research station at Valparai in the Anamalais. I shall also remain grateful for the friendship and inspiration of Ravi Sankaran, R. Rajyashri (Deepa), and Rauf Ali and for having known them when they were alive.

My field research experiences would not have been possible without the permissions and support of various state Forest Departments and officers and the guidance of various people of the forest. In Mizoram, this included C. Ramhluna, Lalramthanga, Lalthanhlua Zathang, V. Lal Fala, Kimthanga, Lakhan (Zara), Thanmawia, Zakhuma, Muankima, and Muanpuia. I am particularly grateful to Kimthanga and Zakhuma, from whom I learnt so much and without whose support much of my work in Mizoram would have been impossible. In Tamil Nadu, in Guindy, Kalakad, and the Anamalais, a number of forest officers supported my work: a set of people too large to list here, including V.R. Chitrapu, V.K. Melkani, V. Ganesan, H. Basavaraju, Rajeev Shrivastava, A. Udhayan, Thangaraj Panneerselvam, Krishnaswami, and many others. For teaching me about the forest and for being my guides and assistants on numerous field trips in the Western Ghats, I thank S. Ganesan, Poovan, G. Murthy, T. Dinesh, Krishnan, A. Silamban, T. Sundararaj, A. Sathish Kumar, M. Jeyapandian, and P. Ganesh. In the Anamalai plantations, I am grateful to Venky Muthiah, Arunkumar Menon, D.G. Hegde, Vivek Ayanna, Mahesh Nair, Oliver Praveenkumar, and several other managers, besides R. Victor J. Ilango and S. Marimuthu for sharing their extensive knowledge and practical experience that informed and influenced my thinking and writing on land use. A special thanks to Venky for his friendship, sage advice, enthusiasm and partnership through the years.

Once again, I thank various publishers for their permission to reproduce, with revision, pieces that appeared earlier in their publications: particularly *The Hindu* group of publications (including *Business Line* and *The Hindu Survey of the Environment*), *Deccan Herald*, *Fountain Ink*, *Orion*, *GEO*, *Outlook Traveller*, *Scroll.in*, *India Together*, *The Telegraph*, *Blackbuck*, *Wildlife Conservation*, *Resonance—Journal of Science Education*, *Saevus*, *Eternal Bhoomi*, *Newslink*, *The Frontier*

*Despatch, Economic and Political Weekly, Focus,* and *WII Newsletter*. At *The Hindu*, a special thanks to the editors who read, edited, and published many of my writings, from op-eds to features to personal essays: G. Ananthakrishnan, Shalini Arun, Divya Gandhi, Nirmala Lakshman, Mukund Padmanabhan, Srinivasan Ramani, Vaishna Roy, and Radhika Santhanam. My thanks are also due to Naresh Fernandes, M. Rajshekar, and Aman Khanna (*Scroll.in*), Gopal Rao and Saurav Kumar (*Fountain Ink*), Kai Friese and Amit Dixit (*Outlook* group), H. Emerson (Chip) Blake and Scott Gast (*Orion*), and Adam Halliday (*The Frontier Despatch*). I also thank #TheBOUBlog and the International Health Policies Blog for allowing republication of my posts.

For their helpful suggestions and guidance towards better writing, my deepest gratitude goes to Peter Kline for his creative nonfiction workshop at Stanford Continuing Studies, Gail Ford of Writers Studio, and Chip Blake of *Orion*. My thanks, too, to Julian Hoffman and my Wildbranch workshop friends and writers, Kimberley Moynahan, Courtney Carlson, Margo Farnsworth, Sarah Boon, and Gavin van Horn, for being there to exchange thoughts on writing and for their helpful comments that improved some of my essays. To Divya, who read with a keen eye and mind, corrected numerous inaccuracies, and helped improve all the essays gathered here; Rohan, who read and edited many essays and commiserated with me endlessly on writing; Suhel for his punctilious eye; Mahesh Rangarajan for his comments and for pointing me to relevant literature; and M. Rajshekar for being a sounding board: a special thanks for helping me with my writing all these years. I thank the book's reviewers and the excellent editors at Oxford University Press India for their interest in my work, their encouragement, and their careful shepherding of this book through publication. I am also very grateful to Sartaj Ghuman for his evocative sketches that reflect the setting and spirit of the essays and for his artwork for the cover.

A number of journalists, writers, scholars, and activists in India, several of whom have written excellent books on conservation and the environment, informed and influenced my thinking and writing over the years. For being an inspiration and for their support when I reached out to them, I thank Ananda Banerjee, Prerna Bindra, Bahar Dutt, Ramachandra Guha, P. Jeganathan, Janaki Lenin, Jay Mazoomdar, Harini Nagendra, Sunita Narain, M. Rajshekar, Mahesh Rangarajan, Nitin Sethi, Kumar Sambhav Shrivastava, and Neha Sinha.

This book is dedicated to my parents and in-laws for the love and extraordinary support they gave as Divya and I embarked on somewhat unconventional paths. My parents, Sitalakshmi Rajagopalan and T.R. Rajagopalan, gave me the freedom to explore the world of nature and the world of books from my schooldays, and were a wellspring of unwavering support and affection over the years. My in-laws, B.P. Mudappa, who is no more, and Chandra Mudappa, always shared our concerns for nature, and gave their love and constant support and impetus to pursue our work. A special thanks to my brother T.R. Ramachandran (Sriram) for being a great birding companion and a benchmark and inspiration, and to Kirti Kandade, Rama and Madhu Jalan, for believing in me and opening their homes to me during my twelve-week writing break in California. In my family, I owe a special debt to Dhanam Sreenivasan and T.R. Santhana Raman for encouraging my writing and interests in nature early on. And finally, to Divya—for being my most thoughtful, consistent, and helpful critic, for all our shared times, journeys, and experiences, for putting up with my endless nonsense, for nudging me towards a greater sensitivity in seeing and in feeling for nature, for your love, and, not the least, for getting me to finish this book—to you, my love and my deepest thanks.

If authors who write about the natural world should also acknowledge their debts to places and the animals and plants that inspired their work, then once again I have a list too long to include here. Still, among places, I must mention Guindy National Park in Chennai city, the Valparai plateau as a countryside landscape in the Anamalais, and Dampa and Kalakad as the wild places that most inspired me and continue to do so. Among wild animals, I would single out the Malabar whistling-thrush, whose song in the predawn chill of the mountains kept me company as I wrote from home. Among plants, I hesitate to list favourites as there are too many of them, from great rainforest trees to tiny water plants, and choose to settle instead for *Coffea arabica*, whose bean and buzz accounts for so much of what you hold in your hands.

To those others I should have acknowledged, but have forgotten to do so, my apologies. With gratitude and a salaam to all, I must admit that any errors, biases, and inaccuracies that remain in this work are wholly my responsibility.